Colloid Science
Principles, methods and applications

Second Edition

Colloid Science Principles, methods and applications

Second Edition

Edited by

TERENCE COSGROVE

School of Chemistry, University of Bristol, Bristol, UK

A John Wiley and Sons, Ltd, Publication

This edition first published 2010

Registered office
John Wiley & Sons Ltd, The Atrium, Southern Gate, Chichester, West Sussex, PO19 8SQ, United Kingdom

For details of our global editorial offices, for customer services and for information about how to apply for permission to reuse the copyright material in this book please see our website at www.wiley.com.

Library of Congress Cataloging-in-Publication Data

Colloid science : principles, methods and applications / edited by Terence Cosgrove. – 2nd ed.
p. cm.
Includes bibliographical references and index.
ISBN 978-1-4443-2019-0 (cloth) – ISBN 978-1-4443-2020-6 (pbk.)
1. Colloids. I. Cosgrove, T. (Terence)
QD549C585 2010
541'.345–dc22

2009045994

A catalogue record for this book is available from the British Library.

ISBN HB: 9781444320190
ISBN PB: 9781444320206

This book is dedicated to the memories of Professor Douglas Everett and Professor Ron Ottewill who were an inspiration to many generations of aspiring Colloid Scientists.

Contents

Preface

This book has grown out of the Spring School in Colloid Science, a one-week course which has been run in the School of Chemistry at Bristol University, during the Easter vacation, every year since 1972. Indeed, this is "the book of the course", since its contents form the basis of the material that is delivered in the lectures. Like the course, this book is primarily intended as a basic introduction to colloid and interface science for those with a first degree in chemistry or physics, or some closely related discipline (e.g. pharmacy, biochemistry), who are working in an industrial R&D laboratory, or indeed a Government or university laboratory, and who need to learn the basics of this subject for their research. Sadly, this is a subject which often receives scant coverage still in many undergraduate courses, despite its great relevance in a wide range of chemical technologies, including pharmaceuticals, agrochemicals and food, personal care and household products, surface coatings, oil and mineral recovery and processing. In addition to these more traditional industries, colloid and interface science underpins many of the new, so-called "nanotechnologies", such as sensors, IT chips, displays, photonics, micro-reactors.

The authors of the various chapters in this book are or have been members of the Colloid Group at Bristol University and are recognised experts in their own areas of the subject. The book begins with an introductory chapter on what a colloid is, what their important features are and this theme is amplified in Chapter 2 by a detailed discussion of the origin of surface charge that is key to the stability of many aqueous colloidal dispersions. In Chapter 3 the issue of colloid stability is addressed taking into account both the electrostatic and dispersion interactions and this theme is expanded in Chapters 9 and 16. In Chapters 4 and 5 two important examples of colloidal systems, surfactants and microemulsions are discussed in depth. Chapter 6 focuses on emulsions which are practical and challenging systems used in everyday life. Polymers are also a key feature in colloid science in that they can form aggregates themselves but are widely used in dispersions to promote or destroy stability and in Chapters 7, 8 and 9 their role is treated in detail. Chapter 10 focuses on another aspect of colloid science and this is the wetting of interfaces and in Chapter 11 some of these ideas are extended to study aerosols and foams which are further examples of colloidal systems. The following three chapters focus on experimental techniques, including rheology in Chapter 12, probably the most basic tool of the colloid scientist, and scattering and imaging methods in Chapters 13 and 14 and 15 for detailed characterisations of colloidal dispersions. Finally in Chapter 16 interfacial forces are explored in depth and some of the methods used to measure them are introduced.

Although this book is based on the course at Bristol it should provide a balanced account of the subject from a practical point of view for anyone requiring an introduction to the important world of colloid and interface science.

Terence Cosgrove and Brian Vincent January 2010

Introduction

Colloid Science at Bristol and The Bristol Spring School

Colloid science has a rich history at Bristol, dating back to the early part of the 20th century when J.W. McBain was appointed to the chemistry staff in 1907. He was subsequently appointed to the first Leverhulme Chair in Physical Chemistry in 1919, largely due to his pioneering work on molecular association in soap solutions (what Hartley later termed "micelles"), an area in which Lord Leverhulme clearly had a strong interest! McBain left for the US (joining Stanford University) in 1926, whereupon W.E. Garner, whose forté was heterogeneous catalysis, took up the Leverhulme Chair. He was succeeded in turn by D.H. Everett (1954) whose speciality was interfacial and colloidal thermodynamics, R.H. Ottewill (1982) with a broad interest in many aspects of colloid science, B. Vincent (1992) whose interests span both the academic and practical aspects of the subject and by T. Cosgrove in 2007.

The first Spring School was conceived by Ron Ottewill who came to Bristol from Cambridge in 1964 at the instigation of Douglas Everett, primarily to set up the M.Sc. Course Advanced Teaching and Research in Colloid and Interface Science. This highly successful course ran for some 30-odd years and only ceased when the introduction of 4-year undergraduate masters degrees in science in the UK made 1-year post graduate courses less relevant. Many leading industrial scientists and academics in the colloid and interface field have passed though Bristol, as Ph.D. students, members of the M.Sc. course, or as participants (around 1000 to date) of the Spring School. As well as Ron Ottewill, other teachers on the first course in 1972 were Aitken Couper, Jim Goodwin, Dudley Thompson and Brian Vincent. The course was initially managed by the Department of Extra-Mural Studies (as it was then called) of the University, and a number of people from that department gave a great deal of help and support in the early days. Special mention must be made of David Wilde and Sue Pringle in this regard.

Since the mid-1990's the Bristol Colloid Centre (BCC) has taken over the management of the course, and it has become one of the range of courses that the BCC offers annually. For example, as well as the Spring School, the BCC also offers more basic courses at the technician level and for those without a first degree in a scientific discipline. The BCC was set up in 1994 by Brian Vincent and Jim Goodwin, with Terry Cosgrove subsequently succeeding Jim as Deputy Director after he retired in 1996. Cheryl Flynn was the first staff member to be appointed followed by Paul Reynolds (as Manager). The staff now number eleven and the management has been taken on by Roy Hughes with Paul being responsible for new business ventures.

The purpose of the BCC is to give research and training support to the large range of industries, in the UK and abroad, which feature colloid and interfacial science and technology in their products or processing. The BCC staff and the academic colloid group work very closely together. From the BCC sprung the DTI/EPSRC funded IMPACT (Innovative Manufacturing and Processing using Applied Colloid Technology) Faraday Partnership in 2001, and on the back of IMPACT came the ACORN (Applied Colloid Research Network) DTI Link programme. The BCC has worked closely alongside IMPACT in producing its own on-line courses in colloid science and technology recently as well as running lab-based formulation courses in conjunction with the RSC where IMPACT supplied distance learning modules. The Spring School Course is complementary to these. In particular, as well as the lectures, there is a strong emphasis in the Spring School on exposure of the participants to a range of basic experimental techniques. It is from these beginnings that this text has emerged.

Although the Spring School course, from the start, has been aimed primarily at industrial scientists, the objective has always been to teach the fundamentals of the subject, with many illustrations of the basic principles taken from real applications in industry. Of course, these basics have not changed over the years, but new ideas do strongly emerge and develop, and the course is frequently updated to accommodate these.

Brian Vincent, Bristol, January 2010

Acknowledgements

We would like to acknowledge the following people who helped in the production of this book.

Yan Zhang who redrew many of the original diagrams and new drawings by Anita Espidel. Pam Byrt who helped with the preparation of the original manuscripts and all the staff of the Bristol Colloid Centre for their support of the project. Edward Elsey is thanked for reading the proof copies and spotting many errant commas and non-sequitors. The authors would especially like to thank their long suffering other halves without whose help and support this book would never have been finished. The Editor would like to thank his wife Maggie for her encouragement and patience during the summer of 2009 in Florida where most of the editing work on the new edition was carried out.

List of Contributors

Bartlett, Paul, School of Chemistry, University of Bristol, UK

Briscoe, Wuge, School of Chemistry, University of Bristol, UK

Cosgrove, Terence, School of Chemistry, University of Bristol, UK

Davis, Sean, School of Chemistry, University of Bristol, UK

Eastman, John, Learning Science Ltd, Bristol, UK

Eastoe, Julian, School of Chemistry, University of Bristol, UK

Fermin, David, School of Chemistry, University of Bristol, UK

Hughes, Roy, Bristol Colloid Centre, University of Bristol, UK

Kwamena, Nana-Owusua A., School of Chemistry, University of Bristol, UK

Reid, Jonathan P., School of Chemistry, University of Bristol, UK

Reynolds, Paul, Bristol Colloid Centre, University of Bristol, UK

Richardson, Robert, Department of Physics, University of Bristol, UK

Riley, Jason, Department of Materials, Imperial College London, UK

van Duijneveldt, Jeroen, School of Chemistry, University of Bristol, UK

Vincent, Brian, School of Chemistry, University of Bristol, UK

1

An Introduction to Colloids

Roy Hughes

Bristol Colloid Centre, University of Bristol, UK

1.1 Introduction

Introductions tend to be dull; scientific ones tend to be rigorously so. This is felt most strongly by readers with some experience of the field in question. To avoid this pitfall we will dwell only briefly on the rich history of the subject and concentrate on introducing simple concepts in colloid science. These are designed as an introduction to the subject but will also provide a small measure of insight for those with more experience, and of course act as the foundation for the following chapters.

Overbeek (1) suggests that the birth pangs of colloid science began in the 1840s with aggregation studies of *pseudo-solutions* in water of sulfur and silver iodide by the toxicologist Francesco Selmi (2). Later came the development of a gold dispersion by Faraday (3). These materials consisted of finely divided mixtures of solid particles (*sols*) dispersed in electrolyte solutions. The term *colloid* was developed by Graham in 1861, from the Greek meaning glue-like, and this name stuck. He also coined the terms *dialysis*, *sol* and *gel*. Graham defined a colloid in terms of its *inability* to pass through a fine membrane. Many of these early colloidal species would have had submicrometre dimensions and some were more or less polymeric in nature. This introduces an important concept, that the dimensions of the species are dominant in defining the colloidal region.

Traditionally the *colloidal domain* is defined as extending over a range of dimensions from a few nanometers to a few tens of micrometres. Operationally in this region a consistent set of physical laws can be applied to effectively describe the behaviour of the materials. At the extremes these laws become less effective descriptors: greater

Colloid Science: Principles, methods and applications, Second Edition Edited by Terence Cosgrove

Table 1.1 Examples of materials inhabiting the colloidal domain

medium(α) ⇒ dispersed phase (β) ⇓	gas (fluid)	liquid	solid
gas (bubbles)	—	foam	solid foam
liquid (droplets)	liquid aerosol	liquid emulsion	solid emulsion
solid (particles)	solid aerosol	sol	solid sol

consideration of molecular and atomic properties is appropriate at lengths of a few nanometres and kinematics and wetting phenomena play a significant role at tens and hundreds of micrometres. We will consider the latter at the end of this chapter. Colloidal materials are composed of at least two phases, one dispersed (denoted as β) in a second phase (denoted as α). Various examples of colloidal systems are given in Table 1.1 and illustrated in Figures 1.1 and 1.2.

One word of caution concerns the use of the term 'phase' in colloid science. Whilst correctly used in Table 1.1, it is often more loosely used to describe the physical state of the system. Some colloidal states do not achieve thermodynamic equilibrium but the state may be so long lived that it is practically difficult or theoretically inconvenient to distinguish it from an equilibrium phase. There are identifiable non-equilibrium states such as colloidal *glasses* and some *creams* and *gels* are locked into a kinetically trapped state for many decades. In Table 1.1 each of the individual 'phases' could be complex, an emulsion may be formed from gel-like particles or equally particles can be dispersed in a gel. Many everyday systems consist of complex mixtures, each with the potential to evolve to a lower energy state. This highlights the need to understand and control stability, a core area in colloid science. Because of the variety of materials that could be used we can appreciate the intellectual challenges that practising commercial scientists experience when faced with the prediction of shelf-life and the control of storage stability.

We have introduced the concept of a timescale for the system and we can also view this from the perspective of the particle. The stochastic (random) Brownian motion experienced by colloidal particles takes its name from Robert Brown and observations made in 1827 on pollen grains. Smoluchowski (4) and Einstein (5) developed the concept into a formal framework. Let us define a time τ, required to move a mean squared distance $\langle x^2 \rangle$. The value

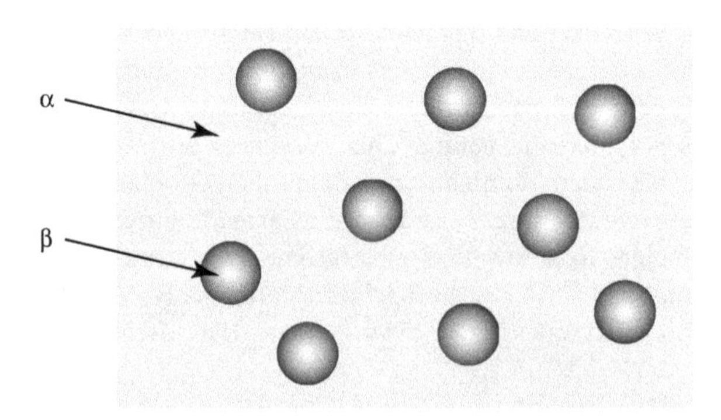

Figure 1.1 Schematic representation of a colloidal system

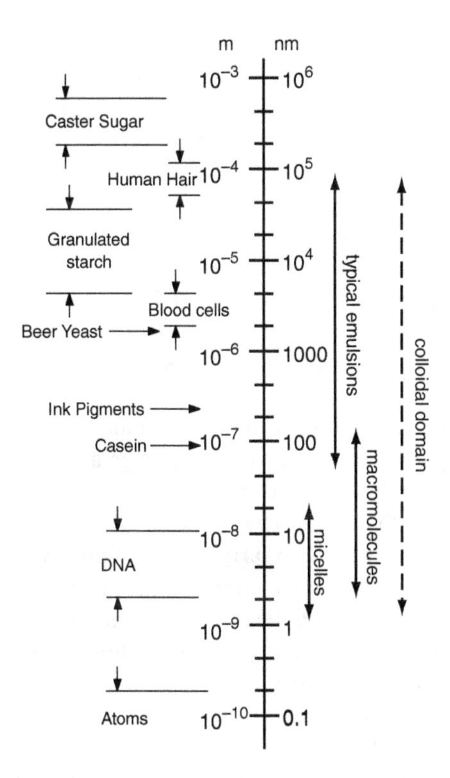

Figure 1.2 *The colloidal domain: the dimensions and typical examples of materials that fall in the colloidal size range*

attained is determined by the diffusion coefficient of the particle, D.

$$D = \frac{\langle x^2 \rangle}{6\tau} \tag{1.1}$$

The diffusion coefficient for spherical particles in a dilute solution is given by the ratio of the thermal energy $k_B T$ and the frictional (Stokes) drag f, on the particles.

$$D = \frac{k_B T}{f} \tag{1.2}$$

$$f = 6\pi\eta_o a \tag{1.3}$$

where k_B is the Boltzmann constant, T is the absolute temperature, η_o is the solvent viscosity and a is the particle radius. We can set the square of the particle radius to be proportional to the mean square distance moved and define a characteristic timescale for a colloid.

$$\tau = \frac{6\pi\eta_o a^3}{k_B T} \tag{1.4}$$

This approach is appropriate for dilute systems under quiescent conditions. It is incomplete in the sense it does not allow for body forces such as gravity or mechanical or thermal convection (the latter a probable factor in Brown's original observations). It also does not allow for concentration effects where interactions between neighbouring colloidal species slows diffusion and increases the effective viscosity, in some cases by many orders of magnitude. We can address this by substituting the viscosity of the material η for the solvent viscosity.

$$\tau = \frac{6\pi\eta a^3}{k_B T} \tag{1.5}$$

The difference between Equations (1.4) and (1.5) can be substantial. For example, an emulsion system such as a hand cream can have a viscosity a million times that of water. It is fair to conclude that a very wide range of timescales are appropriate to the study of colloidal materials.

Many of the early systems investigated at the end of the 19th and start of the 20th century consisted of nano-sized particles but these presented a great challenge. Indeed the nature of small colloidal species, molecules and atoms were controversial and impossible to experimentally visualise at that time. The development of the ultramicroscope in 1903 by Siedentopf and Zsigmondy (6) was an early method of observing individual particles and was an important step in confirming the nature and existence of a colloid. The idea of using microscopy to directly investigate the dynamics of larger particles was grasped by Jean-Baptiste Perrin (7). He supposed that small but observable colloidal particles could provide a model for atoms. By following the distribution of particles under gravity and applying the ideal gas law to the colloid he was able to estimate Avogadro's number, an amazing feat. This approach was an echo of a theme of colloid research in the latter half of the 20th century; that is the analogue of atoms and molecules to colloids. In terms of academic research the discipline divided, to an extent, into two endeavours. One area was the synthesis and characterisation of the systems, the other the drive towards analogues and studying them through theory and experiment.

A critical feature of colloidal particles to both understand and control is the propensity towards aggregation and sedimentation. In order for the systems to remain in suspension the particles should be both small and well dispersed. There exists an attraction between two materials that are alike but separated and dispersed in a different media. In the 1930s the attractive force was notably investigated by Hamaker (8) who provided a consistent description of the van der Waals interaction applied to a pair of colloidal particles. The Hamaker approach demonstrated that particles will tend to strongly aggregate without the presence of an opposing repulsive interaction. The idea of a colloidal repulsive interaction was not novel when in the 1940s a coherent and largely successful description of the pair potential was developed. The DLVO approach, named after Derjaguin, Verwey, Landau and Overbeek (9), summed the attractive and repulsive components to describe the pair interaction energies and forces between a pair of colloidal particles. This is discussed in detail in Chapter 3.

Interestingly, in the 1920s we see a distinction appear between the study of colloids and polymers when Staudinger suggested that the latter consisted of high molecular weight macromolecules. There remains a clear distinction between these fields to this day but many overlapping areas exist and areas of polymer and colloid science operate hand in hand.

Another species which overlaps with the dimensions of the colloidal domain are the microstructures occurring with surface-active agents. As with polymeric materials, the distinction between micellar species and colloids was not initially made. Just prior to the First World War, McBain (10), at the University of Bristol, noted the anomalously high conductivity of detergent molecules. He suggested they associate to form colloidal ions. We would classify these as association colloids; McBain was observing micelles, the aggregation of surfactants into supramolecular species. Since surface-active material is required to form and stabilise colloidal materials it is a subject that has always sat comfortably with colloid science.

Commercial and academic forces demanded increased control of colloidal systems, driving both instrumental development and high-quality synthesis of spherical particles of varying chemistry. Through the 1960s, 1970s and 1980s this was an endeavour undertaken by a number of researchers and University groups with Bristol being one of those at the forefront in synthesis and characterisation. This is well illustrated through the works of Ottewill and workers in Bristol (11). Non-intrusive techniques such as light, X-ray, neutron scattering and nmr reveal realms of order and disorder in colloidal species. Measurements of elasticity (12) and osmotic pressure (13) and theoretical modelling completed the first connections between chemical form, structural order, interparticle forces and macroscopic properties. It is somewhat ironic, at least in a historic sense, that the scientific exploration of the colloidal state showed their richness when compared to many atomic systems. This has generated a wealth of studies towards the close of the 20th century, concentrating on their unique less atomistic-like nature.

Many areas of colloidal physics and chemistry now have well-developed theoretical frameworks. As the 21st century rolls on we are seeing a wider unification of physical and synthetic chemistry with physics. This is coupled with Venn diagram intersection of biology and engineering. There is a drive to use the tools of colloid science to microscopically engineer and synthesise new materials and develop biotechnology. This presents genuinely new and exciting challenges in physics, chemistry, engineering and biology. We can accelerate these developments through a thorough understanding and application of the underlying principles of colloid and surface science.

Of great importance are the practical applications of the science which show no signs of abating. It underpins a number of key aspects of formulation science in colloidal-based products and processes. Some common examples of practical systems are listed in Table 1.2, illustrating how widespread the colloidal form is.

1.2 Basic Definitions

1.2.1 Concentration

The concentration of colloidal species is most appropriately recorded in two forms, as a number density (number of particles per unit volume) or as the fraction of the volume of the system occupied (volume fraction). Volume fraction, ϕ:

$$\phi = nv_\mathrm{p} = n\frac{4}{3}\pi a^3 \tag{1.6}$$

where v_p is the volume of a particle and n the number density of particles. It is expressed above for a dispersion of spherical monodispersed (single-sized) particles. In the laboratory,

Table 1.2 *Colloids in everyday life*

Products	Processes
• surface coatings (paints, video tapes, photographic films)	• clarification of liquids (water, wine, beer)
• cosmetics and personal care (creams, toothpaste, hair shampoo)	• mineral processing (flotation, selective flocculation)
• household products (liquid detergents, polishes, fabric conditioners)	• detergency ('soil' detachment, solubilisation)
• agrochemicals (pesticides, insecticides, fungicides)	• oil recovery (drilling fluids, oil slick dispersal)
• pharmaceuticals (drug delivery systems, aerosol sprays)	• engine and lube oils (dispersion of carbon particles)
• foodstuffs (butter, chocolate ice cream, mayonnaise)	• silting of river estuaries
• pigmented plastics	• ceramic processing ('sol ⇒ gel' processing)
• fire-fighting foams	• road surfacing (bitumen emulsions)
also natural systems, such as:	
• biological cells	
• mists and fogs	

systems are typically prepared in terms of weight or volume. Incorrectly converting between measures of concentration has proved to be a constant source of errors. It is normally supposed that the density is conserved, so that we may write, dispersion density, ρ_T:

$$\rho_T = \rho_o(1-\phi) + \rho_p\phi \tag{1.7}$$

where ρ_p is the density of the particles and ρ_o that of the surrounding medium. From this we can see that by measuring the density of a dispersion, we can establish the concentration of the system providing we know the densities of the individual components. From the previous expressions we can obtain weight fractions, which are often ambiguously described by users, but algebraically clear.

$$\text{weight by volume, } W_v\text{:} \quad W_v = \frac{\text{mass of particles}}{\text{total volume}} = \rho_p\phi = W_m\rho_T \tag{1.8}$$

$$\text{weight by weight, } W_m\text{:} \quad W_m = \frac{\text{mass of particles}}{\text{total mass}} = \frac{\rho_p}{\rho_T}\phi = \frac{W_v}{\rho_T} \tag{1.9}$$

These concentrations refer to concentrations in terms of the total mass or volume of the system. In addition to these measures we can also obtain two other measures of weight fraction in terms of the mass or volume of just the medium used.

$$f_v = \frac{\text{mass of particles}}{\text{volume of medium}} = \frac{W_v}{1-\phi} \tag{1.10}$$

$$f_m = \frac{\text{mass of particles}}{\text{mass of medium}} = \frac{W_m}{1-W_m} \tag{1.11}$$

These measures have been expressed, sometimes confusingly, as a percentage rather than a fraction. The choice of measure is often dictated by the way a sample is prepared, for example when a mass is added to a fluid under mixing, Equation (1.10) or (1.11) is often implied in determining concentrations. It is worth noting that weight by weight measures are dimensionless, weight by volume are not and consequently care must be taken to allow for differences in the units being adopted. The numerical differences between the equations *can* be small but this is only the case at low volume fractions and moderate to low density differences, otherwise beware!

These expressions presuppose the system consists of a single type of particle. In complex multi-component systems or systems where particle size and density can be linked, we should formally calculate the concentrations by integrating or summing the particle size distribution and allow for the volumes occupied by all the components present. There are, as a consequence, a myriad of ways of defining concentrations in multi-component systems. It is important to understand which measure is being used.

1.2.1.1 Size Polydispersity

Colloidal systems tend to show a degree of size polydispersity. The term 'size' is often used to refer to the diameter d of the particles rather than the radius a, although this is not universally the case and is another common cause for calculation errors. Systems can be monomodal, that is have a well-defined mean particle size but have a spread of particle sizes around a modal value. The modal value is the size that occurs most frequently in the system. Distributions may also be multimodal, that is composed of a series of more or less discrete distributions, each region with a given 'modal' value. In natural systems multi-component multimodal behaviour is frequently observed.

Figure 1.3 shows a log/normal particle size distribution in linear and log form. The frequency, on the left-hand axes, is the relative *number* of particles f_i for a given diameter d_i (or radius a_i). The distribution can be characterised by various moments or more

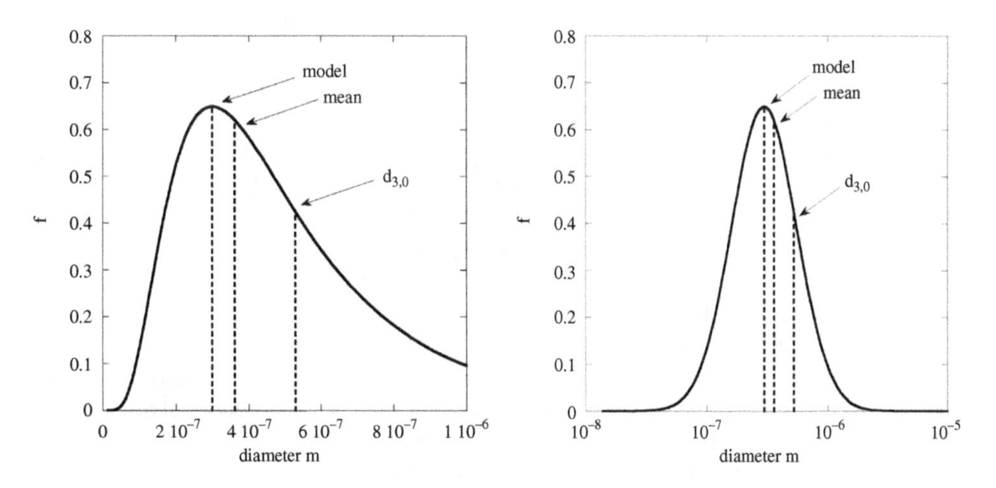

Figure 1.3 *A log/normal distribution displayed on (a) a linear and (b) a logarithmic scale*

conveniently, diameters.

$$d_{m,n} = \left(\frac{\sum_i f_i (d_i)^m}{\sum_i f_i (d_i)^n} \right)^{\frac{1}{m-n}} \quad \text{where} \quad \sum_i f_i = 1 \tag{1.12}$$

The mean particle diameter is given by $\langle d \rangle = d_{1,0} = d$ and is the most common measure of a distribution. It represents the average particle size by number.

$$\langle d \rangle = \sum_i f_i d_i \tag{1.13}$$

Some techniques for particle size measurement and some applications 'weight' towards different averages of the distribution. Consider the Sauter mean value $d_{3,2}$. This is applied in some hydrodynamic applications such as spraying. As droplets are created the ratio of the volume to the surface area is an important feature to characterise, by raising the diameter to powers 3 (volume) and 2 (area) it achieves this end. In general the value of the powers m and n relate to the type of property observed relative to the object. Accordingly $d_{3,0} = d_v$ is the volume-weighted mean and $d_{2,0} = d_a$ the area-weighted mean.

Another nomenclature used to describe particle size distributions arises from the cumulative distribution. For example when sieving particles through a series of meshes of ever reducing size, those retained or *accumulated* on a given mesh are particles with sizes larger than the mesh dimensions. By knowing the mass accumulated on each sieve we can build up a distribution of all the particles below a certain size. The cumulative approach to describing a distribution has become common to a wide range of techniques not least of which is optical diffraction. The measure is often represented as,

$$c_v(j) = 100 \frac{\sum_{i=1}^{i=j} f_i (d_i)^3}{\sum_{i=1}^{i=p} f_i (d_i)^3} \quad \text{where} \quad j \leq p \tag{1.14}$$

for the volume cumulative distribution as a percentage. The volume distribution is apportioned into p discrete sizes. The total distribution is obtained by summing over all the sizes up to p. The total fraction of particles with a diameter equal to, or less than d_j, is obtained by summing up to this size and dividing by the sum of the total distribution. Similar calculations can be performed for both number- and area-weighted values. The accompanying diameter is quoted as $d(0{,}50)$ or $d(50)$, which is the diameter where the cumulative distribution $c_v(j) = 50\%$. This represents the diameter where half the distribution is less than this size and half more. There are numerous variations on this measure for different percentage ranges. Another measure sometimes used is $d(10{,}90)$, which is the average diameter excluding 10% of the smallest and largest particles. Two of these averages are shown on the volume-weighted cumulative distribution in Figure 1.4. It is interesting to note the significant difference between these measures even for a simple log/normal distribution. For the moments studied here we should note that measures of the diameter weighted toward volume provide the largest dimension. This value can prove a very poor indicator of the presence of fine particles in a distribution.

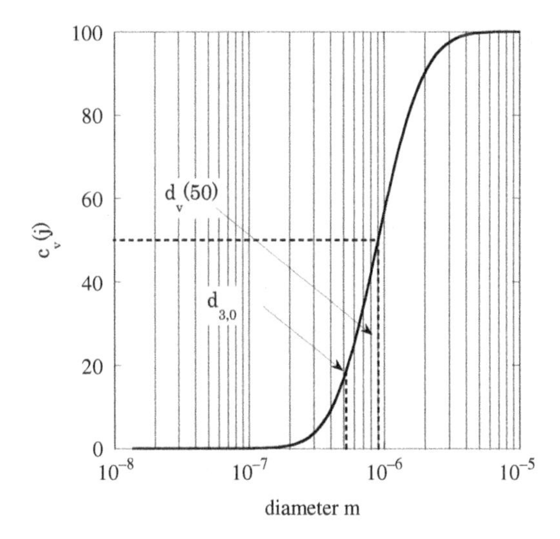

Figure 1.4 *A cumulative volume distribution corresponding to the log/normal distribution shown in Figure 1.3, with the volume average diameter and d(50) compared*

If colloidal particles are being produced via a synthetic route it is common to quote a coefficient of variation. This is a measure of an equivalent peak width of a normal distribution.

$$\nu = \frac{100\sigma}{\langle d \rangle} \quad \text{where} \quad \sigma = \left[\sum_i f_i (d_i - \langle d \rangle)^2 \right]^{1/2} \tag{1.15}$$

where σ is the standard deviation and ν the coefficient of variation (COV), here being expressed as a percentage. Values of a few percent would be considered a near monodisperse system but even 7%–8% is considered good for some more difficult systems.

Much of what has been described above can be applied to emulsions and low-quality (dilute) foams as a characterisation tool. We can include more sophisticated measures for anisotropic particles. Image analysis for larger particles enables maximum and minimum chords to be recorded, perimeter dimensions and areas to be established and all can be expressed in terms of a distribution. One difficulty the experimentalist faces is deciding on the most appropriate measure. For example, for an irregular object should we consider the ratio of the perimeter to the largest length and average this value, or should we alternatively take the ratio of the average largest length to the average perimeter? It is context dependent and choosing the right approach can provide insight, and choosing the wrong one can lead up a 'blind alley'; such characterisation must be used knowledgeably rather than blindly.

The general tools used above apply to polymers but in a subtler manner where variations in chain length (dimension) are demanded by the nature of the polymerisation process. Typically we are rather more bound by techniques and distributions tended to be recorded in terms of weight, number or sometimes viscosity-averaged values, although all relate to the end-to-end dimensions of the chain. These represent different fractions of the distribution. Polydispersity for polymers tends to be recorded in terms of the ratio of the weight to the number average, as outlined in Chapter 6.

1.2.2 Interfacial Area

There are two interfacial areas which influence the behaviour of a dispersion. They are the total area in suspension and the area per unit volume of sample. The total area, A, of the interface in a monodisperse colloidal dispersion in a volume V_T is given by,

$$A = \frac{3\phi}{a} V_T \tag{1.16}$$

As an example, for $V_T = 1\ \mathrm{dm}^{-3}$, $d = 100$ nm and $\phi = 0.1$, then the area A is 6000 m^2, which is the approximate area of a soccer pitch. This highlights the relative importance of the surface area and in fact you will note from Table 1.1 that without an interface you cannot have a colloid. Moreover, it highlights the level of demand required for surface-active material.

Another measure of interfacial area is the specific surface area (SSA). This is the area per unit mass of particles, A_{SSA},

$$A_{SSA} = \frac{3}{\bar{a}\rho_p} \tag{1.17}$$

where $\bar{a}$ is an areal average radius closely related $2\bar{a} = d_{2,3}$. The SSA, a seemingly prosaic measure, has great utility. On a carefully prepared dried dispersion this value can be obtained using gas adsorption and the BET method. The value is also influenced by the distribution of energies for adsorption. The specific surface area gives an indication of particle size through $\bar{a}$ and experimentally it is influenced by the smallest primary particles. This may not be the dispersed particle size but in the absence of pore structures and highly roughened surfaces it is an indication of the size of a characteristic entity. We can assess the relative degree of dispersion by comparing the measured specific surface area with the calculated one. For example, light-scattering data for the distribution of particle sizes in the dispersion can be used to determine the SSA using the expression,

$$A_{SSA} = \frac{\sum_i f_i \pi d_i^2}{\sum_i f_i \rho_p \frac{\pi d_i^3}{6}} \tag{1.18}$$

Large differences between the two measures can be used to detect the presence of fine particles obscured by say the light-scattering technique. Also one frequently encounters 'nanoparticles' which can only be dispersed to the micrometre length scale even though the specific surface area would suggest they are 10–20 nm in dimension. This arises from irreversible aggregation of the particles when they were originally synthesised. It leads to frequent disappointments when acquiring commercial samples. Equally in naturally occurring materials such as minerals the method is able to detect the presence of unexpected quantities of fines.

1.2.3 Effective Concentrations

We have seen that the concentration of the particles is given by their number density. In the absence of any long range repulsive forces, particles will tend to aggregate. In order to mitigate the effects of attractive forces a layer can be adsorbed on the particles, typically polymeric or surfactant. This can provide a measure of stability against aggregation. Whilst the number density of particles remains unaltered the impact of a surface or adsorbed layer increases the effective volume and volume fraction of the particles. If the layer has a thickness δ then the effective volume fraction ϕ_{eff} is given by,

$$\phi_{\mathrm{eff}} = \phi\left(1 + \frac{\delta}{a}\right)^{3} \tag{1.19}$$

where the effective radius of the particle is given by $a+\delta$ as shown in Figure 1.5. This concept whilst simple has great utility. For example, with emulsion systems we are faced with a core droplet, say aqueous, with a radius a, dispersed in an oil phase. It is stabilised by an interfacial layer, say of adsorbed particles (Pickering emulsion). Knowing the volume of the system from the initial phase volumes we can establish the volume fraction of the cores of the droplets. The effective volume fraction will be higher due to the adsorbed particles and this effect becomes more significant the smaller the drop size and the thicker the layer.

The concept of an effective volume fraction not only applies to an actual physical dimension but can also be considered in terms of a 'distance of closest approach' between particles. Consider a polymer latex particle; it often has a surface covered with end-groups formed from the initiator fragments, a good example being carboxylic acid groups. At high pH the protons fully dissociate and this gives rise to a net negative charge on the surface of the particle. The particles will attract cations (counter ions) and repel anions (co-ions). This leads to the development of an ion cloud which shields the surface charge on the particles. It has a characteristic dimension, the reciprocal of which is denoted as κ. The term κ^{-1} is a measure of the effective layer thickness of ions around the particle. If the system is very strongly charged the particles tend to repel one another and this will occur when their surface separation is of the order of κ^{-1}. This is the same as imagining that the particles have a larger

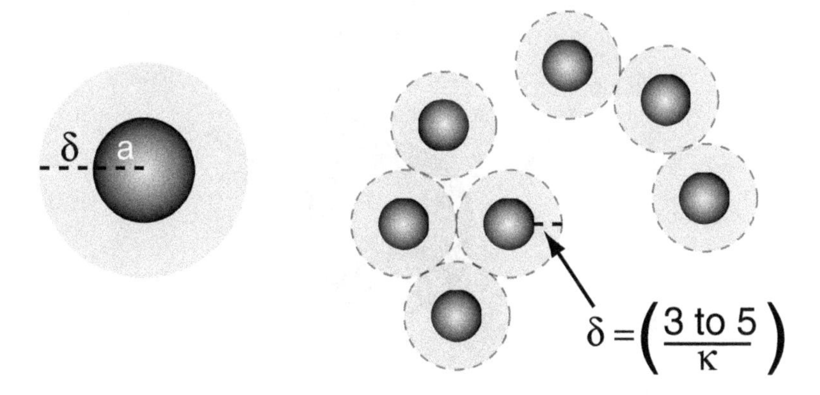

Figure 1.5 The interfacial layer around a particle illustrating the concept of effective volume fraction

radius. We can interpret this dimension as an effective volume fraction. A good rule of thumb for a system where the net interaction is repulsive is given by,

$$\phi_{\text{eff}} = \phi\left(1 + \frac{C}{\kappa a}\right)^3 \tag{1.20}$$

where $C = 3$ to 5.

This volume fraction represents the apparent concentration of the particles. It is as if a shell exists around the particles that another approaching particle is colliding with and bouncing off. The effective volume fraction would be the value needed for a system of non-interacting particles to possess in order to possess the same behaviour as seen in the repulsive system (Figure 1.5). It should be noted that the smaller κa, the greater the effective concentration.

In Chapter 3 we will see the significance of the value of κa, which represents the extent of the ion cloud compared to the radius of the particle. The electrical double layer for a charged particle is illustrated in Figure 1.6, illustrating the ion 'atmosphere' around each charged particle. There is an associated electrostatic potential ψ_0, at the particle surface and an electrokinetically determined zeta potential ζ, an important parameter for determining colloidal stability in many aqueous systems.

1.2.4 Average Separation

We might suppose that the average distance separating the centres of two particles is given by the reciprocal of the cube root of the number density.

$$R = n^{-\frac{1}{3}} = 2a\left(\frac{4\pi}{24}\right)^{\frac{1}{3}}\phi^{-\frac{1}{3}} \approx 2a\left(\frac{0.52}{\phi}\right)^{\frac{1}{3}} \tag{1.21}$$

This equates to the separation seen with the simple cubic packing of spheres. When spheres are packed in this arrangement they occupy 52% of the available volume. This is their maximum packing fraction. Colloid particles can occupy a range of spatial arrangements. They display fluid and ordered 'phases' in addition to long-lived transient states,

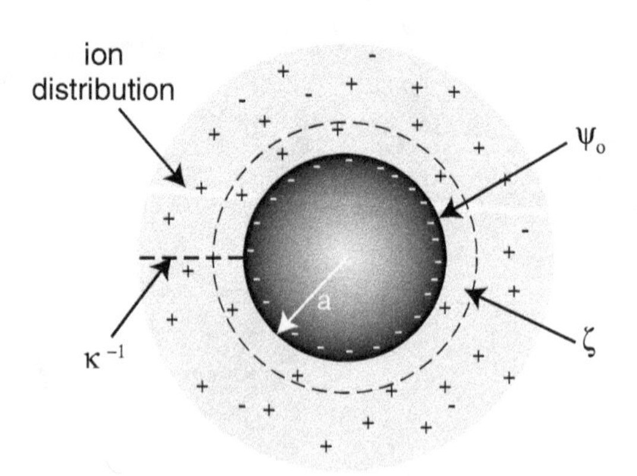

Figure 1.6 A very simple schematic of the electrical double layer around the particle

some glassy in nature. The type of 'phase' observed depends upon the interaction energies between the particles and the preparation path. We can express the average closest separation of particles in terms of a surface separation H,

$$H = 2a\left[\left(\frac{\phi_{\mathrm{m}}}{\phi}\right)^{1/3} - 1\right] \tag{1.22}$$

At this distance there may be one or more particles. For example in a face-centred cubic (fcc) or hexagonal close-packed (hcp) arrangement, the maximum packing fraction is $\phi_{\mathrm{m}} \cong 0.74$. There are 12 particles or *nearest neighbours* (coordination number) at this separation (Figure 1.7). This a structural form with a volume fraction independent number of nearest neighbours. Charged colloidal particles can display this state. However, reality is more complex and if the volume fraction is substantially reduced they can change state and tend to occupy body-centred cubic structures (bcc) with 8 nearest neighbours ($\phi_{\mathrm{m}} \cong 0.68$). Such structures can form in dispersion or be nucleated from surfaces forming two-dimensional arrays. Regular states usually require nearly monodisperse, although not necessarily monomodal, systems to form. The presence of such illustrates the rationale underpinning the concept of colloidal analogues to molecular systems.

Surfactant systems occupy a more varied range of phases than colloidal particles. As the surfactant concentration is increased micelles can arrange themselves into bcc structures but equally can form rods, worm-like structures, lamellar, vesicular and multilayer vesicular structures and a host of other challenging forms. Polymers are, of course, molecular systems and in solution and melt often consist of randomly orientated interpenetrating chains. However, true molecular crystalline zones can develop both in local phase-separating regions or across much larger length scales. We can learn lessons from particulate systems but molecular species present their own chemically specific unique challenges.

The practising particle scientist often encounters less disciplined irregular structures. These can be more fluid-like in arrangement. They can possess a random close-packed (rcp) structure with varying structural order depending upon how the sample is prepared (typically $\phi_{\mathrm{m}} \sim 0.62$–$0.64$). Here the coordination number varies with volume fraction.

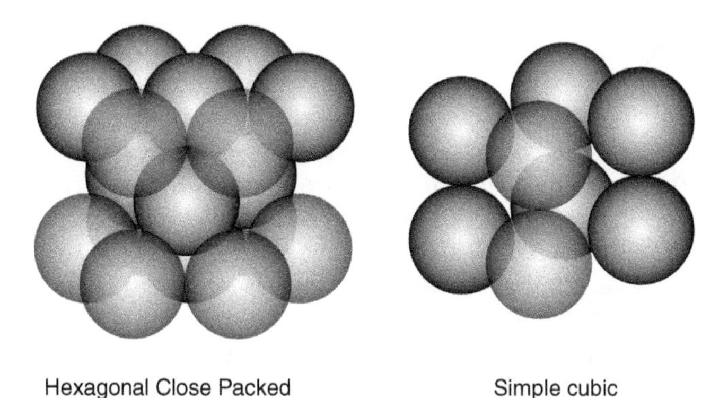

Figure 1.7 *Examples of the maximum packing of ordered structures observed in strongly repulsive colloidal systems*

Aggregated clusters of particles normally possess a consistent nearest neighbour spacing and it is the relative packing and coordination number which will vary with volume fraction. One extreme example are fractal flocs where the coordination number varies from the centre of mass of the floc to the edges. Rarely are large length scale fractal structures encountered, that is one in which the dimensions of the floc are say two orders of magnitude larger than the smallest particle forming the floc. Fractal models can provide a useful tool to consider local arrangements of particles.

Many of the properties of dispersions change once they become concentrated and most commercial products (Table 1.2) are formulated as concentrated dispersions. The location of the boundary between dilute and concentrated is a debatable one but as the effective volume fraction of the system approaches the packing limit of the system, concentrated behaviour is observed. The more strongly the particles interact, the lower the concentration where multi-body interactions occur and thus the higher the effective concentration.

1.3 Stability

The term 'stability' should always be used in context. Colloidal stability relates to the physical state of the system: it is stable if it remains well dispersed. It is often incorrectly assumed that a colloidally stable system is one that remains evenly distributed but this may not be the case in the presence of body forces. In addition, chemical instabilities can drive dispersions into physically unstable states. For example, the oxidation of surface charged groups can reduce repulsive forces or bacterial degradation of polymer stabilisers can lead to changes in stabilising layers.

There is no rigorously applied rule regarding the use of descriptors for the types of particle aggregation that are observed. Unstable colloidal particles will tend to form aggregates. If the attractive forces holding the aggregate together are very large, much greater than the typical body forces used in say stirring, mixing, milling or with ultrasonic probes, then it can be regarded as being permanently aggregated. An irreversible aggregation from the dispersed state is often termed coagulation. The term coagulum is used to describe *en masse* the resulting aggregates. The term flocculation is often used to refer to weaker, more or less reversible aggregation seen with weakly attractive forces. The term precipitation is loosely used but often reserved to describe the phase change to a solid from a molecularly dispersed system when it is subjected to a perturbation. The resulting solid can be amorphous, glassy, a single crystal or polycrystalline. When precipitating in solution the solid phase is often finely divided. In the absence of a stabilising influence the species will aggregate. The resulting precipitate can be a floc or a dispersion depending upon the environment in which it is formed. The nucleation and growth of small, well-defined and well-dispersed crystals is of great value in many agrochemical and pharmaceutical applications.

Electrostatic interactions are not the most versatile method to impart stability (Chapter 8). By adsorbing a layer on the particle either using a polymer or surfactant, a steric hindrance can be built into the pair interaction energy. In order to prevent very close approach the polymer can be tuned to give rise to highly unfavourable interactions as chains on neighbouring particles interpenetrate in a Brownian collision. The addition of a polymer can be detrimental to colloidal stability. If a polymer is used which has a strong preference for the particle surfaces it can 'bridge' particles together leading to fairly weak aggregates at low levels of polymer dosing. Non-adsorbing polymers can osmotically drive particles to an aggregated state.

Generally the concentration of surfactant or polymer required to impart stability depends upon the ability of the particle to adsorb. This can be achieved through a chemisorption process via a locking on reaction between chemical groups on the stabiliser and the particle. This can also occur through physisorption where molecular forces bind molecules to the particle surface in a variety of conformations depending upon their structure. The corresponding adsorption isotherm can also take a variety of forms. An ideal form was proposed by Langmuir (Chapter 7) where the adsorbed amount increases as the concentration in solution is increased until the surface 'saturates' and the amount adsorbed on the surface reaches a constant value. Provided we maintain our system at this surfactant concentration we should be on the plateau region of the isotherm. In practical systems the form of the isotherm is frequently not known or is complicated by the presence of other materials. One common practice is to maintain the concentration of material as a weight percent, p_l, of the total mass of the particles.

$$\mathrm{p_l} = 100\frac{\text{mass of stabilliser}}{\text{mass of particles}} = 100\left(\frac{C_\mathrm{m}}{f_\mathrm{v}}\right) \tag{1.23}$$

where C_m is the concentration of stabiliser in the volume of the dispersing medium (*not* the total volume). In order to maintain a constant percentage of stabiliser (p_l), as the concentration of particles is increased (f_v), then the concentration of stabiliser (C_m) must also increase. Typical values of p_l vary in the region of 3%–5%. At moderate and high particle concentration this can lead to high levels of surface adsorption, very complex condensed surface layers and a large excess of non-adsorbed stabiliser. This is rarely desirable or commercially cost-effective.

1.3.1 Quiescent Systems

The balance of the long-range attractive forces (Hamaker) with electrostatic repulsive forces was the basis of DLVO theory (Figure 1.8). Providing the net interaction energy is sufficiently repulsive the particles cannot collide and stick together. We consider this energy relative to $k_\mathrm{B}T$, the thermal energy. So particles moving with Brownian motion will collide with average energies of the order of a few $k_\mathrm{B}T$. Thus we might imagine that a few $k_\mathrm{B}T$ would be sufficient to impart stability. This is not the case as there is a distribution of collision speeds and energies. Thus we must elaborate this approach a little. If the total energy of interaction (repulsion) between the particles is V_r then the number of collisions that would lead to the particles sticking may be given by a distribution based upon thermal and repulsive energies.

$$N_\mathrm{rel} = \exp\left(-\frac{V_\mathrm{r}}{k_\mathrm{B}T}\right) \tag{1.24}$$

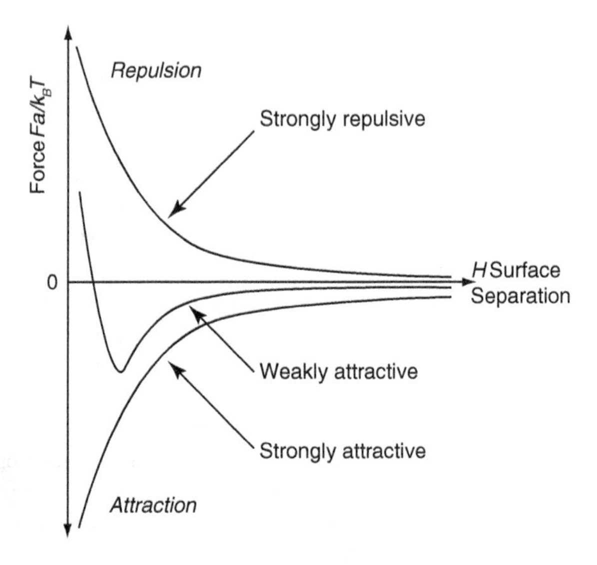

Figure 1.8 *The interaction forces between pairs of colloidal particles showing strongly repulsive, strongly attractive and weakly attractive interactions*

Here $N_{\rm rel}$ represent the relative number of particles that aggregate. Let us consider a system which is repulsion dominated. A repulsion energy of $2.5k_{\rm B}T$ results in about 8% of the collisions giving rise to aggregation, whereas $25k_{\rm B}T$ results in less than 2×10^{-9} % of the collisions giving rise to aggregation. The greater the repulsion the more stable the system. In practical terms a barrier of $25k_{\rm B}T$ gives a shelf life to the material of several months.

1.3.2 Sedimentation or Creaming

In the presence of the Earth's gravitational field an additional body force will act on the particles. They will tend to sediment if they are more dense than the medium or cream if they are less dense. The sedimentation/creaming velocity $v_{\rm s}$ is obtained by equating the Stokes drag with the gravitational force.

$$v_{\rm s} = \frac{2}{9}\frac{\Delta\rho g a^2}{\eta_0} \tag{1.25}$$

where $\Delta\rho$ is the density difference between the particle and the medium and g is the gravitational constant. This is for a dilute system. From this we can note that the larger the particle the greater the tendency to sediment and of course the smaller the Brownian motion which acts to counter this tendency. Colloidally stable particles will still sediment even though the attractive forces between them are small. Aggregates will tend to sediment faster, *unless* they form a fully space-filling structure. The irony here is that systems which are unstable and weakly attractive can produce percolating, that is connected, structures which are stable against sedimentation. Typically this occurs when the attractive energy between the particles at contact is about $-10k_{\rm B}T$ and is usually achieved using adsorbed polymers or surfactants.

1.3.3 Shearing Flows

The application of shear forces can disrupt aggregates and increase both the number of collisions between particles and the energy of those collisions. As a consequence, particles can collide with sufficient force to overcome the colloidal repulsive forces and form aggregates. So shearing fields and body forces in general can both disperse and aggregate systems depending upon the nature of those forces.

The simplest approach to this challenge is to consider the ratio of the diffusive process to the convective process. This is denoted by the Peclet number, *Pe*.

$$Pe = \frac{6\pi a^3 \sigma_0}{k_B T} \tag{1.26}$$

The symbol σ_0 represents the shear stress. This is the body force per unit area applied to the system in a shearing flow field. The Peclet number is a dimensionless quantity and is a measure of the balance of convective flow to the Brownian movement of the system. We can classify the response of the system as follows:

$Pe > 1$	$Pe = 1$	$Pe < 1$
convection	transition	diffusion

At high stresses convection dominates, and in this region aggregates form or break. We can assess this response by equating the Peclet number with the dimensionless force of interaction between the particles as,

$$Pe = |\bar{F}| \quad \text{where} \quad \bar{F} = \frac{Fa}{k_B T} \tag{1.27}$$

where F is the colloid interaction force between the particles and $\bar{F}$ is the force in dimensionless form. So for example when $\bar{F}$ is attractive (negative) and the Peclet number is much greater in magnitude than this value, the aggregates will be disrupted. The concept embodied in Equation (1.27) can be used as a good comparative tool for investigating how flow influences stability.

1.3.4 Other Forms of Instability

We can think about other forms of instability that influence the colloidal state; that is mechanisms by which the distribution of particle size and the homogeneity of the system can change.

Notwithstanding changes in entropy, within any size distribution of say latex particles, crystals, bubbles or droplets, the formation of larger particles tend to be energetically favoured over the smaller ones. The surface free energy change when the area of the interface between the particles and the medium changes is given by ΔG_{12},

$$\Delta G_{12} = \gamma_{12} \Delta A \tag{1.28}$$

where γ_{12} is the interfacial tension and ΔA is the change in area. It is energetically favourable for the total area of the particles to reduce and thus their size to increase. This is associated with the decrease in Laplace pressure Δp with increasing particle radius.

$$\Delta p = \frac{2\gamma_{12}}{a} \tag{1.29}$$

As a result if there is a degree of solubility, the particle size distribution changes with smaller particles being lost and larger particles increasing in size. The size distribution progressively moves to larger particle sizes. This process is known as Ostwald ripening (14). It occurs when there is a mechanism by which change is possible. For example when large and small emulsion droplets collide there is an opportunity to share some of the core material, a process by which the volume of the droplets can change. It is most noticeable with small particle sizes. For solid particles there needs to be a small but finite solubility of the solid material in the continuous phase to allow molecules to move from smaller to larger particles. It is analogous to the small particles being more 'soluble' than their larger cousins.

The reduction in surface free energy through the merging, or coalescence, of droplets or bubbles is a process driving instability in foams and emulsions. The coalescence of two droplets, for example, will reduce the total area and energy. This can be opposed by developing a strong viscoelastic barrier between the drops. As the droplets approach, the liquid layer developing between them drains. This will displace the stabilising molecules at the interface which in turn generates an opposing interfacial tension gradient. This is the Gibbs Marangoni effect and is another stabilising mechanism, albeit one that is difficult to control.

Little has been discussed so far about instabilities observed with polymers and surfactant systems as they lie beyond the scope of this chapter. These are molecular systems and so prone to all the changes associated with state variables such as temperature and pressure. Also by changes in salt, pH, solvency and so forth, we can indirectly influence other state variables such as entropy, enthalpy and internal energy. This can lead to precipitation, conformation changes and various complex phases. Aggregated states can be observed with large vesicular structures formed by surfactant multi-layers. They can be visualised as a series of onion skins around a central cavity. These entities possess some of the features of surfactant systems and some of those seen with colloidal particulates. Equally, polymer molecules can be lightly crosslinked to form microgels, a molecular cluster with some particle-like properties. Colloids cover a broad spectrum of behaviour.

1.4 Colloid Frontiers

The ubiquitous nature of colloid science is such that it is inevitable that we stray from the colloid path and delve into other fields. Thus it is useful for the colloid scientist to know when to think about their system in a different way. As this chapter draws to a close it is appropriate to end where we began and consider the boundaries of the colloidal range.

The calculations using the approach of Hamaker show that van der Waals forces will tend to cause two objects to 'stick together'. The level of attraction on close approach is extremely large. This would suggest that separating this book, assuming you are reading it in paper form, from its resting surface would require some considerable force. In reality other features become significant. The surface roughness, which would be of at least colloidal dimensions, reduces the number of points of contact, weakening the attraction. At the smaller length scale the atomic repulsions and the nature of the interaction would weaken the attraction too. The important forces are also different on the macroscopic length scale; we are less concerned about the Brownian motion of the book and more about kinematics and inertia. The same influences will apply to particles provided they are large enough. There are several ways in which we might consider this boundary between the macroscopic and microscopic.

A particle moving through a fluid experiences a hydrodynamic drag. At low velocities the Stokes drag is dominant but as the velocity of the particle increases the inertia forces become increasingly significant. The balance of the viscous drag to the inertial forces defines the *Reynolds number Re*, for the *particle*.

$$Re = \frac{2\rho_p a v}{\eta_0} \tag{1.30}$$

where v is the velocity. Once Re exceeds about 0.1 the drag is no longer purely Stokesian in form and inertia begins to become increasingly significant. It occurs most readily for large dense particles. This is a hydrodynamic boundary where we can consider a characteristic type of colloidal behaviour is lost.

Another boundary definition concerns the transition between capillary and colloid pair interaction forces such as that described by DLVO. As an example consider a hydrated floc of hydrophilic particles dispersed in a hydrophobic medium (Figure 1.9). The floc is made up of a collection of particles with a van der Waals force acting between them. If the particles are damp enough and large enough a water bridge also exists between them, creating a region of fluid curvature.

This arises from the Laplace pressure (15), Δp either side of the oil water interface. We can express this in the more general Young–Laplace form.

$$\Delta p = \gamma_{OW}\left(\frac{1}{r_1} + \frac{1}{r_2}\right) \tag{1.31}$$

where γ_{OW} is the interfacial tension between the oil and water and r_1 and r_2 are the principle radii of curvature of the surface. In order to disrupt the flocs it is necessary to overcome both

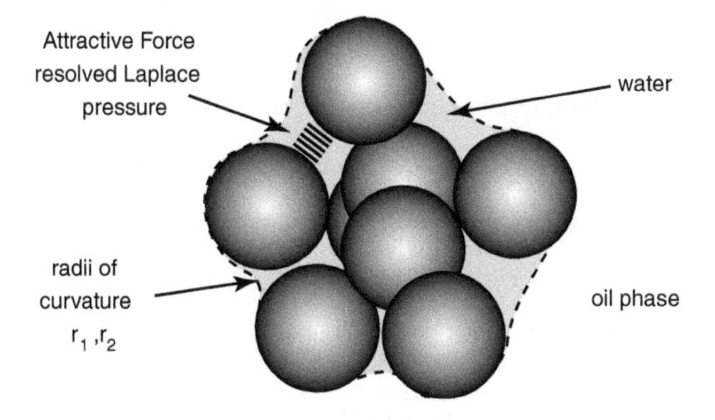

Figure 1.9 *A schematic showing the aqueous bridges and resolved interaction forces in a hydrated aggregate*

the capillary and van der Waals forces. This is when surfactants can aid dispersion as they can 'wet out' the flocs by lowering interfacial tension and lubricating the motion of one particle past another. Capillary forces are large and significant for large particles; grains of sand on a beach are more effectively held together by the forces associated with wetting and capillary action than a resolved pair interaction potential (16). The boundary between microscopic pair interactions and the dimensions where wetting forces are important is moot and an area of active research. We can estimate that this and the role of inertia become significant at a few tens of micrometres in size.

When we examine the other extreme, the small particle limit, we are examining the transition region between the microscopic and atomisitic. A rather arbitrary boundary, as all these boundaries tend to be, can be defined in terms of the number of atoms that sit at the surface of the particle relative to the number of molecules in the bulk of the particle. Suppose we consider a small ball-like compact molecule such as citric acid precipitating out to form colloidal particles. Depending upon the assumptions we make, we can perform some simple calculations on this system. For a 20 nm diameter particle that is composed of a little more than 3000 molecules we find that 14 % of these molecules are at the surface. At 2 nm we have a particle made of a few tens of molecules and all are in direct contact with the surface. Somewhere in the region of these two sizes the bulk properties of the material are influenced by the particle dimensions. For example, size restrictions for semiconducting particles can lead to shifts in their quantum states, changes in electronic states and optical properties. Even at larger diameters changes in behaviour can be seen. For example as the particle size of a crystalline solid decreases, below its typical bulk grain boundary dimensions, a transition occurs. Larger polycrystalline colloids of the same material can show differing dielectric properties to their small single crystal counterparts. In this region, whilst we may consider our system as possessing colloidal properties, care must be taken over the application of colloidal ideas. Ultimately the good colloid scientist is less interested in labels and more interested in tackling the challenges and using the tools appropriate for the task. The rest of this text lays some of the foundations to enable the reader to do just that.

References

Historical References

(1) Overbeek, J. Th. G. (1982) Colloidal Dispersions, ed. Goodwin, J. W. The Royal Society of Chemistry, London, p. 1.
(2) Selmi, F. (1845) Nuovi Ann. Science Natur. Di Bologna 2: 24, 225.
(3) Faraday, M. (1857) Phil. Trans. Roy. Soc. 147: 145.
(4) Smoluchowski, M. (1906) Zur kinetischen Theorie der Brownschen Molekularbewegung und der Suspensionen. Annalen der Physik 21: 756–780.
(5) Einstein, A. (1956) Investigations on the Theory of Brownian Movement. Dover, New York. ISBN 0-486-60304-0.
(6) Siedentopf, H. Zsigmondy, R. (1903) Ann. Physik 4: 10.
(7) Perrin, J. (1909) Ann. de Chem. et de Phys. 8: 18.
(8) Hamaker, H. C. (1937) Physica 4: 1058.
(9) Derjaguin, B. V., Landau, L. D. (1944) Acta Physiochim. URSS 14: 633–662;Verwey, E. J. W. Overbeek, J. Th. G. (1948) Theory of the Stability of Lyophobic Colloids. Elsevier, Holland: (2000) Dover reprint.

(10) McBain, J. W. (1913) Trans. Farad. Soc. 9: 99; McBain, J. W. (1944) Colloid Chemistry, Vol. 5, ed. Alexander, J. Reinhold, New York, p. 102.
(11) Ottewill, R. H., Rennie, A. R.(eds.) (1992) Modern Aspects of Colloidal Dispersions. Kluwer Academic Publishers, Netherlands.
(12) Goodwin, J. W., Hughes, R. W., Partridge, S. J., Zukoski, C. F. (1986) J. Chem. Phys. 85: 559.
(13) Vrij, A., Jansen, J. W., Dhont, J. K. G., Pathmamanoharan, C. Kops-Werkhoven, M. M., Fijnaut, H. M. (1983) Farad. Disc. 76: 19.
(14) Ostwald, W. O. (1917) An Introduction to Theoretical and Applied Colloid Chemistry. John Wiley & Sons, Ltd, New York.
(15) Pierre-Simon, Marquis de Laplace (1806) Mécanique Céleste, Supplement to the 10th edition.
(16) Rumpf, H. (1962) The Strength of Granules and Agglomerates. Agglomeration. Knepper, W. A. Ed. AIME, Interscience, New York, p. 379.

Background Introductory Texts

Goodwin, J. (2004) Colloids and Interfaces with Surfactants and Polymers: An Introduction. Wiley-VCH.
Pashley, R. M. (2004) Applied Colloid and Surface Chemistry, John Wiley & Sons, Ltd.
Norde, W. (2003) Colloids and Interfaces in Life Sciences, Marcel Dekker Inc.
Lyklema, J. H. (2005) Fundamentals of Interface Colloid Science. Academic Press, Vols 1–4.
Hunter, R. J. (2000) Foundations of Colloid Science, Oxford University Press.
Evans, D. F. Wennestrom, H. (1998) The Colloidal Domain: Where Physics, Chemistry and Biology Meet. John Wiley & Sons, Ltd.
Morrison, I. D., Ross, S. (2002) Colloidal Dispersions: Suspensions, Emulsions and Foams. John Wiley & Sons, Ltd.

2

Charge in Colloidal Systems

David Fermin[a] *and Jason Riley*[b]

[a]*School of Chemistry, University of Bristol, UK*
[b]*Department of Materials, Imperial College London, UK*

2.1 Introduction

Particles in any suspension experience attractive van der Waals interactions which can promote reversible or irreversible aggregation. To prepare stable colloidal suspensions it is necessary to introduce interactions between particles that oppose the van der Waals attraction. One method of achieving this is to charge the particles, the surface charge resulting in a repulsive interparticle force (1). Methods of charging particles can be classified under four broad headings: ionisation of surface groups; ion adsorption; non-symmetric ion dissolution; and isomorphous ion substitution. In an electrolyte the solvated ions surround the particles and shield their surface charge. The distribution of counter-ions in the vicinity of a charged surface may be described using Stern–Gouy–Chapman theory in which the potential at the surface is dropped across two layers; a compact inner layer and a diffuse outer layer. As two particles approach each other their diffuse layers will overlap and the resultant repulsive force may outweigh the attractive van der Waals attraction, thus rendering the suspension stable. The distribution of ions in the diffuse layer is dependent upon the concentration of the electrolyte, the formal charge of the ions, the solvent and the potential at the boundary between the compact inner layer of ions and the diffuse outer layer of ions. The potential at this interface is often equated to the zeta (ζ) potential; that is the potential at the shear plane between particle and solvent under flow. The ζ potential is determined in electrokinetic experiments; techniques in which relationships between the current or voltage and the relative flow of the two phases in the suspension are measured.

Colloid Science: Principles, methods and applications, Second Edition Edited by Terence Cosgrove

Hence, the scientist wishing to prepare stable colloids must have an appreciation of the origins of surface charge, understand how ions are distributed at the particle/liquid interface and be familiar with methods of measuring the ζ potential.

2.2 The Origin of Surface Charge

Many methods of preparing colloidal suspensions yield particles that possess charge at their surface. The surface charge may be further modified by altering the environment, for example changing the pH or adding an ionic surfactant. There are four generic mechanisms by which a surface immersed in a liquid may attain a charge; these are summarised in Figure 2.1.

2.2.1 Ionisation of Surface Groups

Particles that possess suitable chemical functionality can become charged as a result of the ionisation of surface groups. In aqueous solutions pH is commonly used to control the degree and nature of the ionisation. For example, metal oxides may become charged as a result of the protonation or deprotonation of surface groups. The pH of the isoelectric point of titania is 5.8, i.e. at a pH of 5.8 the ζ potential of titania is zero. At a pH less than 5.8 the titania is positively charged:

$$\mathrm{Ti{-}OH + H^+ \rightarrow Ti{-}OH_2{}^+}$$

and at a pH greater than 5.8 the oxide is negatively charged:

$$\mathrm{Ti{-}OH + OH^- \rightarrow Ti{-}O^- + H_2O}$$

Similarly, proteins may acquire charge as a result of the ionisation of carboxyl and amino groups to $-COO^-$ and $-NH_3{}^+$ respectively.

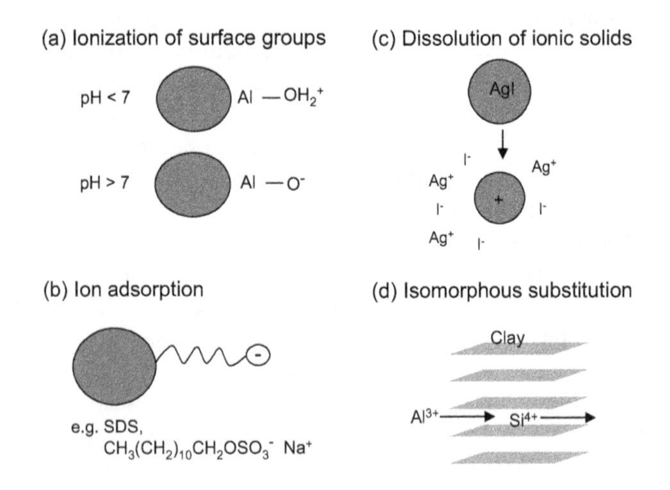

Figure 2.1 *The methods of charging a solid surface immersed in electrolyte*

2.2.2 Ion Adsorption

If the bulk material cannot be ionised, ionic surfactants may be added to generate charge-stabilised suspensions. For example, particles of carbon black on which anionic surfactants are adsorbed may be suspended in water. This is the basis of inks, with the proprietary mixture of surfactants used termed a dispersant. Increases in pH may result in protonation of the anionic surfactants of the dispersant and lead to instability of the inks.

2.2.3 Dissolution of Ionic Solids

Silver halide sols underpin photographic film technology and have played an important role in the development of our understanding of charged colloids. Silver halides are sparingly soluble salts, for example, the solubility product of silver iodide ($K_{SP} = a_{Ag^+} a_{I^-}$) in pure water is 8.5×10^{-17}. If the dissolution of Ag^+ and I^- ions are unequal then at equilibrium the sol will contain charged silver iodide particles. Thus, in the presence of excess iodide ions the particles will be negatively charged and with excess silver ions positively charged particles will be obtained.

2.2.4 Isomorphous Substitution

The replacement of one atom by another of similar size in a crystal lattice is termed isomorphous substitution. Clays are naturally occurring colloidal particles of size less than 2 μm. The building blocks used to construct the wide variety of clays found in nature are silica and alumina sheets. The silica sheets consist of linked silicon–oxygen tetrahedra with four oxygen atoms surrounding each central silicon. The alumina sheets are formed from octahedra, each central aluminium being surrounded by six oxygens or hydroxyls. In simple clays such as kaolinite, a sheet of alumina octahedra shares apical oxygens with a sheet of silica tetrahedra. The tetrahedra–octahedra bilayers are then bound together by a combination of van der Waals forces and hydrogen bonding. Clays become charged as a result of amorphous substitution. The substitution of Si^{4+} by Al^{3+} in the tetrahedral layers or of Al^{3+} by Mg^{2+}, Zn^{2+} or Fe^{2+} in the octahedral layers leads to a net negative charge.

2.2.5 Potential Determining Ions

When discussing the influence of ions on colloid stability, it is important to differentiate between inert ions and potential-determining ions. *Potential-determining ions are species which by virtue of their electron distribution between the solid and liquid phase determine the difference in potential between these phases* (IUPAC). The *potential*, sometimes termed the *inner-potential*, of a phase determines the electrical work done in taking a test charge from infinity to a point inside the phase. In the case of silver iodide sols, described above, it is apparent that the surface charge density, and hence the potential, changes on addition of silver or iodide ions. Hence, for this system Ag^+ and I^- ions are potential-determining ions. It is of note that with respect to metal oxides, protons are potential-determining ions and thus a change in pH will result in a change in the surface charge. Inert ions do not change the charge density at the surface of particles but may influence the interfacial potential difference by virtue of their local distribution.

2.3 The Electrochemical Double Layer

If a positively charged surface is placed in an electrolyte containing inert ions then simple electrostatics indicates that cations will be repelled from and anions attracted to the interface. Electroneutrality will be attained when the electrolyte layer near the interface possesses a net negative charge of equal magnitude to the surface charge of the solid material. The structure of the charged atmosphere of electrolyte, commonly known as the *electrochemical double layer*, is characterised by a potential drop across the solid/liquid interface which is dependent on the concentration and nature of the ionic species.

2.3.1 The Stern–Gouy–Chapman (SGC) Model of the Double Layer

The classical description of the ionic distribution in the vicinity of a charged surface underpinning the SGC model is depicted in Figure 2.2. This model considers the solvent as a

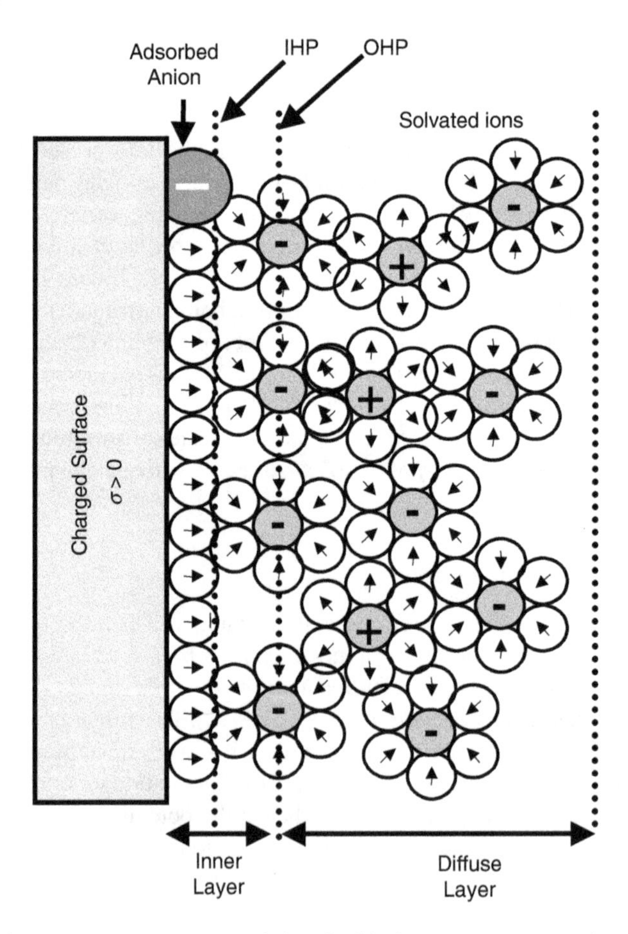

Figure 2.2 *Schematic representation of the double layer structure at the solid/electrolyte interface according to the SGC model*

dielectric continuum and the ionic species as non-interacting point charges. In the absence of ionic species *in direct contact* with the surface, the surface charge is effectively counterbalanced by ions located across two distinctive regions, the *Helmholtz* and *diffuse layers*. As discussed in this section, the contribution of each of the layers to the total interfacial charge density is dependent on the concentration of the ionic species in solution as well as the relative permittivity of the solvent.

The Helmholtz layer contains two planes commonly referred to as the inner (IHP) and outer Helmholtz planes (OHP). The former corresponds to the plane in which specifically adsorbed species are located as well as solvent molecules in direct contact with the surface. Specifically adsorbed species are those molecules which possess a solvation layer that is affected by the interaction with the surface. The OHP defines the plane of closest approach of the fully solvated ions (non-specifically adsorbed). Ionic species located in this layer are not considered in thermal motion, therefore the stored charge in this layer is only dependent on the structure and dielectric properties of the medium in contact with the charged surface. In the absence of specifically adsorbed ions, the potential drop across the Helmholtz layer ($\Delta\phi_H$) is determined by the charge density at the OHP (σ_H) and the distance between the OHP and the surface (d_H),

$$\Delta\phi_H = \frac{\sigma_H d_H}{\varepsilon\varepsilon_0} \tag{2.1}$$

where ε is the relative permittivity of the solvent and ε_0 the permittivity of free space. The opposing charges at the surface and OHP can be viewed as two plates of a capacitor of capacitance C_H,

$$C_H = \frac{\varepsilon\varepsilon_0}{d_H} \tag{2.2}$$

It is also useful to define the capacitance as the derivative of the charge with respect to the potential difference (*differential capacitance*). To a first approximation, the capacitance of the Helmholtz layer is independent of the applied potential. However, experimental analysis described in the next section demonstrates that this approximation is not strictly valid at the metal/electrolyte interface. This is essentially due to the fact that the dipoles as well as the structure of the hydration layer within the IHP are affected by the magnitude of the interfacial electric field (2).

Beyond the OHP, the surface charge is counterbalanced by a dynamic ionic atmosphere commonly referred to as the diffuse or Gouy-Chapman layer. Considering an infinite flat plane with a charge σ_{OHP}, the concentration of the ionic species 'i' (c_i) along the axis perpendicular to the surface (x) can be described in terms of the Boltzmann distribution:

$$c_i(x) = c_i(\text{aq})\exp\left(-\frac{z_i e[\phi(x)-\phi(\text{aq})]}{k_B T}\right) \tag{2.3}$$

where the index 'aq' denotes the value at the bulk of the electrolyte solution, z_i is the ionic charge and e is the elementary charge. In order to correlate the electrical potential at a distance x from the OHP with the local charge density (ρ_e), the Poisson equation can be invoked:

$$\rho_e(x) = -\varepsilon\varepsilon_0\left(\frac{d^2\phi}{dx^2}\right) = \sum_i \frac{z_i e}{N_A} c_i(x) \tag{2.4}$$

Combining equations (2.3) and (2.4) for a symmetric electrolyte ($z_{anions} = z_{cations} = z$), it follows:

$$\tanh\left[\frac{F}{4RT}(\phi(x)-\phi(aq))\right] = \exp(-\kappa x)\tanh\left[\frac{F}{4RT}(\phi(OHP)-\phi(aq))\right] \tag{2.5}$$

where the parameter κ is the reciprocal of the *Debye length*:

$$\kappa = \left(\frac{2z^2e^2N_A c(aq)}{\varepsilon\varepsilon_0 k_B T}\right)^{\frac{1}{2}} \tag{2.6}$$

Assuming $ze(\phi(OHP)-\phi(aq)) \ll k_B T$, Equation (2.5) can be simplified to a single exponential decay:

$$\phi(x) = (\phi(OHP)-\phi(aq))\exp(-\kappa x) = \Delta_{aq}^{OHP}\phi\exp(-\kappa x) \tag{2.7}$$

The potential distribution across the diffuse layer as evaluated from Equations (2.5) and (2.7) are compared in Figure 2.3a. The so-called Debye–Huckel approximation (Equation 2.7) underestimates to a certain extent the potential profile in comparison to the analytical solution (Equation 2.5). The difference in both solutions is slightly more significant as the bulk electrolyte concentration decreases. However, Equation (2.7) does give a realistic potential decay for the magnitude of the surface potentials typically encountered in colloid science.

The Debye lengths for various concentrations of z:z aqueous electrolyte at 25 °C (as estimated from Equation 2.6) are summarised in Table 2.1. This parameter plays a key role in the stability of colloids in solution as it reflects the electrostatic repulsion associated with

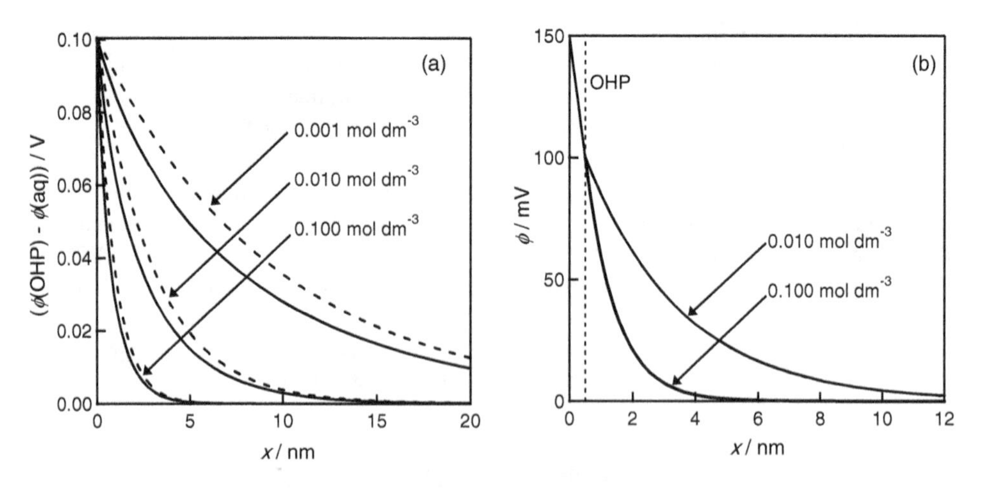

Figure 2.3 Electrostatic potential distribution across the diffuse layer (a) and the overall interfacial region (b) for various electrolyte concentrations

Table 2.1 *The dependence of the Debye length on electrolyte concentration*

C(aq)/(mol dm^{-3})	Debye length for z^+:z^- electrolyte/nm			
	1:1	1:2/2:1	2:2	1:3/3:1
10^{-1}	1	0.6	0.5	0.4
10^{-2}	3	1.8	1.5	1.2
10^{-3}	10	5.6	4.8	3.9
10^{-4}	30	18	15	12
10^{-5}	100			

overlapping diffuse layers. As the concentration of the electrolyte solution increases, the decay of the electrostatic potential is sharper. As a consequence, the colloidal particles will approach each other closer at higher electrolyte concentration. This behaviour becomes more pronounced as the charge of the ionic species (z) increases.

The overall interfacial potential drop including the Helmholtz and diffuse layers is illustrated in Figure 2.3b. As there are no ions present between the OHP and the metal surface, the electrostatic potential falls linearly in this region. From the OHP, the potential decreases exponentially towards the bulk of the electrolyte solution. The charge associated with the diffuse layer (σ_d) can be obtained from Equations (2.4) and (2.5), yielding:

$$\sigma_d = \frac{2c(\text{aq})zeN_A}{\kappa}\sinh\left(\frac{ze\Delta_{\text{aq}}^{\text{OHP}}\phi}{2k_BT}\right) \tag{2.8}$$

While the charge stored at the Helmholtz layer (σ_H) has a linear dependence with the potential drop across this layer, σ_d exhibits a more complex dependence on the electrostatic potential as a result of the ionic distribution in the diffuse layer. As discussed for the Helmholtz layer, the capacitance of the diffuse layer (C_d) can be expressed in terms of σ_d:

$$C_d = \frac{2c(\text{aq})z^2e^2N_A}{\kappa k_BT}\cosh\left(\frac{ze\Delta_{\text{aq}}^{\text{OHP}}\phi}{2k_BT}\right) \tag{2.9}$$

The potential dependence of C_d for various electrolyte concentrations is illustrated in Figure 2.4. The minimum in the parabolic capacitance voltage curves corresponds to the potential at which the charge at the OHP and the diffuse layer is effectively zero. In the absence of specifically adsorbed ions, this potential is defined as the *potential of zero charge* (pzc). The decrease in the Debye length with increasing concentration of the electrolyte (Figure 2.3 and Table 2.1) manifests itself with the increase in C_d.

Considering that the total capacitance (C_T) of the double layer can be defined as a series combination of the Helmholtz and diffuse layer capacitances, it follows:

$$C_T = \frac{C_dC_H}{C_d + C_H} \tag{2.10}$$

According to Equation (2.10), the smallest capacitance will have the largest contribution to C_T. Figure 2.5 displays the total capacitance as a function of the potential across the interface

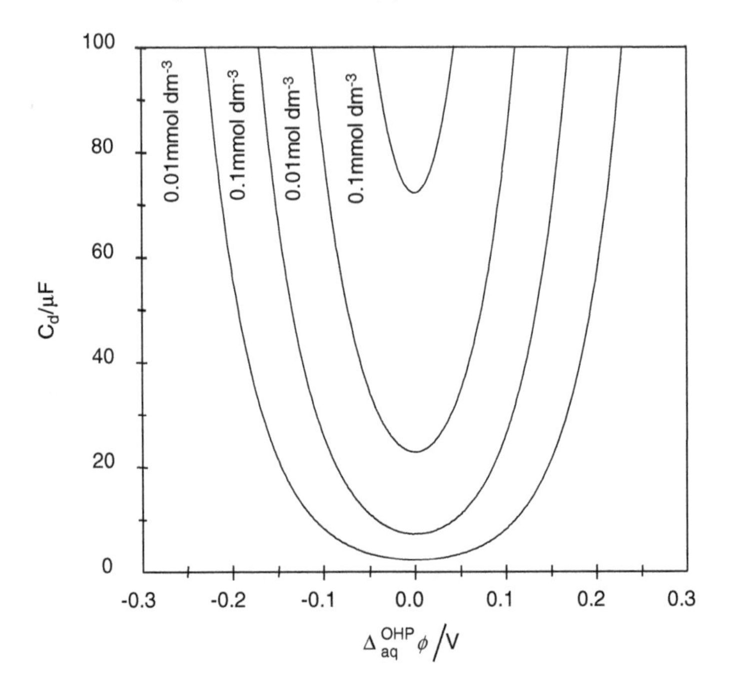

Figure 2.4 *Diffuse layer capacitance as a function of the potential difference across the diffuse layer for various concentrations of electrolyte in aqueous solution at room temperature*

($\Delta\phi$) and aqueous electrolyte concentration. Taking $C_H = 20\,\mu\text{F cm}^{-2}$, it can be seen that the contribution of C_d is limited to potentials close to the pzc. The advantage of expressing the potential distribution across the double layer in terms of the differential capacitance will become clear when discussing experimental observations at mercury electrodes.

2.3.2 The Double Layer at the Hg/Electrolyte Interface

The relationship between surface charge, interfacial potential difference and electrolyte composition can be experimentally investigated at the metal/electrolyte interface. Historically, the mercury electrode in aqueous electrolyte has been considered a model metal/electrolyte system; the interface between the two liquids being atomically smooth and the continual renewal of surface minimising contamination. Further, and perhaps most importantly, the mercury/aqueous electrolyte interface is ideally polarisable, i.e. charge carriers are not transferred across the interface, over a wide range of potential differences. Hence, electrochemical studies at this interface have been central to our current understanding of the double layer.

The two-electrode assembly typically employed for studying the relationship between surface charge and interfacial potential is shown in Figure 2.6, where ϕ_i is the potential of phase i. The experimentally applied potential, $E(\phi_{Cu1} - \phi_{Cu2})$, is partitioned across all the interfaces of the system:

$$E = (\phi_{Cu1} - \phi_{Hg}) + (\phi_{Hg} - \phi_{aq}) + (\phi_{aq} - \phi_{s,ref}) + (\phi_{s,ref} - \phi_{ref}) + (\phi_{ref} - \phi_{Cu2}) \qquad (2.11)$$

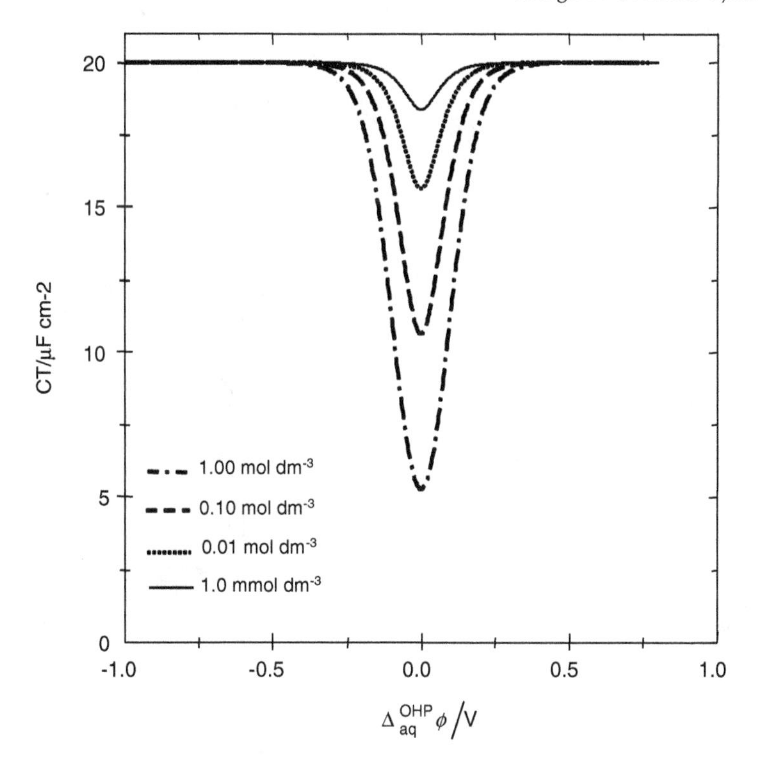

Figure 2.5 *The total differential capacitance of the double layer as a function of the potential difference across the interface for various concentrations of electrolyte in aqueous solution at room temperature*

Careful experimental design ensures that as the applied potential is altered the potential drop across the reference electrode remains constant and changes to both liquid junction potentials and metal junction potentials are negligible. Hence, for the simple experimental apparatus displayed in Figure 2.6 a change to the applied potential of magnitude ΔE results in an equivalent change in the potential difference across the mercury/electrolyte interface, i.e.,

$$\Delta E = \Delta(\phi_{\mathrm{Hg}} - \phi_{\mathrm{aq}}) = \Delta_{\mathrm{aq}}^{\mathrm{Hg}}\phi \tag{2.12}$$

The surface charge at a particular applied potential can be obtained by measuring either the surface tension of the mercury drop or the integrated capacitance. For the dropping mercury electrode system shown in Figure 2.6, the surface tension can be calculated from the time required for a drop of maximum size to form, t_{max}. Immediately prior to the instant at which the mercury drop detaches from the capillary the forces of gravity and surface tension are equal, i.e.,

$$m_{\mathrm{Hg}} g t_{\mathrm{max}} = 2\pi r_{\mathrm{c}} \gamma \tag{2.13}$$

where m_{Hg} is the mass flow rate of mercury, g is the standard acceleration of gravity, r_{c} is the radius of the capillary and γ the mercury surface tension. γ is related to the surface charge

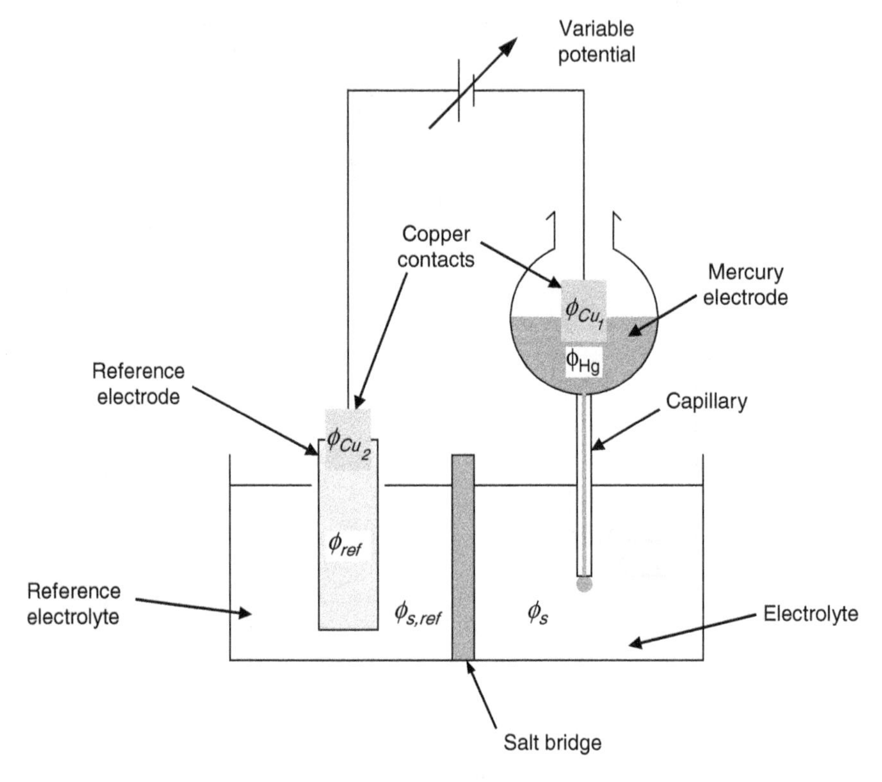

Figure 2.6 *The dropping mercury electrode*

density of the metal, σ_m, by the Lippman equation:

$$\sigma_m = -\frac{d\gamma}{dE} \tag{2.14}$$

which is derived by equating the work done charging the electrode, σdE, to the change in surface free energy, $-d\gamma$. As mentioned in the previous section, the derivative of the surface charge with the applied potential corresponds to the total capacitance at the double layer (C_T). At the Hg/electrolyte interface, C_T can be obtained with a high degree of accuracy employing techniques such as electrochemical impedance spectroscopy at a hanging mercury drop electrode (4). It follows that the relationship between surface tension, surface charge and differential capacitance is:

$$-\frac{d^2\gamma}{dE^2} = \frac{d\sigma_m}{dE} = C_T \tag{2.15}$$

Equation (2.15) establishes that the interfacial charge can be estimated by integration of C_T over a potential range, taking the charge as zero at the minimum of C_t (pzc). In addition, the surface tension will exhibit a maximum at the pzc.

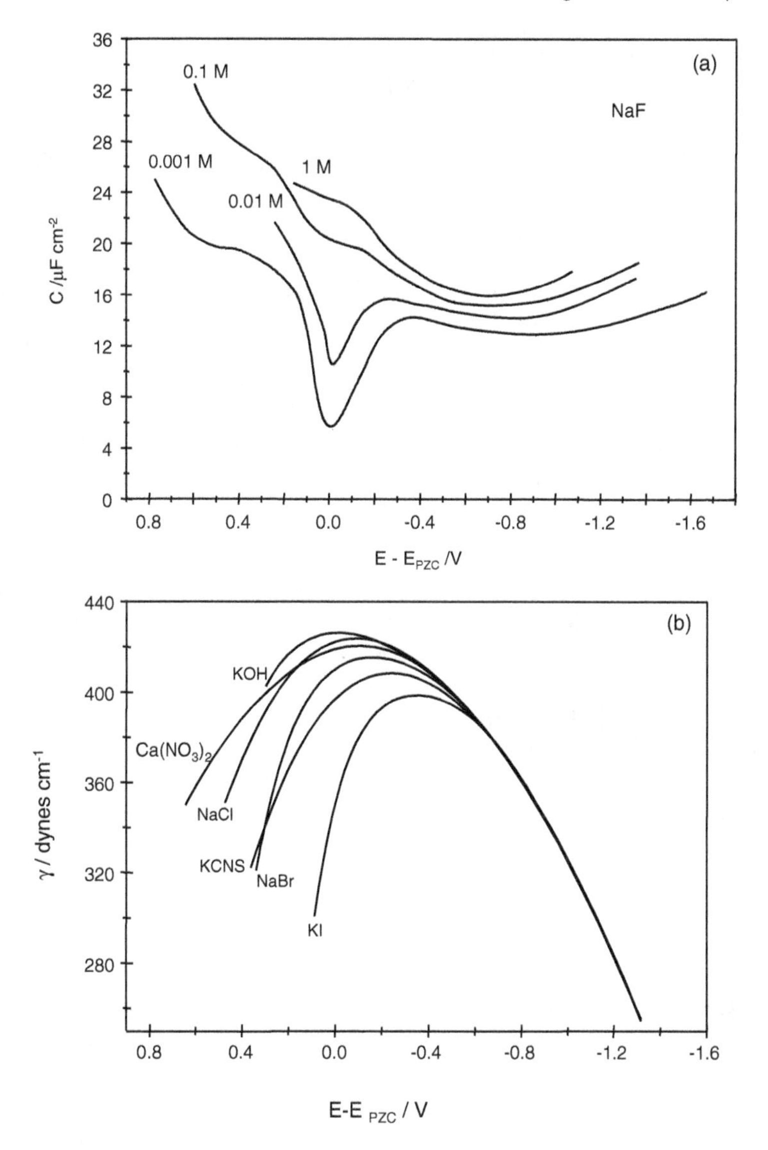

Figure 2.7 *The differential capacitance (a) and surface tension (b) of the Hg electrode versus applied potential for various electrolytes. (Reprinted with permission from Ref. (3)). Copyright (1947) American Chemical Society*

Figure 2.7 shows the potential dependence of the capacitance and surface tension at the mercury electrode in the presence of various electrolyte solutions (3). The key points to note from the experimental data are:

(a) at low electrolyte concentration the capacitance displays a minimum at the pzc,
(b) at high electrolyte concentrations or high potential C_T shows minimal dependence on potential,

(c) the surface tension goes through a maximum near the pzc
(d) at positive potentials with respect to the pzc the surface tension at a particular potential varies with ion type.

These observations are essentially consistent with the SGC model of the potential distribution across the electrochemical double layer. The sharp potential dependence of the capacitance around the pzc at low concentrations of NaF indicates that C_d (Equation 2.9) has the strongest contribution to C_T. On the other hand, C_H (Equation 2.2) dominates the total capacitance far from the pzc and at high electrolyte concentrations. The main qualitative differences between Figures 2.5 and 2.7a arises from a variety of aspects including ion–ion interactions, interactions between the solvent molecules and the charged surface and so forth (2,5). It can also be seen in Figure 2.7b that the maximum of the surface tension is affected by the nature of the anion present. As mentioned in the next section, the specific interaction of anions with the charged surface can significantly affect the potential distribution across the interface.

2.3.3 Specific Adsorption

In Section 2.3.1, it was considered that the plane of closest approach of the solvated ions is the OHP. However, some ions, as a result of favourable interactions with the surface, may lose all or some of the molecules in the solvation shell and hence move towards the inner Helmholtz plane (IHP). The IHP is formerly defined as the average distance of closest approach of the specifically adsorbed ions.

It is generally considered that sodium and fluoride ions do not specifically adsorb at the mercury electrode. This is consistent with the fact that the potential of minimum C_T (pzc) in Figure 2.7a is effectively independent of the concentration of NaF. In Figure 2.7b, the limited dependence of γ with type of electrolyte at negative potentials indicates that the various cations do not exhibit a significant affinity to the mercury surface. On the other hand, larger anions do exhibit specific adsorption at the Hg surface at potentials positive to the pzc. The trends in Figure 2.7b show that the extent of specific anion adsorption not only depends on the properties of the anion but also on the applied potential. The thermodynamic relationships developed in Chapter 4 provide access to the surface excess of the ionic species at a given electrode potential from the concentration dependence of γ.

A schematic of cation adsorption at a silica interface is shown in Figure 2.8. It is worth noting that the adsorption of Al^{3+} ions is so favourable that overcompensation of the surface charge occurs and even though the silica is negatively charged it is necessary for the diffuse layer to contain an excess of negative ions. This charge overcompensation phenomenon can be conveniently used for the fabrication of multilayered materials at the nanoscale (6).

Two examples of the so-called electrostatic adsorption of charge colloids and polymers are displayed in Figure 2.9. Silicon surfaces spontaneously form a thin oxide layer which is negatively charged in aqueous solution at neutral pH (Figure 2.8). The exposure of the surface to a solution containing poly-L-lysine promotes the electrostatic adsorption of the polycation, generating a film less than 1 nm thick. Charge overcompensation self-limits the growth of the thin polycationic film. Negatively charged colloidal Au nanoparticles can be adsorbed on the poly-L-lysine film as shown in the scanning electron micrograph in

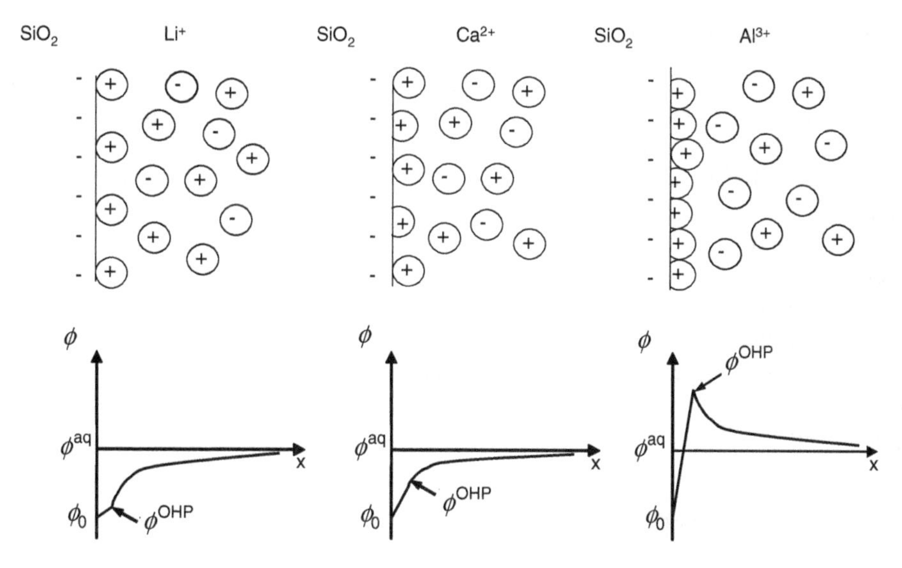

Figure 2.8 Schematics of the SiO_2 electrolyte interface in the presence of specifically adsorbing ions

Figure 2.9a. Electrostatic repulsion among the colloids ensures that the nanoparticles do not aggregate at the surface, allowing the spontaneous adsorption of a single layer of nanoparticles.

Two-dimensional films of DNA can be formed at mica surfaces in the presence of Mg^{2+} in saline solutions, as shown in Figure 2.9b. The topographic image obtained with an atomic force microscope (AFM) demonstrates that Mg^{2+} ions are able to overcompensate the negative charges at the mica surface, allowing the adsorption of DNA.

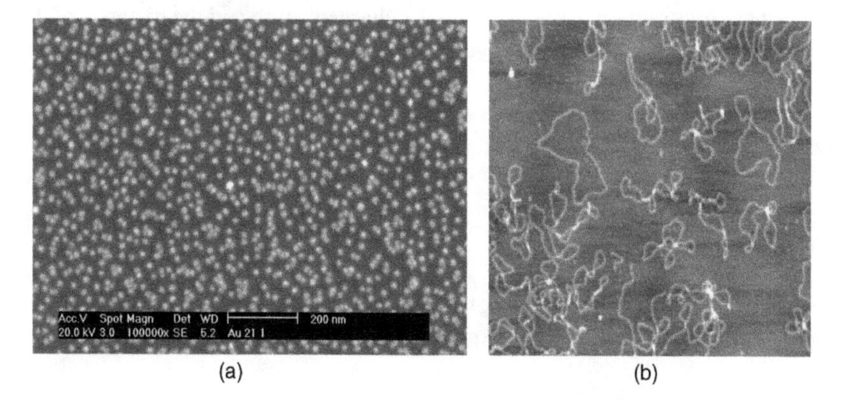

Figure 2.9 *Assembly of charged colloids and polymers employing electrostatic forces at charged surfaces. (a) Scanning electron micrograph of a 2D assembly of 20 nm Au nanoparticles electrostatically adsorbed at a SiO_2 surface modified with poly-L-lysine; (b) 2 μm × 2 μm AFM image of plasmid DNA adsorbed on mica in the presence of Mg^{2+} ions*

2.3.4 Interparticle Forces

A Hg electrode has been considered for illustrating the ion distribution at the interface between a charged phase and electrolyte. We now turn our attention back to charged colloidal particles. At a charged colloid there will be a non-uniform distribution of counter-ions; with specifically adsorbed ions, ions 'stuck' at the IHP and a diffuse layer in which counter-ions are in excess. The fact that the ionic atmosphere counter balances the colloid charge begs the question, 'how does the charging of the particles infer stability on a colloidal suspension?' This issue will be discussed in detail in Chapter 3. Here we note that as the particles approach each other repulsive interactions occur as their diffuse layers begin to overlap. The distribution of ions in the diffuse layer depends on $\Delta_{\text{aq}}^{\text{OHP}}\phi$ and the $c(\text{aq})$. This suggests that it is these two parameters that must be measured and controlled in order to formulate stable suspensions. The remainder of this chapter is concerned with methods of approximating $\Delta_{\text{aq}}^{\text{OHP}}\phi$.

2.4 Electrokinetic Properties

So far in describing ion distribution at charged interfaces it has been assumed that both bulk phases are static. As a result of the surface charge, interesting effects occur when one or both of the bulk phases are set in motion. The study of these phenomena is termed electrokinetics. In an electrokinetic experiment either; the solid phase is static and the electrolyte flows, the electrolyte is static and the solid moves or both phases are in motion. There are two experimental strategies in electrokinetic experiments, either the motion is controlled and an electrical property is measured, e.g. *streaming current and streaming potential* measurement, or the electrical field is controlled and motion monitored, e.g. *electro-osmotic, electrophoretic* and *electrosonic* experiments.

When discussing the relative motion of the two phases in electrokinetics it is useful to define a *shear plane*, the effective location of the solid/liquid interface. Above, the ions whose centre lies on the OHP were described as being 'stuck' to the interface. Under flow conditions these ions remain in contact with the charged phase and the distance of the shear plane from the phase boundary is an hydrated ion diameter, as shown in Figure 2.10. The *zeta potential*, ζ, is the potential at the shear plane and may be experimentally determined using electrokinetic methods. It is often assumed, even though the shear plane is at least an hydrated ion diameter from the interface and the outer Helmholtz plane is an hydrated ion radius, that ζ is equivalent to $\Delta_{\text{aq}}^{\text{OHP}}\phi$.

2.4.1 Electrolyte Flow

The Reynolds number, *Re*, for the flow of electrolyte, of density ρ and viscosity η, through a capillary of radius a is given by the expression

$$Re = \frac{2\rho v_{\text{m}} a}{\eta} \tag{2.16}$$

where v_{m} is the mean flow rate. If *Re* is less than *ca.* 1000 laminar flow will occur. A parabolic flow profile is established with the laminae in contact with the walls having zero

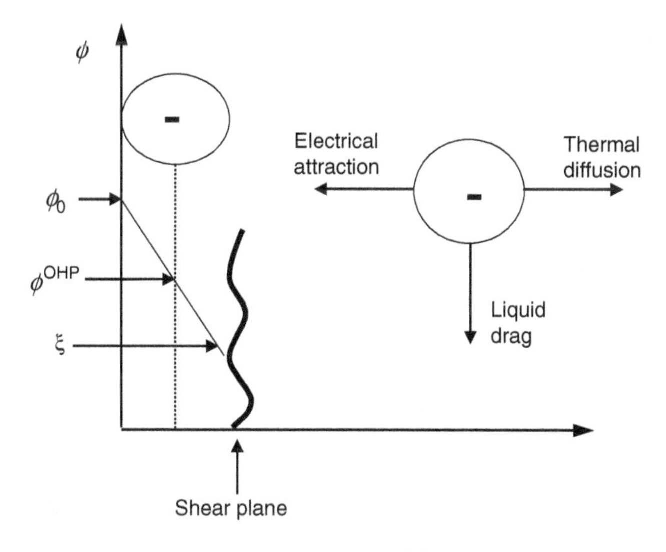

Figure 2.10 An illustration of the shear plane

velocity and that at the centre the maximum velocity, as shown in Figure 2.11. If the capillary walls are charged then whilst the ions at the OHP are stationary those in the diffuse layer will flow. As the diffuse layer possesses a net charge this flow leads to a *streaming current*. As the current is proportional to the charge of the diffuse layer, which in turn depends on ζ, a study of the streaming current as a function of flow rate allows ζ to be determined. To obtain expressions relating streaming current to flow rate the system may be reduced to a single lamina of charge σ_d flowing at a distance $1/\kappa$ from the interface. Although it is possible to measure streaming currents it is more common to monitor the *streaming potential*. The streaming potential is related to the streaming current by Ohm's law.

2.4.2 Streaming Potential Measurements

To measure a streaming potential the electrolyte is made to flow past the surface of interest, as shown in Figure 2.12. The streaming potential is measured using polarisable electrodes set perpendicular to the direction of flow. For materials of high dimension a capillary is

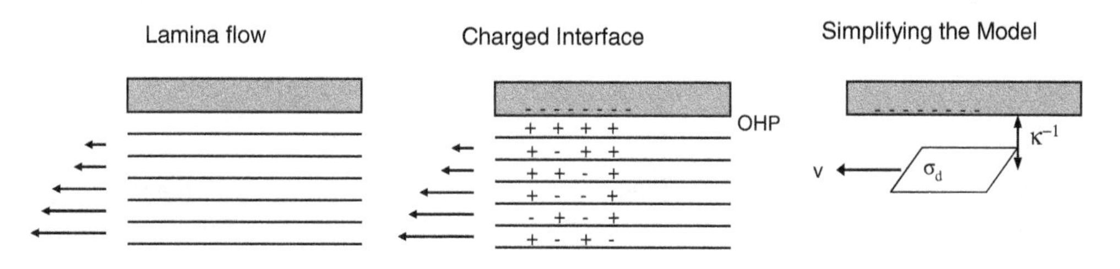

Figure 2.11 Charge transport under laminar flow conditions

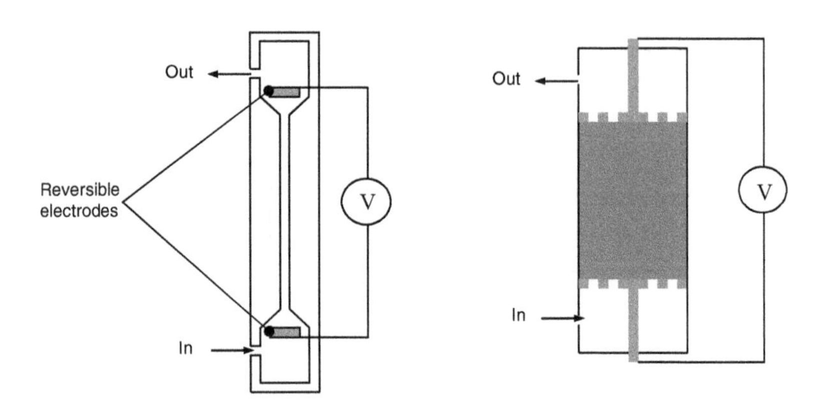

Figure 2.12 *Experimental apparatus for determining streaming potentials*

formed and the solution pumped through the cell. For materials of low dimension, e.g. powders and fibres, a cell consisting of porous electrodes may be employed. In such cells the streaming potential E_S is:

$$E_S = \frac{\varepsilon\varepsilon_0 \Delta p}{c(\mathrm{aq})\eta\Lambda}\zeta \quad (2.17)$$

where Δp, the pressure across the cell, and η, the electrolyte viscosity, describe the hydrodynamics and Λ, the molar conductivity of the solution, stems from application of Ohm's law.

2.4.3 Electro-osmosis

If flowing a solution through a capillary leads to a streaming current and a streaming potential then applying a potential E across a capillary should lead to a flow of solution. The phenomenon of solution flow when a potential is applied is termed *electro-osmosis* and is now often employed to pump solutions in microfluidic systems (7). Two experiments that employ electro-osmosis to determine the ζ potential of the surface of a capillary are show in Figure 2.13. In dynamic electro-osmosis measurements, Figure 2.13a, a potential is applied and the resultant solution flow monitored by measuring the velocity of the bubble. The ζ potential is calculated using the expression:

$$\zeta = v_b \frac{a_b^2}{a_c^2} \frac{i\eta\Lambda}{\varepsilon\varepsilon_0} \quad (2.18)$$

where v_b is the bubble velocity, a_b is the radius of the capillary containing the bubble, i is the current and a_c is the radius of the capillary across which the potential is applied. In equilibrium electro-osmosis measurements, Figure 2.13b, the pressure required to stop the

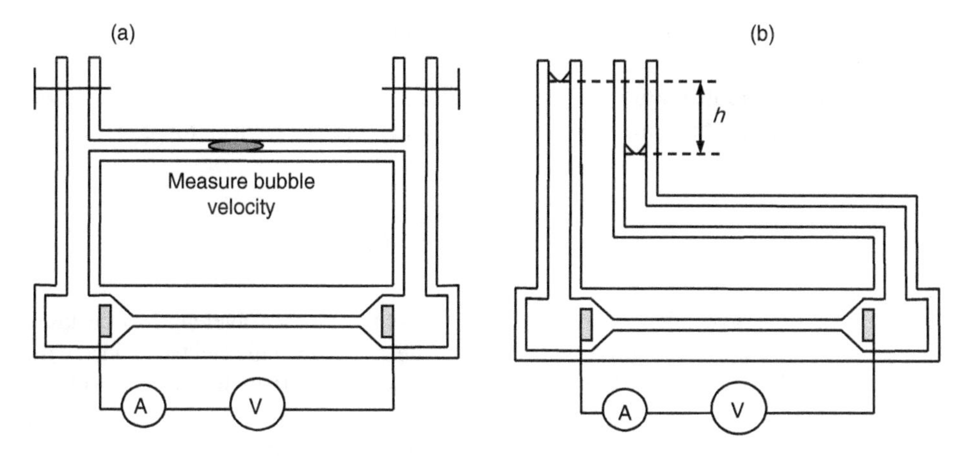

Figure 2.13 Experimental apparatus for determining the electro-osmotic pressure

potential induced flow is measured. The electro-osmotic pressure is related to the ζ potential by the expression:

$$\zeta = h\rho g \frac{i}{E^2} \frac{1}{8\pi\varepsilon\varepsilon_0\Lambda} \tag{2.19}$$

where ρ is the electrolyte density and g the standard acceleration of gravity.

2.4.4 Electrophoresis

In electrophoresis, the colloidal particles are the mobile phase and the electrolyte is stationary. Hence, electrophoretic methods are commonly employed when information on the charge of colloidal particles is required. Conceptually it is a very simple technique; optical methods are employed to monitor the velocity of the colloid particles, v_p, in an electric field. However, the fact that if the walls of a cell are charged fluid will flow when an electric field is applied means that maintaining a stationary electrolyte layer is experimentally challenging. This problem may be overcome by treating the walls of the electrophoresis cell to prevent charging. Alternatively, a closed cell is employed and the optical field is focused on particles in the stationary lamina, which is where the potential-induced flow along the cell walls is cancelled by the reverse flow of solution through the centre of the cell. The depth of the stationary solvent lamina is dependent on cell geometry: for a cylindrical cell it is at $0.146d$ and for a rectangular cell $0.2d$, where d is the cell diameter or depth. The velocity of some colloid particles may be measured directly under a microscope. However, techniques in which particle velocity is calculated by determining the change in light scattering when the particles move in an electric field are more commonly employed. Light scattering is discussed in detail in Chapter 12, here it is considered only in relation to laser Doppler electrophoresis (LDE) and phase analysis light scattering (PALS), techniques that may be employed to determine the velocity of particles in electric fields.

LDE (8) is based on the fact that when laser radiation of frequency ν_0 is scattered by a particle that is moving in an electric field the frequency of the scattered light ν_s is Doppler shifted. The magnitude of the Doppler shift $\delta\nu$ is directly proportional to the particles' velocity in the direction of the scattering vector. Hence, for quasi-elastic scattering:

$$\delta\nu = \frac{2nv_{\mathrm{p}}}{\lambda_0}\sin\left(\frac{\theta}{2}\right)\cos\varphi \tag{2.20}$$

where n is the refractive index of the suspension medium, λ_0 is the wavelength of the laser, θ is the scattering angle and φ is the angle between the direction of the electric field and the scattering vector. To determine the magnitude of the Doppler shift, which even for a highly charged mobile particle moving in a high electric field is only a few tens of Hz, the scattered beam is added to a reference beam. A schematic of the apparatus employed in LDE is displayed in Figure 2.14; note that the scattering vector $\boldsymbol{Q}$ lies along the direction of the electric field, i.e. the angle φ is zero. The photocurrent measured at the detector has an AC component at the beat frequency $|\nu_0 - \nu_s|$, thus the Doppler shift may be determined and the particle velocity calculated.

The maximum fields employed in LDE experiments are typically of the order of 10 V cm^{-1} and are square wave modulated at a frequency of 1 Hz. The form of the field is limited by electrode polarisation and Joule heating of the suspensions. If the charge on the colloidal particles is low then their velocity in the maximum applied electric field

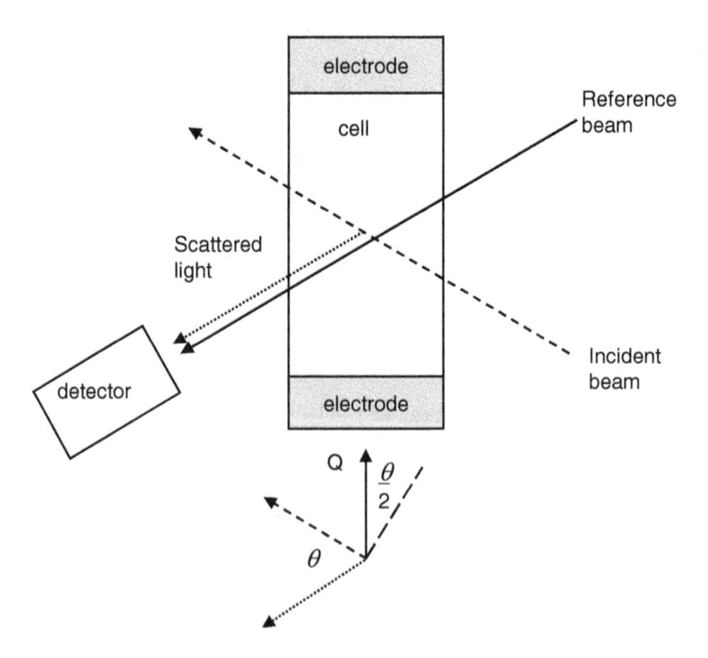

Figure 2.14 *A schematic of the principle features of the apparatus for laser Doppler electrophoresis*

will be small and, given that the electric field can only be applied for short periods, the Doppler shift in frequency may be immeasurable. Hence, when determining the velocity of particles that exhibit low ζ potential (e.g. in solutions of high ionic strength, suspensions in low polarity solvents or measurements near the isoelectric point), LDE may not have the sensitivity required. Similarly, the application of LDE to studies of highly viscous media is limited. To measure the velocity of slow-moving particles PALS (9) was developed. PALS is closely related to LDE but uses a more sophisticated method of comparing the reference and scattered beams and as a result can routinely measure mobilities a 1000 times lower than the limit of the LDE technique. A PALS experiment involves phase modulation of the reference beam and modulation of the electric field. In a simplified PALS experiment the reference beam is phase modulated at the frequency of the scattered beam in zero field. The phase shift between the scattered and reference beams are compared. When no field is applied the phase shift will be constant. When a field is applied the phase shift will change with time. Experimentally, the relative phase shift, the difference between the phase shift at time t and the constant phase shift in the absence of an electric field, is recorded as a function of time. The relative phase shift can be determined with high accuracy as it may be measured over a high number of cycles. Typically in a PALS experiment a sinusoidal field of frequency in the range 5 Hz–60 Hz is applied to the sol and the relative phase shift monitored (Figure 2.15). The amplitude of the sinusoidally modulated relative phase shift allows the velocity of the particles to be determined. Corrections for the drift velocity of the particles in zero field may be made. Using the PALS technique particle velocities

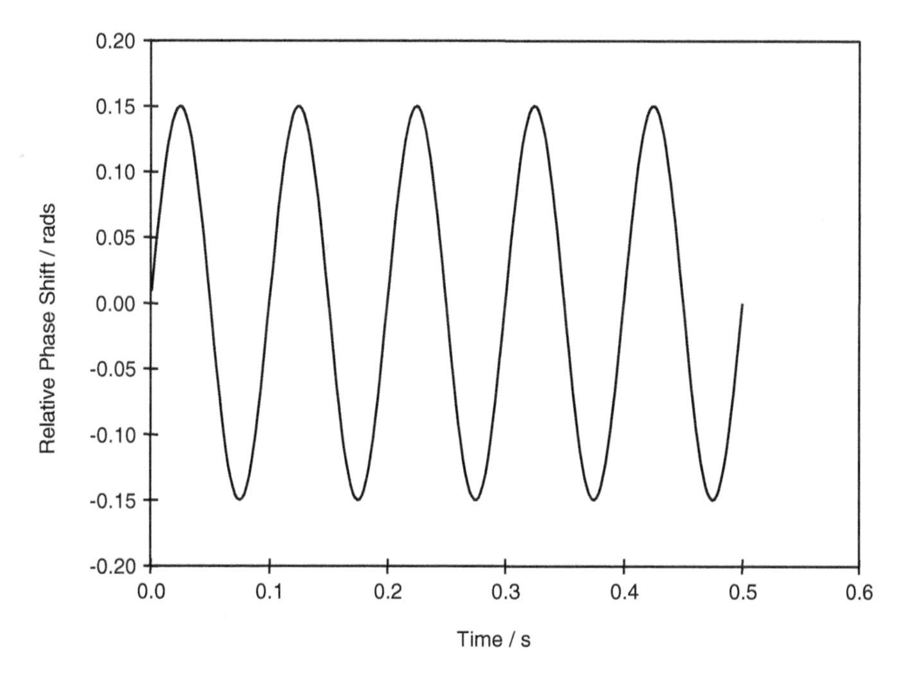

Figure 2.15 The raw data obtained in a PALS experiment when the applied electric field is modulated at a frequency of 10 Hz

equivalent to Doppler shifts as low as 0.001 Hz may be determined. Hence PALS is the method of choice to determine the velocity of particles in systems where the ζ potential is low or the viscosity is high.

To relate the velocity of the colloid particles to the electric field it is necessary to consider all the forces acting. The motion due to the electric field will be opposed by the viscous drag on the particle, the viscous drag on the ionic atmosphere and the electrostatic force that results from distortion of the diffuse layer. The relative importance of these forces depends on the dimensionless quantity κa which is the ratio of the radius of curvature of the particle to the double layer thickness. When κa is small ($\ll 1$) the charged particle may be treated as a point charge and the Huckel equation

$$\zeta = \frac{v_p}{E}\frac{3\eta}{2\varepsilon\varepsilon_0} \tag{2.21}$$

relates the particle velocity to the ζ potential. Generally, the Huckel equation is only valid for studies of particles suspended in non-aqueous media of low conductance. When κa is large the particle–electrolyte interface may be treated as a flat sheet and the Smoluchowski equation applies:

$$\zeta = \frac{v_p}{E}\frac{\eta}{\varepsilon\varepsilon_0} \tag{2.22}$$

2.4.5 Electroacoustic Technique

A final method of determining the ζ potential is to use electroacoustics (10). When a high frequency, approximately 1 MHz, AC field is applied to a colloidal sol both the particles and the diffuse layer are set in motion. As the inertia of particles differs to that of the diffuse layer, the velocity/field transfer function differs for the two components. Pressure waves result and from measurements of the sound the ζ potential may be determined. A notable advantage of the electroacoustic method is that it may be employed with optically dense systems where light-scattering techniques are not applicable.

References

(1) Attard, P. (2001) Recent advances in the electric double layer in colloid science. Curr. Opin. Colloid Interface Sci., 6: 366–371.
(2) Bockris, J. O'M., Reddy, A. K. N. (2000) Modern Electrochemistry, 2nd edn. Plenum Publishers, New York.
(3) Grahame, D. C. (1947) The electrical double layer and the theory of electrocapillarity. Chem. Rev., 41: 441–501.
(4) Bard, A. J., Faulkner, L. R. (2001) Electrochemical Methods: Fundamentals and Applications. John Willey & Sons, Ltd., New York.
(5) Delahay, P. (1965) Double Layer and Electrode Kinetics. Interscience Publishers, New York.
(6) Decher, G. (1997) Fuzzy nanoassemblies: toward layered polymeric multicomposites. Science, 277: 1232.
(7) Schasfoort, R. B. M., Schlautmann, S., Hendrikse, J., van der Berg, A. (1999) Field-effect fluid control for microfabricated fluidic networks. Science, 286: 942–945.

(8) Uzgiris, E. E. (1981) Laser Doppler methods in electrophoresis. Prog. Surface Sci., 10: 53–164.
(9) McNeil-Watson, F., Tscharnuter, W., Miller, J. (1998) A new instrument for the measurement of very small electrophoretic mobilities using phase analysis light scattering (PALS). Colloids Surfaces A, 140: 53–57.
(10) Hunter, R. J. (1998) Recent developments in the electroacoustic characterisation of colloidal suspensions and emulsions. Colloids Surfaces A, 141: 37–65.

3

Stability of Charge-stabilised Colloids

John Eastman

Learning Science Ltd, Bristol, UK

3.1 Introduction

One of the important aspects of the study of colloidal dispersions is understanding their stability so that we can manipulate the state of dispersions for particular applications.

Charge stabilisation is one means by which this can be achieved and we can manipulate the stability through changes in the chemical environment such as salt concentration, ion type and pH.

We must understand how this works both in quiescent systems and when external fields, such as gravitational or shear fields, are present.

So what do we mean by stability? Well, this very much depends on the circumstances which are being considered. We can define stability in colloidal systems either in terms of their tendency to aggregate or in terms of their tendency to sediment under the action of gravity.

In this chapter we will focus on the stability to aggregation and look at the factors which control this stability. We will study this by considering the interaction between two representative particles in the system. By considering what happens when two particles come together (during a Brownian collision) we can predict the stability of the whole system by looking at the form of the colloidal pair potential.

Colloid Science: Principles, methods and applications, Second Edition Edited by Terence Cosgrove

3.2 The Colloidal Pair Potential

The pair potential is the total potential energy of interaction between two colloidal particles as the separation or distance between them is varied. Formally it is a free energy, and we calculate it by simply summing the various components that we can identify.

The calculation is normally done for two particles in isolation, i.e. at infinite dilution. In a concentrated system (a condensed phase), multi-body interactions should be accounted for, and then we would refer to the potential of mean force. However, we get an adequate estimation of the total potential in a concentrated system simply by the summation of the interaction from the nearest neighbours.

It is this interaction energy that governs the stability of colloidal dispersions and which we effectively manipulate whenever we make a change to a formulation.

The components of the pair potential that we are most interested in for charged colloids are those due to the attractive van der Waals forces and the repulsive force between similarly charged particles.

3.2.1 Attractive Forces

Molecules with a permanent dipole will attract similar molecules as the dipoles align. They will also induce a dipole in an adjacent neutral atom or molecule and cause an attraction. This is relatively easy to understand; however, the motion of the electrons in *any* atom cause rapidly fluctuating dipoles. This leads to the London dispersion interaction as the oscillating dipoles become coupled. Even neutral atoms have a fluctuating dipole due to the motion of the electrons around the nucleus. It is energetically more favourable for adjacent atoms to be oscillating in unison. This is the interaction which we recognise from the non-ideal behaviour of the inert gases.

This interaction is non-directional, so that when large assemblies of atoms are considered, different dipolar orientations do not cancel each other. Colloidal particles are of course large assemblies of atoms and hence the van der Waals forces from the London dispersion interaction act between particles to cause attraction.

The early calculations were due to Hamaker and de Boer (1). The route is to sum the interaction of one atom in a particle with each atom in the adjacent particle (Figure 3.1). That interaction is then summed over all the atoms in the first particle.

The result is a long-range interaction – much longer range than the interaction between two isolated atoms. The range of the interaction is comparable with the radii of colloidal particles.

The attractive potential energy is directly proportional to a particle radius (a), a material constant – the Hamaker constant (A) and is inversely proportional to distance of separation (h).

$$V_{\mathrm{A}} = -\frac{A}{12}\left[\frac{1}{x(x+2)} + \frac{1}{(x+1)^2} + 2\ln\frac{x(x+2)}{(x+1)^2}\right] \quad \text{where } x = \frac{h}{2a} \tag{3.1}$$

When the particle separation is small ($h \ll 2a$) this reduces to a simple form of

$$V_{\mathrm{A}} = -\frac{Aa}{12h} \tag{3.2}$$

and a full derivation is given in Chapter 16.

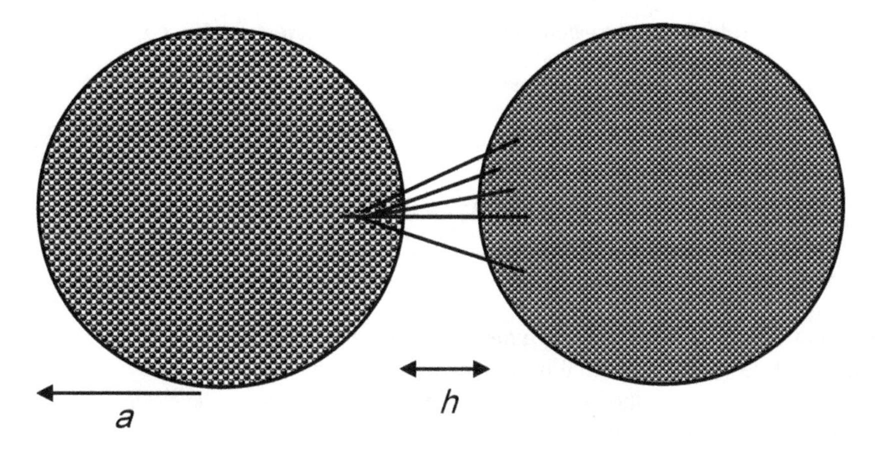

Figure 3.1 London forces between atoms in two adjacent colloidal particles

The Hamaker constant is a function of both the electronic polarisability and the density of the material. When particles are immersed in a medium the attraction between particles is weakened as there is an attraction with the medium also. The combined or composite Hamaker constant (A) can be estimated as the geometric mean of that of the particle (A_{particle}) and that of the medium (A_{medium}) with respect to their values in vacuum, and it is this that should be used in the calculation of the attractive potential.

$$A = \left(\sqrt{A_{\text{particle}}} - \sqrt{A_{\text{medium}}}\right)^2 \tag{3.3}$$

Hamaker constants have values in the range of 10^{-20} J, and a selection of values are given in Table 3.1.

3.2.2 Electrostatic Repulsion

Electrical repulsion is an important stabilising mechanism for particles dispersed in aqueous solutions or moderate polarity liquids such as ethylene glycol.

The diffuse part of the electrical double layer extends in solution over distances characterised by the Debye length ($1/\kappa$). In practice we use the experimentally accessible

Table 3.1 Hamaker constants for various materials

Particles	Hamaker constant (J/10^{-20})	Media	Hamaker constant (J/10^{-20})
poly(tetrafluorethylene)	3.8	water	3.7
poly(methyl methacrylate)	7.1	pentane	3.8
poly(styrene)	7.8	ethanol	4.2
silica (fused)	6.5	decane	4.8
titanium dioxide	19.5	hexadecane	5.1
metals (Au, Ag, Pt, etc.)	~40	cyclohexane	5.2

Table 3.2 *The extent of the double layer thickness as a function of electrolyte concentration*

NaCl concentration	Double layer thickness
30 mM	2 nm
10 mM	3 nm
1 mM	10 nm
0.1 mM	30 nm

zeta potential (see Section 3.3.3) as a measure of the electrical potential at the Stern layer. The rate of decay of this potential is governed by the reciprocal of the Debye length and is commonly referred to as the double layer thickness. This defines the extent to which the ionic atmosphere, which is different from the bulk ionic medium, extends from the particle surface (see Table 3.2 and Table 2.1).

When two particles approach each other, the ionic atmospheres overlap (Figure 3.2) and the local ion concentration midway between the particles can be estimated by summing the contributions from each particle. The difference in this local mid-point ion concentration and that in the bulk results in an osmotic pressure acting to force the particles apart. Integration of this force with respect to distance gives us the energy.

When two charged particles come together there are two extreme cases which we can envisage. If the ion adsorption equilibrium is maintained then either the surface charge remains constant and the surface potential compensates (constant charge) or the surface potential remains constant and the surface charge density changes to compensate (constant potential). Hogg, Healy and Fuerstenau (2) derived expressions which enable us to calculate the interaction between non-identical spheres under both constant charge and constant

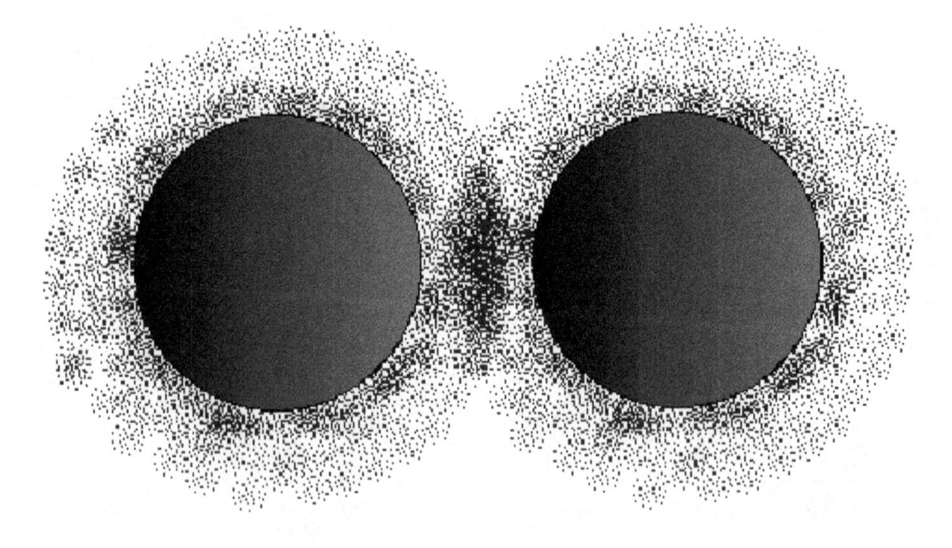

Figure 3.2 *The overlap of electrical double layers on adjacent particles*

potential conditions.

$$V_R^{\psi} = \frac{a_1 a_2(\psi_{0_1}^2 + \psi_{0_2}^2)}{4(a_1 + a_2)}\left[\frac{2\psi_{0_1}\psi_{0_2}}{\psi_{0_1}^2 + \psi_{0_2}^2}\ln\left(\frac{1+\exp(-\kappa h)}{1-\exp(-\kappa h)}\right) + \ln(1+\exp(-2\kappa h))\right] \quad (3.4)$$

$$V_R^{\sigma} = \frac{a_1 a_2(\psi_{0_1}^2 + \psi_{0_2}^2)}{4(a_1 + a_2)}\left[\frac{2\psi_{0_1}\psi_{0_2}}{\psi_{0_1}^2 + \psi_{0_2}^2}\ln\left(\frac{1+\exp(-\kappa h)}{1-\exp(-\kappa h)}\right) + \ln(1-\exp(-2\kappa h))\right] \quad (3.5)$$

These reduce to the basic expressions:

$$V_R^{\psi} = \frac{\varepsilon a \psi_0^2}{2}\ln(1+\exp(-\kappa h)) \quad (3.6)$$

$$V_R^{\sigma} = \frac{\varepsilon a \psi_0^2}{2}\ln(1-\exp(-\kappa h)) \quad (3.7)$$

for the interaction between identical particles with a radius a. The expressions are valid in the regime where κa, the product of the Debye constant and the particle radius, is greater than 10.

For conditions where κa is less than 3 the general expression is

$$V_R = 2\pi\varepsilon a \psi_\delta^2 \exp(-\kappa h) \quad (3.8)$$

3.2.3 Effect of Particle Concentration

Whenever we add charged colloidal particles to a liquid we do two things:

- add counter-ions with each particle
- reduce the solution volume available to the ions.

Both of these factors become important as the particle concentration increases and when the background electrolyte concentration is low. We can expand the expression for κ from Chapter 2 (Equation 2.6) to take account of these extra effects. The expanded expression is:

$$\kappa^2 = \frac{e^2 z^2}{\varepsilon k_B T}\frac{2n_0 + \dfrac{3\sigma_\delta \phi}{ae}}{1-\phi} \quad (3.9)$$

where z is the counter-ion valency, n_0 the concentration of counter-ions in solution (added electrolyte), a is the particle radius and e is the formal charge on an electron.

We recognise the first part of the expression from Equation (2.6), but now we have the expression $\frac{3\sigma_\delta \phi}{ae}$ which takes into account the counter-ions which are carried by the particle through the surface charge density σ_δ. We can see that this expression becomes important when the particles are small and the volume fraction ϕ and surface charge density are high. The denominator of $(1-\phi)$ takes into account the volume taken up by the particles in the dispersion.

This effect is only important when the background electrolyte levels are low and where we have high concentrations of small highly charged particles. We can see the effect in

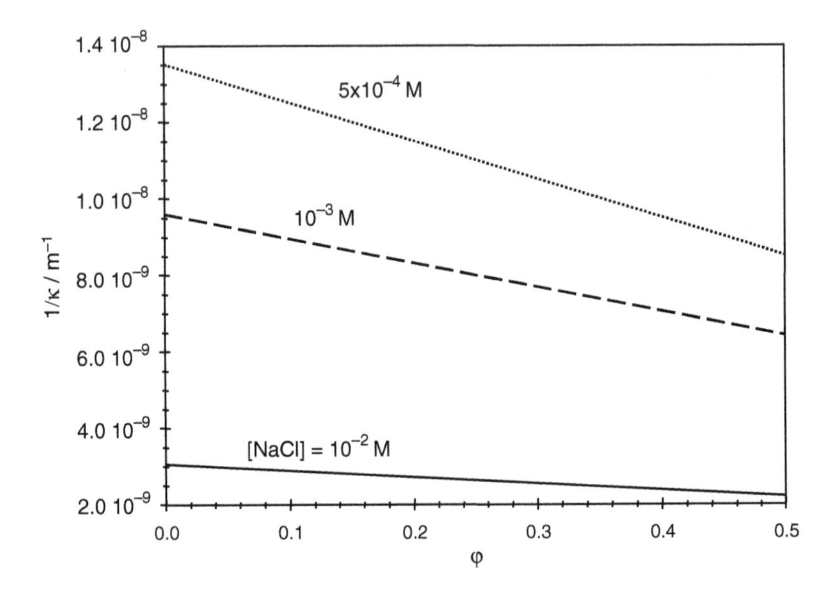

Figure 3.3 *The effect of the counter-ions associated with the particle surface on the Debye length, 1/κ*

Figure 3.3, which shows the effect for different background electrolytes as a function of particle volume fraction for 85 nm radius particles with a surface charge density of $0.15\,\mu C\,cm^{-2}$.

3.2.4 Total Potential

The linear addition of the electrostatic and dispersion potentials is the basis of the DLVO theory for colloid stability (3, 4). When we add the attractive potential to the repulsive electrostatic potential we have the typical curve for charge-stabilised colloidal particles.

$$V_T = V_A + V_R \tag{3.10}$$

This curve has a number of interesting and important features. The shape of the curve is the consequence of the addition of the exponential decay of the repulsive term and the more steeply decaying one-over-distance relationship of the attractive term.

This linear superposition leads to a maximum in the curve, as seen in Figure 3.4, and is known as the primary maximum. It is this maximum in the pair potential which provides the mechanism for stability of charged colloidal particles. It creates an effective activation energy for aggregation. As two particles come together they must collide with sufficient energy to overcome the barrier provided by the primary maximum.

It is important to realise that this barrier to aggregation only provides kinetic stability to a dispersion. The thermodynamic drive is towards an aggregated, phase-separated state. We can say that the larger the barrier the longer the system will remain stable.

Note that the potential is plotted in units of k_BT. These thermal energy units help us relate the height of the maximum with the energy of a Brownian collision, which will be of the order of $1.5\,k_BT$.

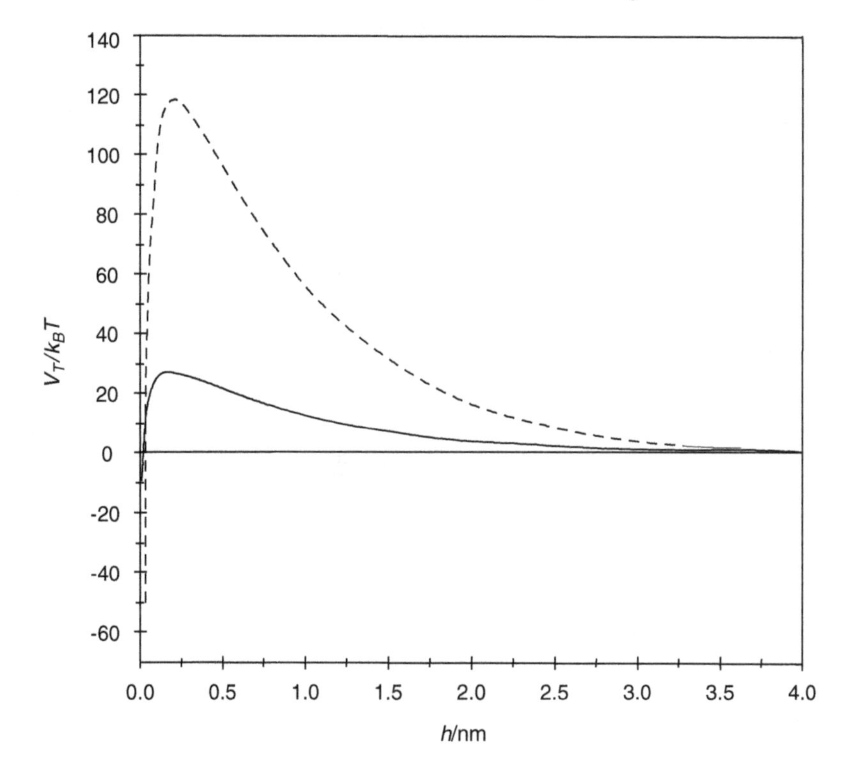

Figure 3.4 *Examples of a total interaction potential curve for two charge-stabilised systems*

Therefore, in order to pose a suitable barrier to aggregation this primary maximum must be at least 1.5 k_BT. In practice we need to manipulate the system so that the primary maximum is at least 20 k_BT in order to achieve a level of stability which can be relied upon over an extended period of time.

3.3 Criteria for Stability

We need to define how the various factors that we put into our systems affect the stability so that we can define threshold values.

We need to consider:

- the effect of ion type and concentration
- the value of the zeta potential
- the effect of particle size.

3.3.1 Salt Concentration

The example given in Figure 3.5 shows the curve for titanium dioxide particles of 100 nm radius and with a zeta potential of –50 mV at different concentrations of sodium chloride.

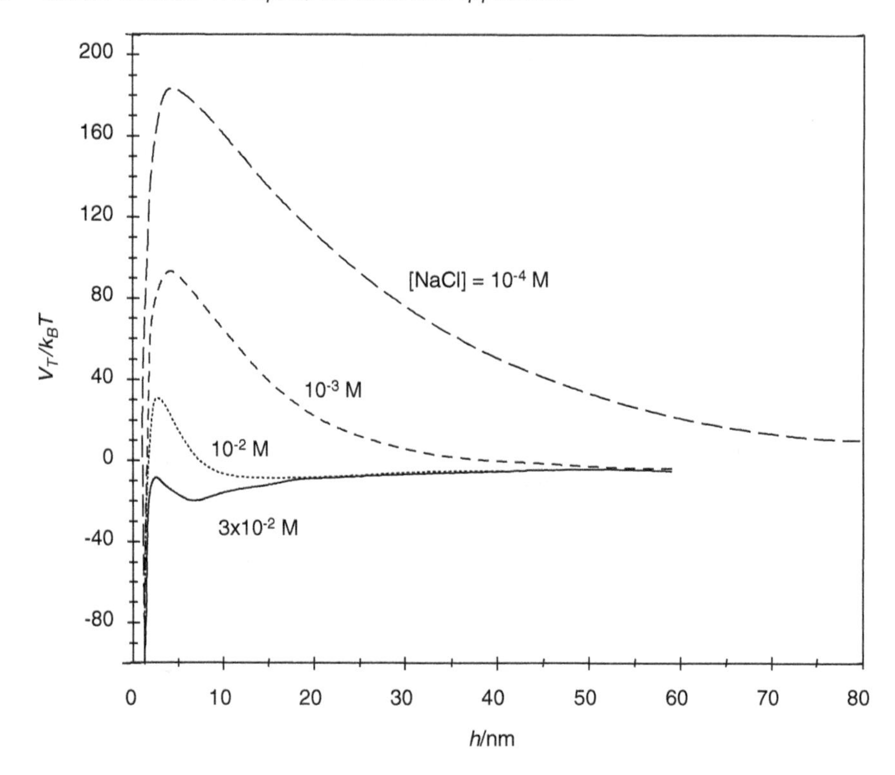

Figure 3.5 *The effect of salt concentration on the shape of the total interaction potential curve*

The points to note are that:

- in each case there is a steep rise to the primary maximum at small separations
- at larger separations there is a long repulsive tail, most notable at lower electrolyte concentrations
- the range of the tail reduces as the electrolyte concentration increases (in line with the decrease in the Debye length)
- the height of the maximum decreases with increasing electrolyte concentration.
- at some point (in this case around 10^{-2} M NaCl) a significant energy minimum develops since the van der Waals dispersion term is insensitive to the electrolyte changes; the attractive minimum is known as the secondary minimum
- as the primary maximum falls to just a few k_BT, a significant fraction of colliding particles will collide with at least that energy and stick
- as the primary maximum falls to below zero (above 3×10^{-2} M NaCl in this case), all collisions will lead to aggregation as there is no barrier.

3.3.2 Counter-ion Valency

The counter-ions are the dominant ions in the Stern and diffuse layers and hence the stability is more sensitive to the counter-ion type than the co-ion type. The valency of the counter-ions is in fact of major importance in determining the stability of charged colloids.

We can estimate a critical coagulation concentration (c.c.c.) from the pair potential exercise by taking the condition that when there is no primary maximum barrier the inter-particle force is also zero and so the coagulation of the particles will be diffusion controlled (all collisions result in coagulation).

Early observations noted a 6th power dependence of z on the c.c.c., and this was formalised in the Shultz–Hardy rule (5–7).

$$c.c.c. \propto \frac{1}{z^n} \tag{3.11}$$

- $n=6$ for high potentials (unusual for coagulation due to ion adsorption)
- $n=2$ for low potentials (the more common occurrence).

With many systems the adsorption of ions and resultant decrease in the Stern potential and hence zeta potential (see also Chapter 2.2) results in a power less than 6; however, a trivalent counter-ion such as Al^{3+} is 6 times as effective a coagulant as Na^+.

Figure 3.6 shows some experimentally determined values of the c.c.c. for two types of dispersions. Both systems show the marked sensitivity to the counter-ion valency predicted by the Shultz–Hardy rule. The more dense AgI particles have a higher Hamaker constant than the polystyrene and hence are more easily aggregated and have lower values of the c.c.c. at each valency of the counter-ion. The line plotted shows the expected 6th power dependence.

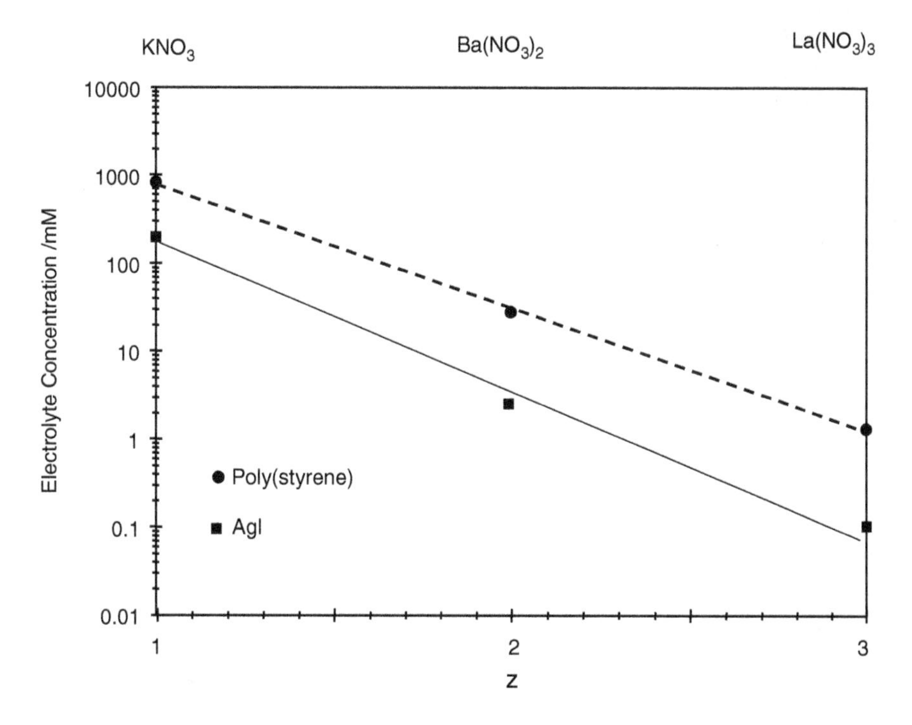

Figure 3.6 *Critical coagulation concentrations for two different colloidal systems using three different electrolytes*

It is important to note that the pH of the systems containing the tri-valent ion was held at pH = 4. It is common for tri-valent ions such as Al^{3+}, Fe^{3+} and L^{3+} to form large hydrated complexes with higher valencies than 3, at pH values greater than 4. Hence at pH 7 or 8, aluminium chloride is an even more effective coagulant than at pH 3 or 4.

3.3.3 Zeta Potential

The zeta potential is a characteristic that is often experimentally accessible. We can expect it to be close to the value of the Stern potential and hence it is often used in the calculation of the pair potential. It may be possible to determine the surface charge density through a titration procedure and to then calculate the Stern potential, but this is experimentally laborious.

In the expression for the electrostatic repulsion the surface potential appears as the square, and so it is a key parameter in estimating the primary maximum.

$$V_R = 2\pi\varepsilon a\psi_\delta^2 \exp(-\kappa h) \tag{3.12}$$

In the example given in Figure 3.7 for polymer latex particles in 1 mM NaCl, we need a value in excess of 20 mV to produce a stable dispersion.

When the zeta potential is –25 mV the value of V_{max} is $\sim 40\, k_B T$. When we double that to –50 mV we can see that $V_{max} \sim 160\, k_B T$. So as expected, since V_R is related to the square of the zeta potential, a doubling of the zeta potential leads to a quadrupling of the value of V_{max}.

In this example, when the zeta potential reduces to less than –20 mV the value of V_{max} drops below $20\, k_B T$ and significant aggregation will occur.

3.3.4 Particle Size

Both the attractive and repulsive contributions are proportional to the particle radius. At small sizes the value of V_T is directly proportional to the particle size. However, at large sizes the value of V_T has a more complicated variation.

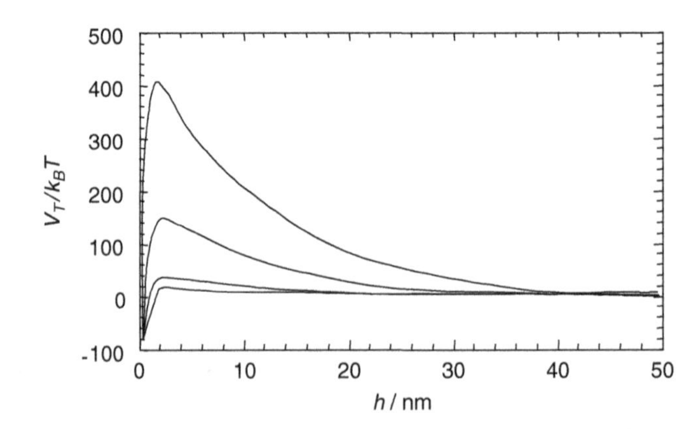

Figure 3.7 The effect of zeta potential on the shape of the total interaction potential curve for a polystyrene latex: from the bottom to top the zeta potentials are –20 mV, –25 mV, –50 mV and –80 mV

In all cases a larger particle radius leads to a higher energy barrier, in other words electrostatic stability increases with increasing particle radius (all other factors remaining constant). For small particle sizes (< 100 nm radius) the primary maximum is directly proportional to the radius. However, the relationship breaks down at larger sizes and the height of the primary maximum increases at a lower rate.

The shape of the curve of the attractive interaction is the important point here. Another feature of this is that for particles with large radii, the attraction often dominates again at long range giving rise to a secondary minimum at distances of the order of 5–10 nm. This attraction is manifest as weak but reversible aggregation.

We can make a distinction here between two types of aggregation. Coagulation is the rapid aggregation that happens in the absence of a primary maximum and leads to a strong irreversible aggregated structure. Flocculation is a reversible aggregation that occurs in a secondary minimum as described. Flocculation is reversible on the addition of energy to the system, usually the application of a shear field by shaking, stirring or other mechanical process.

3.4 Kinetics of Coagulation

The rate of coagulation is used either directly or indirectly to determine the c.c.c. For example if the rate is monitored, we find that it increases as the electrolyte concentration is increased until it reaches a plateau value. If we just add particles to different electrolyte solutions in a series of tubes, then we can check for the onset of aggregation after a fixed time. This may be 5 minutes, or we may choose a little longer, but it is still the faster rate that we notice.

As we have already established that electrostatically stabilised dispersions are kinetically stable and not thermodynamically stable, the key factor is the kinetics. If the rate is so slow that we don't detect a significant change during our period of use, we would consider that to be adequately stable.

3.4.1 Diffusion-limited Rapid Coagulation

$$J_{\mathrm{p}} = D \bullet 4\pi r^2 \bullet \frac{\mathrm{d}N}{\mathrm{d}r} \tag{3.13}$$

Recall that the diffusion constant is in terms of the flux through a unit area per second (J_{p}). We can calculate the flow through a spherical surface around a reference particle (Figure 3.8). This gives us a differential equation that is easily solved to give the number of collisions with that reference particle. Of course each particle is itself such a particle and so the total number of collisions is just that for one particle multiplied by the total particle number. We must divide by 2 as particle A colliding with particle B is the same collision as particle B colliding with particle A.

We can allow for the fact that all particles are in motion by using the sum of the diffusion constants of the two colliding particles. As we are assuming them all to be the same size, we multiply by 2. As each collision results in coagulation the coagulation rate is simply the collision rate.

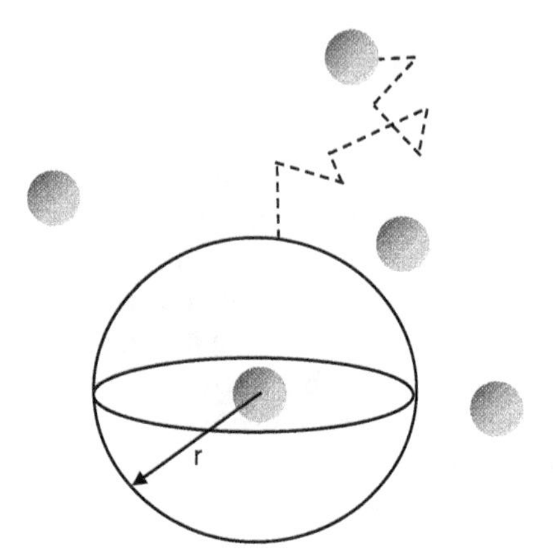

Figure 3.8 *Theoretical spherical surface of influence surrounding a reference particle*

We may write the rate constant in terms of the diffusion constant and the particle radius. The diffusion constant is:

$$k_{\mathrm{D}} = 8\pi Da \tag{3.14}$$

and the half-life is

$$t_{1/2} = \frac{3\eta}{4k_{\mathrm{B}}TN_{\mathrm{p}}} \tag{3.15}$$

where η is the viscosity. We may write the half-life for the second-order reaction in terms of the reciprocal of the particle number. The graph in Figure 3.9 shows how the half-life decreases with size and concentration. (For a given volume fraction the number increases as the size decreases.)

3.4.2 Interaction-limited Coagulation

When there is an energy barrier to prevent the particles coming together the rate is slowed, as only a fraction of the particles collide with sufficient energy to exceed the height of the barrier and stick. This is known as reaction-limited aggregation with a rate constant k_{R} in contrast to the diffusion-limited rate constant k_{D}.

Fuchs (8) defined the stability ratio, W, as the ratio of the rate constants so that the higher the stability ratio the slower the rate.

$$W = \frac{k_{\mathrm{D}}}{k_{\mathrm{R}}} \tag{3.16}$$

To a good approximation the reaction-limited diffusion rate constant is proportional to the Boltzmann factor which gives the fraction of particles at any instant with energy in excess of the primary maximum. The rate of aggregation drops rapidly with the increase in the energy barrier, so that 10 to 20 $k_{\mathrm{B}}T$ gives us reasonable kinetic stability.

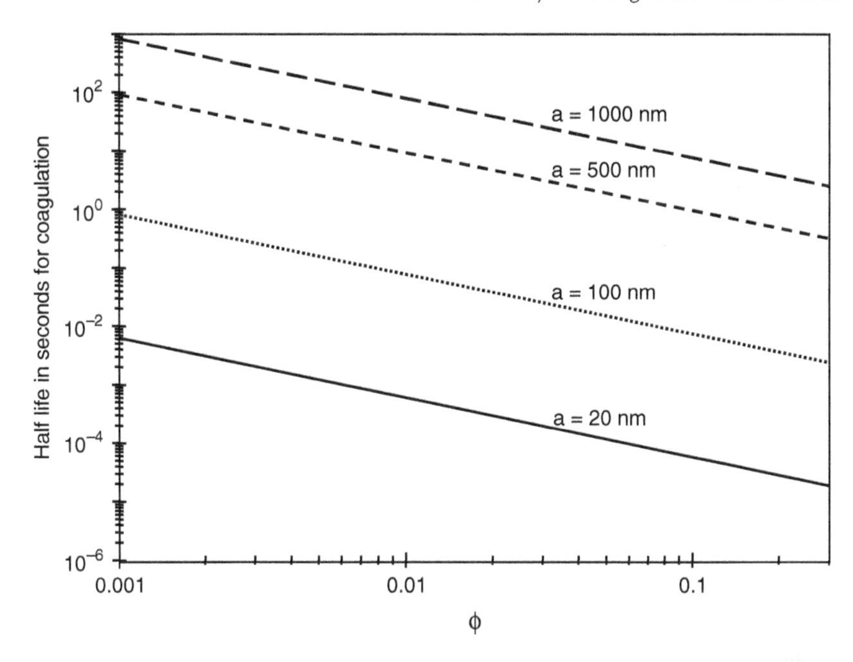

Figure 3.9 *Half-life for coagulation of a series of dispersions of different particle sizes as a function of volume fraction: from top to bottom, 1000 nm, 500 nm, 100 nm and 20 nm radius*

Overbeek (4) has shown that a reasonable approximation to the stability ratio is obtained using the value of the primary maximum.

$$W = \frac{1}{2\kappa a} \exp\left(\frac{V_{\max}}{k_B T}\right) \tag{3.17}$$

from which we can show that

$$k_R \approx 16\pi\kappa D a^2 \exp\left(\frac{-V_{\max}}{k_B T}\right) \tag{3.18}$$

3.4.3 Experimental Determination of c.c.c.

The critical coagulation concentration (c.c.c.) is the salt concentration at which there is a change from aggregation which is limited by the presence of a primary maximum to aggregation which has no barrier. At this point there is a distinct change in the coagulation rate. Measurement of the coagulation rate can be made in two ways:

- directly – measuring the number of particles as a function of time by particle counting (best for large particles)
- indirectly – by light scattering (best for small particles).

In each case we are observing the change in the number of aggregates with time. This tells us about the loss of primary particles from the system.

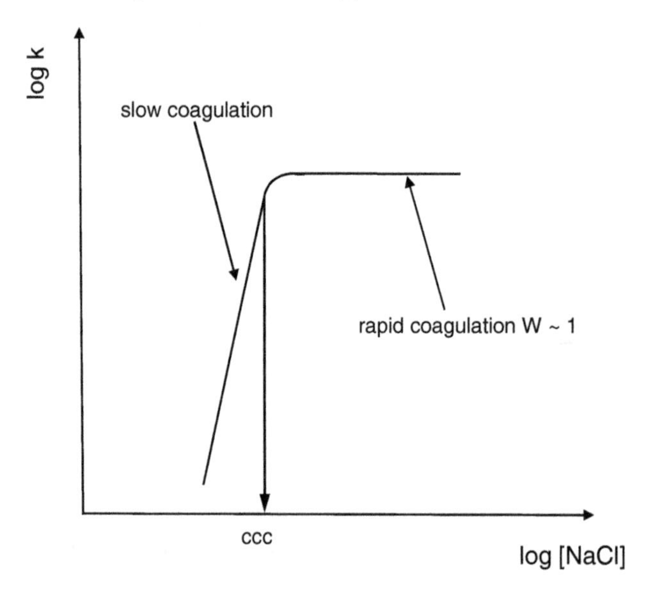

Figure 3.10 *Rate of coagulation against salt concentration indicating the critical coagulation concentration (c.c.c.)*

The simplest experiment is to make up a series of tubes with different electrolyte concentrations and observe at which concentration aggregation becomes apparent after, say, 5 or 10 minutes. It can be quantified if the tubes are lightly centrifuged and a spectrophotometer is used to measure the percent transmission of the supernatant as a measure of the number of particles in suspension.

More precise determination can be made if the rate of aggregation itself is measured. At higher electrolyte concentrations the rate increases to the plateau value (as seen in Figure 3.10), representing the fast or diffusion-limited rate.

Note that the analysis of the rate constant in terms of the diffusion of single particles is strictly the initial rate. As we progress into the coagulation process the particle number changes and the mechanism changes to one where the large, much less mobile aggregates get larger by adding singlets. Accurately describing the rates can then be quite complex.

3.5 Conclusions

In this chapter we have explored the basic theory surrounding the stability of systems containing charged colloidal particles. The balance between the van der Waals interaction and the repulsion between the electrical double layers surrounding charged particles can be controlled to provide an energetic barrier to the coagulation of particles.

We have seen that charge stabilisation can only be effective where significant surface charge can be achieved and so is normally limited to systems in polar solvents.

The stability achieved is only a kinetic stability to coagulation and in the absence of any other stabilising mechanism these systems will eventually coagulate. However, in the stability ratio we have a method for evaluating the rate of coagulation compared with the theoretical rate when there is no barrier to aggregation.

References

(1) Hamaker, H. C. (1937) Physica, 4: 1058.
(2) Hogg, R., Healy, T. W., Fuerstenau, D. W. (1966) Trans. Faraday Soc., 62: 1638.
(3) Derjaguin, B. V., Landau, L. (1941) Acta Physicochim. (URSS), 14: 633.
(4) Verwey, E. J., Overbeek, J. T. G. (1948) Theory of the Stability of Lyophobic Colloids. Elsevier, Amsterdam.
(5) Schulze, J. (1882) pr. Chem., 25: 471.
(6) Schulze, J. (1883) pr. Chem., 27: 320.
(7) Hardy, W. B. (1900) Proc. Roy. Soc., 66a: 110.
(8) Fuchs, N. (1934) Z. Phys., 89: 736.

4

Surfactant Aggregation and Adsorption at Interfaces

Julian Eastoe

School of Chemistry, University of Bristol, UK

4.1 Introduction

A major group of colloidal systems, also classified as lyophilic, is that of the so-called *association colloids* (1). These are aggregates of *amphiphilic* (both oil and water-loving') molecules that associate in a dynamic and thermodynamically driven process that may be simultaneously a molecular solution and a true colloidal system. Such molecules are commonly termed surfactants', a contraction of the term *surface-active agents*. Surfactants are an important and versatile class of chemicals. Due to their dual nature, they are associated with many useful interfacial phenomena, e.g. wetting, and as such are found in many diverse industrial products and processes.

4.2 Characteristic Features of Surfactants

Surface-active agents are organic molecules that, when dissolved in a solvent at low concentration, have the ability to adsorb (or locate) at interfaces, thereby altering significantly the physical properties of those interfaces. The term interface' is commonly employed here to describe the boundary in liquid/liquid, solid/liquid and gas/liquid systems, although in the latter case the term 'surface' can also be used. This adsorption behaviour can

Colloid Science: Principles, methods and applications, Second Edition Edited by Terence Cosgrove

be attributed to the solvent nature and to a chemical structure for surfactants that combines both a polar and a non-polar (amphiphilic) group into a single molecule. To accommodate for their dual nature, amphiphiles therefore sit' at interfaces so that their lyophobic moiety keeps away from strong solvent interactions while the lyophilic part remains in solution. Since water is the most common solvent, and is the liquid of most academic and industrial interest, amphiphiles will be described with regard to their hydrophilic and hydrophobic moieties, or 'head' and 'tail' respectively.

Adsorption is associated with significant energetic changes since the free energy of a surfactant molecule located at the interface is lower than that of a molecule solubilised in either bulk phase. Accumulation of amphiphiles at the interface (liquid/liquid or gas/liquid) is therefore a spontaneous process and results in a decrease of the interfacial (surface) tension. However, such a definition applies to many substances: medium- or long-chain alcohols are surface active (e.g. *n*-hexanol, dodecanol) but these are not considered as surfactants. True surfactants are distinguished by an ability to form oriented monolayers at the interface (here air/water or oil/water) and, most importantly, self-assembly structures (micelles, vesicles) in bulk phases. They also stand out from the more general class of surface-active agents owing to emulsification, dispersion, wetting, foaming or detergency properties.

Both adsorption and aggregation phenomena result from the hydrophobic effect (2); i.e. the expulsion of surfactant tails from water. Basically this originates from water–water intermolecular interactions being stronger than those between water–tail. Finally, another characteristic of surfactants, when their aqueous concentration exceeds approximately 40%, is an ability to form liquid crystalline phases (or lyotropic mesophases). These systems consist of extended aggregation of surfactant molecules into large organised structures.

Owing to such a versatile phase behaviour and diversity in colloidal structures, surfactants find application in many industrial processes, essentially where high surface areas, modification of the interfacial activity or stability of colloidal systems are required. The variety of surfactants and the synergism offered by mixed-surfactant systems (3) also explains the ever-growing interest in fundamental studies and practical applications. Listing the various physical properties and associated uses of surfactants is beyond the scope of this chapter. However, a few relevant examples are presented in the following section, giving an idea of their widespread industrial use.

4.3 Classification and Applications of Surfactants

4.3.1 Types of Surfactants

Numerous variations are possible within the structure of both the head and tail group of surfactants. The head group can be charged or neutral, small and compact in size, or a polymeric chain. The tail group is usually a single or double, straight or branched hydrocarbon chain, but may also be a fluorocarbon, or a siloxane, or contain aromatic group(s). Commonly encountered hydrophilic and hydrophobic groups are listed in Tables 4.1 and 4.2 respectively.

Since the hydrophilic part normally achieves its solubility either by ionic interactions or by hydrogen bonding, the simplest classification is based on surfactant head group type,

Table 4.1 *Common hydrophilic groups found in commercially available surfactants*

Class	General structure
Sulfonate	$R\text{-}SO_3^-\ M^+$
Sulfate	$R\text{-}OSO_3^-\ M^+$
Carboxylate	$R\text{-}COO^-\ M^+$
Phosphate	$R\text{-}OSO_3^-\ M^+$
Ammonium	$R_xH_yN^+X^-$ ($x = 1\text{-}3$, $y = 4\text{-}x$)
Quaternary ammonium	$R_4N^+X^-$
Betaines	$RN^+(CH_3)_2CH_2COO^-$
Sulfobetaines	$RN^+(CH_3)_2CH_2CH_2\ SO_3^-$
Polyoxyethylene (POE)	$R\text{-}OCH_2CH_2(OCH_2CH_2)_nOH$
Polyols	Sucrose, sorbitan, glycerol, ethylene glycol, etc
Polypeptide	$R\text{-}NH\text{-}CHR\text{-}CO\text{-}NH\text{-}CHR'\text{-}CO\text{-}\ldots\text{-}CO_2H$
Polyglycidyl	$R\text{-}(OCH_2CH[CH_2OH]CH_2)_n\text{-}\ldots\text{-}OCH_2CH[CH_2OH]CH_2OH$

with further subgroups according to the nature of the lyophobic moiety. Four basic classes therefore emerge as:

- the anionics and cationics, which dissociate in water into two oppositely charged species (the surfactant ion and its counter-ion)
- the non-ionics, which include a highly polar (non-charged) moiety, such as poly(ethylene oxide) ($-OCH_2CH_2O-$) or polyol groups
- the zwitterionics (or amphoterics), which combine both a positive and a negative group.

Table 4.2 *Common hydrophobic groups used in commercially available surfactants*

Group	General structure	
Alkyls	$CH_3(CH_2)_n$	$n = 12–18$
Olefins	$CH_3(CH_2)_nCH{=}CH_2$	$n = 7–17$
Alkylbenzenes	$CH_3(CH_2)_nCH_2$–(benzene ring)	$n = 6–10$, linear or branched
Alkylaromatics	$CH_3(CH_2)_nCH_3$ on naphthalene ring with substituents R, R	$n = 1–2$ for water soluble, $n = 8$ or 9 for oil soluble surfactants
Alkylphenols	$CH_3(CH_2)_nCH_2$–(benzene ring)–OH	$n = 6–10$, linear or branched
Polyoxypropylene	$CH_3CH(X)CH_2O(CH(CH_3)CH_2)_n$	n = degree of oligomerisation, X = oligomerisation initiator
Fluorocarbons	$CF_3(CF_2)_nCOOH$	$n = 4–8$, linear or branched, or H-terminated
Silicones	$CH_3O(Si(CH_3)_2O)_nCH_3$	

Table 4.3 *Structural features and examples of new surfactant classes*

Classes	Structural characteristics	Example
Catanionic	Equimolar mixture of cationic and anionic surfactants (no inorganic counterion)	*n*-dodecyltrimethylammonium *n*-dodecyl sulfate (DTADS) $C_{12}H_{25}$ $(CH_3)_3$ $N^{+-}O_4S$ $C_{12}H_{25}$
Bolaform	Two charged headgroups connected by a long linear polymethylene chain	Hexadecanediyl-1,16-bis(trimethyl ammonium bromide) Br^- $(CH_3)_3$ N^+– $(CH_2)_{16}$– $N^+(CH_3)_3$ Br^-
Gemini (or dimeric)	Two identical surfactants connected by a spacer close to or at the level of the headgroup	Propane-1,3-bis(dodecyldimethyl ammonium bromide) C_3H_6 -1,3-bis$[(CH_3)_2$ N^+ $C_{12}H_{25}$ $Br^-]$
Polymeric	Polymer with surface active properties	Copolymer of isobutylene and succinic anhydride $H_3C-[C(CH_3)_2-CH_2]_n-CH_2CH(CH_2COOH)-C(=O)-NH-CH_2CH_2OH$
Polymerisable	Surfactant that can undergo homo-polymerisation or copolymerisation with other components of the system	11-(acryloyloxy)undecyltrimethyl ammonium bromide $CH_2=CH-C(=O)-O-(CH_2)_{11}-N^+(CH_3)_3$ Br^-

With the continuous search for improving surfactant properties, new structures have recently emerged that exhibit interesting synergistic interactions or enhanced surface and aggregation properties. These novel surfactants have attracted much interest, and include the catanionics, bolaforms, gemini (or dimeric) surfactants, polymeric and polymerisable surfactants (4, 5). Characteristics and typical examples are shown in Table 4.3. Another important driving force for this research is the need for enhanced surfactant biodegradability. In particular, for personal care products and household detergents, regulations (6) require high biodegradability and non-toxicity of each component present in the formulation.

A typical example of a double-chain surfactant is sodium bis(2-ethylhexyl)sulfosuccinate, often referred to by its American Cyanamid trade name Aerosol-OT, or AOT. Its chemical structure is illustrated in Figure 4.1, along with other typical double-chain compounds within the four basic surfactant classes.

4.3.2 Surfactant Uses and Development

Surfactants may be from natural or synthetic sources. The first category includes naturally occurring amphiphiles such as the lipids, which are surfactants based on glycerol and are vital components of the cell membrane. Also in this group are the so-called soaps', the first recognised surfactants (7). These can be traced back to Egyptian times; by combining animal and vegetable oils with alkaline salts a soap-like material was formed, and this was used for treating skin diseases, as well as for washing. Soaps remained the only source of natural detergents from the 7th till the early 20th century, with gradually more varieties becoming available for shaving and shampooing, as well as bathing and laundering. In 1916, in response to a World War I-related shortage of fats for making soap, the first

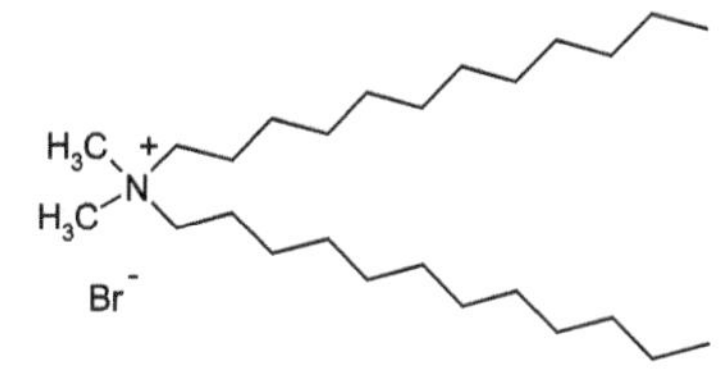

Cationic: *n*-didodecyldimethylammonium bromide (DD AB)

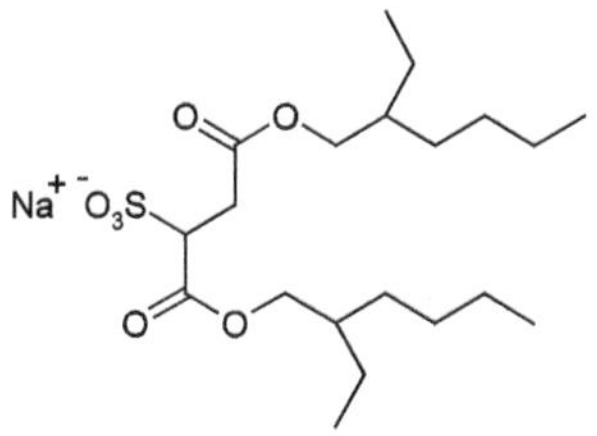

Anionic: Sodium bis(2-ethylhexyl) sulfosuccinate (Aerosol-OT or AOT)

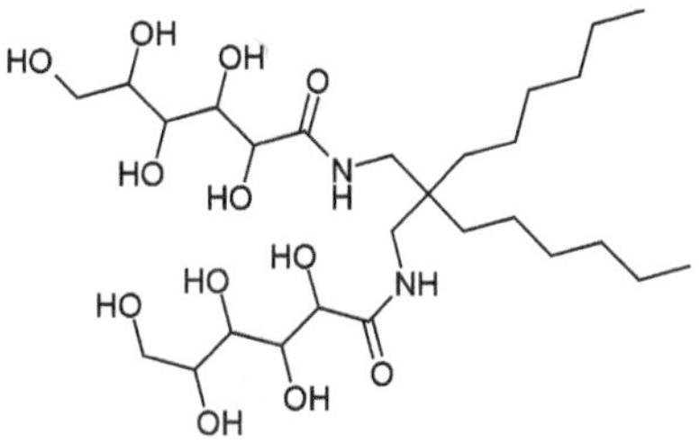

Non-ionic: di(hexyl) glucamide (di-(C6-Glu))

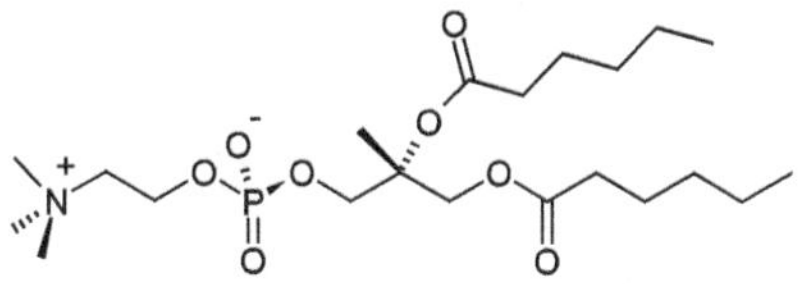

Zwitterionic: di-hexylphosphatidylcholine ((diC6)PC)

Figure 4.1 *Chemical structure of typical double-chain surfactants*

synthetic detergent was developed in Germany. Known today simply as detergents, synthetic detergents are washing and cleaning products obtained from a variety of raw materials.

Nowadays, synthetic surfactants are essential components in many industrial processes and formulations (8–10). Depending on the precise chemical nature of the product, the properties of, for example, emulsification, detergency and foaming may be exhibited in varying degree. The number and arrangement of the hydrocarbon groups together with the nature and position of the hydrophilic groups combine to determine the surface-active properties of the molecule. For example C_{12} to C_{20} is generally regarded as the range covering optimum detergency, whilst wetting and foaming are best achieved with shorter chain lengths. Structure–performance relationships and chemical compatibility are therefore key elements in surfactant-based formulations, so that much research is devoted to this area.

Amongst the different classes of surfactants, anionics are often used in applications, mainly because of the ease and low cost of manufacture. They contain negatively charged head group, e.g. carboxylates ($-CO_2^-$), used in soaps, sulfate ($-OSO_3^-$), and sulfonates ($-SO_3^-$) groups. Their main applications are in detergency, personal care products, emulsifiers and soaps.

Cationics have positively charged head groups, e.g. trimethylammonium ion ($-N(CH_3)_3^+$) – and are mainly involved in applications related to their absorption at surfaces. These are generally negatively charged (e.g. metal, plastics, minerals, fibres, hairs and cell membranes) so that they can be modified upon treatment with cationic surfactants. They are therefore used as anticorrosion and antistatic agents, flotation collectors, fabric softeners, hair conditioners and bactericides.

Non-ionics contain groups with a strong affinity for water due to strong dipole–dipole interactions arising from hydrogen bonding, e.g. ethoxylates ($-(OCH_2CH_2)_mOH$). One advantage over ionics is that the length of both the hydrophilic and hydrophobic groups can be varied to obtain maximum efficiency in use. They find applications in low-temperature detergents and emulsifiers.

Zwitterionics constitute the smallest surfactant class due to their high cost of manufacture. They are characterised by excellent dermatological properties and skin compatibility. Because of their low eye and skin irritation, common uses are in shampoos and cosmetics.

4.4 Adsorption of Surfactants at Interfaces

4.4.1 Surface Tension and Surface Activity

Due to the different environment of molecules located at an interface compared to those from either bulk phase, an interface is associated with a surface free energy. At the air–water surface for example, water molecules are subjected to unequal short-range attraction forces and, thus, undergo a net inward pull to the bulk phase. Minimisation of the contact area with the gas phase is therefore a spontaneous process, explaining why drops and bubbles are round. The surface free energy per unit area, defined as the *surface tension* (γ_0), is then the minimum amount of work (W_{min}) required to create new unit area of that interface (ΔA), so $W_{min} = \gamma_0 \times \Delta A$. Another, but less intuitive, definition of surface tension is given as the force acting normal to the liquid–gas interface per unit length of the resulting thin film on the surface.

A surface-active agent is therefore a substance that at low concentrations adsorbs, thereby changing the amount of work required to expand that interface. In particular, surfactants can significantly reduce interfacial tension due to their dual chemical nature. Considering the air–water boundary, the force driving adsorption is unfavourable hydrophobic interactions within the bulk phase. There, water molecules interact with one another through hydrogen bonding, so the presence of hydrocarbon groups in dissolved amphiphilic molecules causes distortion of this solvent structure apparently increasing the free energy of the system. This is known as the hydrophobic effect (11). Less work is required to bring a surfactant molecule to the surface than a water molecule, so that migration of the surfactant to the surface is a spontaneous process. At the gas–liquid interface, the result is the creation of new unit area of surface and the formation of an *oriented surfactant monolayer* with the hydrophobic tails pointing out of, and the head group inside, the water phase. The balance against the tendency of the surface to contract under normal surface tension forces causes an increase in the surface (or expanding) pressure π, and therefore a decrease in surface tension γ of the solution. The surface pressure is defined as $\pi = \gamma_0 - \gamma$, where γ_0 is the surface tension of a clean air–water surface.

Depending on the surfactant molecular structure, adsorption takes place over various concentration ranges and rates, but typically, above a well-defined concentration – the critical micelle concentration (CMC) – micellisation or aggregation takes place. At the CMC, the interface is at (near) maximum coverage and to minimise further free energy, molecules begin to aggregate in the bulk phase. Above the CMC, the system then consists of an adsorbed monomolecular layer, free monomers and micellised surfactant in the bulk, with all these three states in equilibrium. The structure and formation of micelles will be briefly described in Section 4.6. Below the CMC, adsorption is a dynamic equilibrium with surfactant molecules perpetually arriving at, and leaving, the surface. Nevertheless, a time-averaged value for the surface concentration can be defined and quantified either directly or indirectly using thermodynamic equations (see Section 4.4.2).

Dynamic surface tension – as opposed to the equilibrium quantity – is an important property of surfactant systems as it governs many important industrial and biological applications (12–15). Examples are printing and coating processes where an equilibrium surface tension is never attained, and a new area of interface is continuously formed. In any surfactant solution, the equilibrium surface tension is not achieved instantaneously and surfactant molecules must first diffuse from the bulk to the surface, then adsorb, whilst also achieving the correct orientation. Therefore, a freshly formed interface of a surfactant solution has a surface tension very close to that of the solvent, and this dynamic surface tension will then decay over a certain period of time to the equilibrium value. This relaxation can range from milliseconds to days depending on the surfactant type and concentration. In order to control this dynamic behaviour, it is necessary to understand the main processes governing transport of surfactant molecules from the bulk to the interface. This area of research therefore attracts much attention and recent developments can be found in the references (16–18). However, in the present chapter equilibrium surface tension will now be considered.

4.4.2 Surface Excess and Thermodynamics of Adsorption

Following on the formation of an oriented surfactant monolayer, a fundamental associated physical quantity is the *surface excess*. This is defined as the concentration of surfactant molecules in a surface plane, relative to that at a similar plane in the bulk. A common thermodynamic treatment of the variation of surface tension with composition has been derived by Gibbs (19).

An important approximation associated with this Gibbs adsorption equation is the 'exact' location of the interface. Consider a surfactant aqueous phase α in equilibrium with vapour β. The interface is a region of indeterminate thickness τ across which the properties of the system vary from values specific to phase α to those characteristic of β. Since properties within this real interface cannot be well defined, a convenient assumption is to consider a mathematical plane, with zero thickness, so that the properties of α and β apply right up to that dividing plane positioned at some specific value X. Figure 4.2 illustrates this idealised system.

In the definition of the Gibbs dividing surface XX′ is arbitrarily chosen so that the surface excess adsorption of the solvent is zero. Then the surface excess concentration of component i is given by

$$\Gamma_i^\sigma = \frac{n_i^\sigma}{A} \tag{4.1}$$

where A is the interfacial area. The term n_i^σ is the amount of component i in the surface phase σ over and above that which would have been in the phase σ if the bulk phases α and β had extended to the surface XX′, without any change of composition. Γ_i^σ may be positive or negative, and its magnitude clearly depends on the location of XX′.

Now consider the internal energy U of the total system consisting of the bulk phases α and β:

$$\begin{aligned} U &= U^\alpha + U^\beta + U^\sigma \\ U^\alpha &= TS^\alpha - PV^\alpha + \textstyle\sum_i \mu_i n_i^\alpha \\ U^\beta &= TS^\beta - PV^\beta + \textstyle\sum_i \mu_i n_i^\beta \end{aligned} \tag{4.2}$$

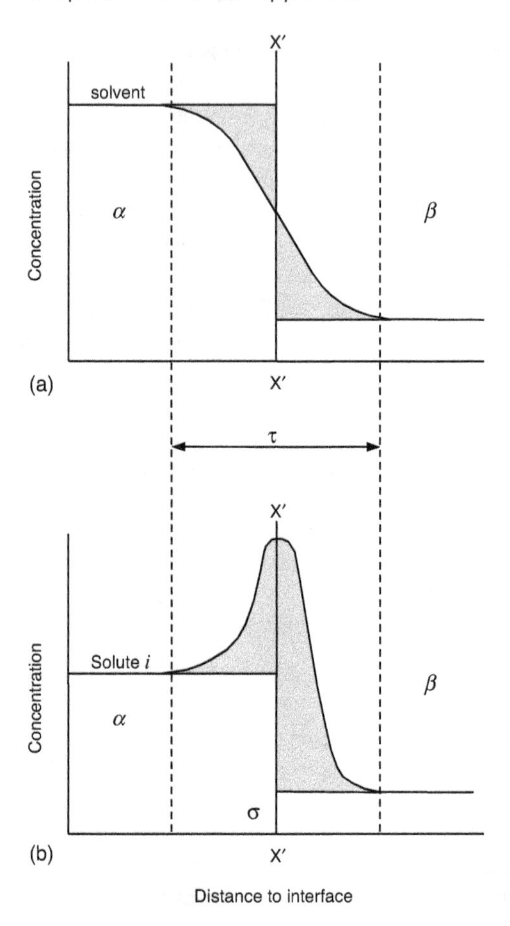

Figure 4.2 *In the Gibbs approach to defining the surface excess concentration Γ, the Gibbs dividing surface is defined as the plane in which the solvent excess concentration becomes zero (the shaded area is equal on each side of the plane) as in (a). The surface excess of component i will then be the difference in the concentrations of that component on either side of that plane (the shaded area) (b)*

The corresponding expression for the thermodynamic energy of the interfacial region σ is

$$U^{\sigma} = TS^{\sigma} + \gamma A + \sum_{i} \mu_i n_i^{\sigma} \tag{4.3}$$

For any infinitesimal change in T, S, A, μ, n, differentiation of Equation (4.3) gives

$$\mathrm{d}U^{\sigma} = T\mathrm{d}S^{\sigma} + S^{\sigma}\mathrm{d}T + \gamma dA + Ad\gamma + \sum_{i} \mu_i dn_i^{\sigma} + \sum_{i} n_i^{\sigma} d\mu_i \tag{4.4}$$

For a small, isobaric, isothermal, reversible change the differential total internal energy in any bulk phase is

$$\mathrm{d}U = T\mathrm{d}S - P\mathrm{d}V + \sum_{i} \mu_i \mathrm{d}n_i \tag{4.5}$$

Similarly, for the differential internal energy in the interfacial region

$$dU^\sigma = TdS^\sigma + \gamma dA + \sum_i \mu_i dn_i^\sigma \tag{4.6}$$

Subtracting Equation (4.6) from Equation (4.4) leads to

$$S^\sigma dT + Ad\gamma + \sum_i n_i^\sigma d\mu_i = 0 \tag{4.7}$$

Then at constant temperature, with the surface excess of component i, Γ_i^σ, as defined in Equation (4.1), the general form of the Gibbs equation is

$$d\gamma = -\sum_i \Gamma_i^\sigma d\mu_i \tag{4.8}$$

For a simple system consisting of a solvent and a solute, denoted by the subscripts 1 and 2 respectively, then Equation (4.8) reduces to

$$d\gamma = -\Gamma_1^\sigma d\mu_1 - \Gamma_2^\sigma d\mu_2 \tag{4.9}$$

Considering the choice of the Gibbs dividing surface position, i.e. so that $\Gamma_1^\sigma = 0$, then Equation (4.9) simplifies to

$$d\gamma = -\Gamma_2^\sigma d\mu_2 \tag{4.10}$$

where Γ_2^σ is the solute surface excess concentration.

The chemical potential is given by

$$\mu_i = \mu_i^0 + RT \ln a_i$$

so at constant temperature

$$d\mu_i = \text{const} + RTd \ln a_i \tag{4.11}$$

where μ_i^0 is the standard chemical potential of component i.

Therefore applying to Equation (4.10) gives the common form of the Gibbs equation for non-dissociating materials (e.g. non-ionic surfactants)

$$d\gamma = -\Gamma_2^\sigma RT \, d \ln a_2 \tag{4.12}$$

or

$$\Gamma_2^\sigma = -\frac{1}{RT}\frac{d\gamma}{d \ln a_2} \tag{4.13}$$

For dissociating solutes, such as ionic surfactants of the form R^-M^+ and assuming ideal behaviour below the CMC, Equation (4.12) becomes

$$d\gamma = -\Gamma_R^\sigma d\mu_R - \Gamma_M^\sigma d\mu_M \tag{4.14}$$

If no electrolyte is added, electroneutrality of the interface requires that $\Gamma_R^\sigma = \Gamma_M^\sigma$. Using the mean ionic activities so that $a_2 = (a_R a_M)^{1/2}$ and substituting in Equation (4.14) gives the Gibbs equation for 1 : 1 dissociating compounds:

$$\Gamma_2^\sigma = -\frac{1}{2RT}\frac{d\gamma}{d \ln a_2} \tag{4.15}$$

If swamping electrolyte is introduced (i.e. sufficient salt to make electrostatic effects unimportant) and the same gegenion M^+ as the surfactant is present, then the activity of M^+ is constant and the pre-factor becomes unity, so that Equation (4.13) is appropriate.

For materials that are strongly adsorbed at an interface such as surfactants, a dramatic reduction in interfacial (surface) tension is observed with small changes in bulk phase concentration. The practical applicability of this relationship is that the relative adsorption of a material at an interface, its surface activity, can be determined from measurement of the interfacial tension as a function of solute concentration. Note that in Equations (4.13) and (4.15), for dilute surfactant systems, the concentration can be substituted for activity without loss of generality.

Figure 4.3 shows a typical decay of surface tension of water on increase in surfactant concentration, and how the Gibbs equation (Equation 4.13 or 4.15 is used to quantify

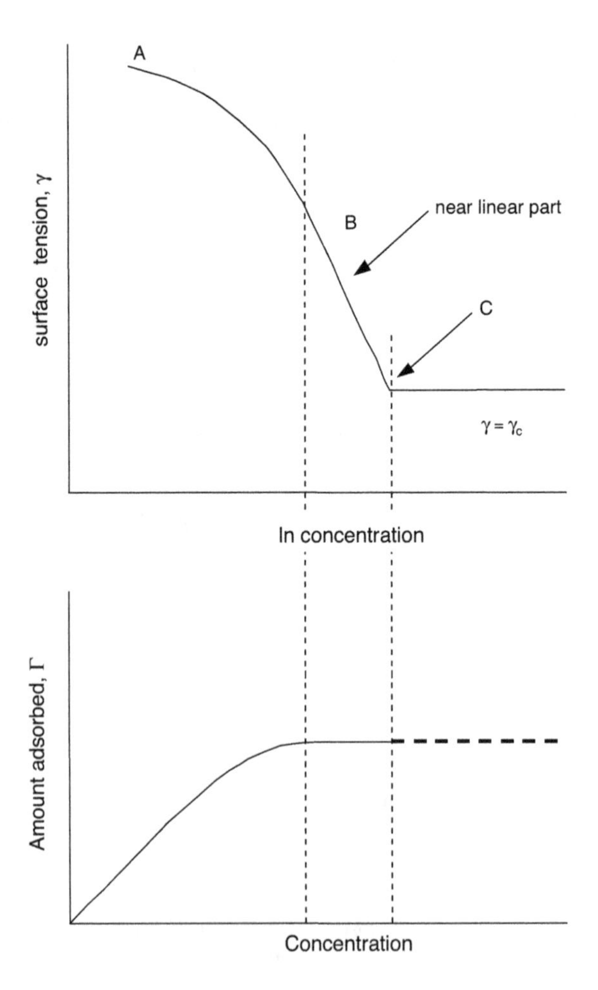

Figure 4.3 Determination of the interfacial adsorption isotherm from surface tension measurement and the Gibbs adsorption equation

adsorption at the surface. At low concentrations a gradual decay in surface tension is observed (from the surface tension of pure water i.e. 72.5 mN m^{-1} at 25 °C) corresponding to an increase in the surface excess of component 2 (region A to B). Then at concentrations close to the CMC, the adsorption tends to a limiting value so the surface tension curve may appear to be essentially linear (region B to C). However, in practice, for most surfactants in the pre-CMC region the γ/ln c is curved so that the local tangent $-d\gamma/d \ln c$ is proportional to Γ_2^σ via Equation (4.13) or (4.15). For single-chain, pure surfactants typical values for Γ_2^σ at the CMC are in the range 2–4 × 10^{-6} mol m^{-2}, with the associated limiting molecular areas being from 0.4–0.6 nm^2.

The value for the Gibbs pre-factor in the case of ionic surfactants has been a matter of discussion (e.g. references 20–23). Of particular concern is the question whether, in the case of ionics, complete dissociation occurs giving rise to a pre-factor of 2, or a depletion layer in the sub-surface could be present so that a somewhat lower pre-factor could be expected. Recent detailed experiments combining tensiometry and neutron reflectivity, which enables direct measurement of the surface excess (as detailed in Chapter 12), have confirmed the use of a pre-factor of 2 in the case of 1:1 dissociating ionic surfactants (24).

Although the Gibbs equation is the most commonly used mathematical relation for adsorption at liquid–liquid and liquid–gas interfaces, other adsorption isotherms have been proposed such as the Langmuir (25), the Szyszkowski (26) and the Frumkin (27) equations. The Gibbs equation itself has been simplified by Guggenheim and Adam with the choice of a different dividing plane and where the interfacial region is considered as a separate bulk phase (of finite volume) (28).

4.4.3 Efficiency and Effectiveness of Surfactant Adsorption

The performance of a surfactant in lowering the surface tension of a solution can be discussed in terms of (i) the concentration required to produce a given surface tension reduction and (ii) the maximum reduction in surface tension that can be obtained regardless of the concentration. These are referred to as the surfactant *efficiency* and *effectiveness* respectively.

A good measure of the *surfactant adsorption efficiency* is the concentration of surfactant required to produce a 20 mN m^{-1} reduction in surface tension. At this value the surfactant concentration is close to the minimum concentration needed to produce maximum adsorption at the interface. This is confirmed by the Frumkin adsorption equation (4.16), which relates the reduction in surface tension (or surface pressure π) and surface excess concentration.

$$\gamma_0 - \gamma = \pi = -2.303RT\,\Gamma_m \log\left(1 - \frac{\Gamma_1}{\Gamma_m}\right) \tag{4.16}$$

The maximum surface excess generally lies in the range 1–4.4 × 10^{-6} mol m^{-2} (29): solving Equation (4.16) indicates that when the surface tension has been reduced by 20 mN m^{-1}, at 25 °C, the surface is 84–99.9% saturated. The negative logarithm of such concentration, pC_{20}, is then a useful quantity since it can be related to the free energy change $\Delta G^{\ominus}$ involved in the transfer of a surfactant molecule from the interior of the bulk liquid phase to the interface. The surfactant adsorption efficiency thus relates to the structural groups in the molecule via the standard free energy change of the individual groups (i.e. free energies of transfer of methylene, terminal methyl and head groups). In particular, for a given homologous series of straight-chain surfactants in water, $CH_3(CH_2)_n$–M, where M is the hydrophilic head group and n is the number of methylene units in the chain, and when the systems are at $\pi = 20\ \mathrm{mN\ m^{-1}}$, the standard free energy of adsorption is

$$\Delta G^{\ominus} = n\,\Delta G^{\ominus}\,(\text{-CH}_2\text{-}) + \Delta G^{\ominus}(\mathrm{M}) + \Delta G^{\ominus}\,(\text{CH}_3\text{-}) \tag{4.17}$$

Then the adsorption efficiency is directly related to the length of the hydrophobic chain (the hydrophilic group remains the same), viz.

$$-\log (C)_{20} = \mathrm{p}C_{20} = n\left[\frac{-\Delta G^{\ominus}(\text{-CH}_2\text{-})}{2.303RT}\right] + \text{constant} \tag{4.18}$$

$\Delta G^{\ominus}(\mathrm{M})$ is considered as a constant and it is assumed that Γ_m does not differ significantly with increasing chain length, and that activity coefficients are unity. The efficiency factor $\mathrm{p}C_{20}$ therefore increases linearly with the number of carbon atoms in the hydrophobic chain. This is also described by Traube's rule (30) (Equation 4.19).

$$\mathrm{Log}\, C_s = B - n\,\mathrm{Log}\, K_T \tag{4.19}$$

C_s is the surfactant concentration, B is a constant, n is the chain length within a homologous series and K_T is Traube's constant. For hydrocarbon straight-chain surfactants K_T is usually around 3 (31) or by analogy to Equation (4.18) is given by

$$\frac{C_n}{C_{n+1}} = K_T = \exp\left[\frac{-\Delta G^{\ominus}(\text{-CH}_2\text{-})}{2RT}\right] \tag{4.20}$$

For compounds having a phenyl group in the hydrophobic chain it is equivalent to about three and one-half normal $\text{-CH}_2\text{-}$ groups.

The larger $\mathrm{p}C_{20}$ the more efficiently the surfactant is adsorbed at the interface and the more efficiently it reduces surface tension. The other main factors that contribute to an increase in surfactant efficiency are summarised below:

- a straight alkyl chain as the hydrophobic group, rather than a branched alkyl chain containing the same number of carbon atoms
- a single hydrophilic group situated at the end of the hydrophobic group, rather than one (or more) at a central position
- a non-ionic or zwitterionic hydrophilic group, rather than an ionic one.

For ionic surfactants, this can be driven by a reduction in the effective charge by (a) use of a more tightly bound (less hydrated) counter-ion and (b) increase in ionic strength of the aqueous phase.

The choice of $20\ \mathrm{mN\ m^{-1}}$ as a standard value of surface tension lowering for the definition of adsorption efficiency is convenient but somewhat arbitrary, and is not valid for systems where surfactants differ significantly in maximum surface excess or when the surface pressure is less than $20\ \mathrm{mN\ m^{-1}}$. Pitt *et al.* (32) circumvented this problem by defining $\Delta\gamma$ as half the surface pressure at the CMC.

The performance of a surfactant can also be discussed in terms of *effectiveness of adsorption*. This is usually defined as the maximum lowering of surface tension γ_{min} regardless of concentration, or as the surface excess concentration at surface saturation Γ_m

since it represents the maximum adsorption. γ_{min}, and Γ_m, are controlled mainly by the critical micelle concentration, and for certain ionics by the solubility limit or Krafft temperature T_k, which will be described briefly in Section 4.5.1. The effectiveness of adsorption is an important factor in determining such properties as foaming, wetting and emulsification, since Γ_m through the Gibbs adsorption equation gives a measure of the interfacial packing.

The efficiency and effectiveness of surfactants do not necessarily run parallel, and it is commonly observed – as shown by Rosen's extensive data listing (29) – that materials producing significant lowering of the surface tension at low concentrations (i.e. they are more efficient) have smaller Γ_m (i.e. they are less effective). In determining surfactant efficiency the role of the molecular structure is primarily thermodynamic, while its role in effectiveness is directly related to the relative size of the hydrophilic and hydrophobic portions of the adsorbing molecule. The area occupied by each molecule is determined either by the hydrophobic chain cross-sectional area, or the area required for closest packing of head groups, whichever is greater. Therefore, surfactant films can be tightly or loosely packed resulting in very different interfacial properties. For instance, straight chains and large head groups (relative to the tail cross section) favour close, effective packing, while branched, bulky, or multiple hydrophobic chains give rise to steric hindrance at the interface. On the other hand, within a series of single straight-chain surfactants, increasing the hydrocarbon chain length from C_8 to C_{20} will have little effect on adsorption effectiveness (29).

4.5 Surfactant Solubility

In aqueous solution, when all available interfaces are saturated, the overall energy reduction may continue through other mechanisms. Depending on the system composition, a surfactant molecule can play different roles in terms of aggregation (formation of micelles, liquid crystal phases, bilayers or vesicles, etc). The physical manifestation of one such mechanism is crystallisation or precipitation of surfactant from solution – that is, bulk-phase separation. While most common surfactants have a substantial solubility in water, this can change significantly with variations in hydrophobic tail length, head group nature, counter-ion valence, solution environment and most, importantly, temperature.

4.5.1 The Krafft Temperature

As for most solutes in water, increasing temperature produces an increase in solubility. However, for surfactants, which are initially insoluble, there is often a temperature at which the solubility suddenly increases very dramatically. This is known as the Krafft point or Krafft temperature, T_K, and is defined as the intersection of the solubility and the CMC curves, i.e. it is the temperature at which the solubility of the monomeric surfactant is equivalent to its CMC at the same temperature. This is illustrated in Figure 4.4. Below T_K surfactant monomers only exist in equilibrium with the hydrated crystalline phase, and above T_K micelles are formed providing much greater surfactant solubility.

The Krafft point of ionic surfactants is found to vary with counter-ion (33), alkyl chain length and chain structure. Knowledge of the Krafft temperature is crucial in many

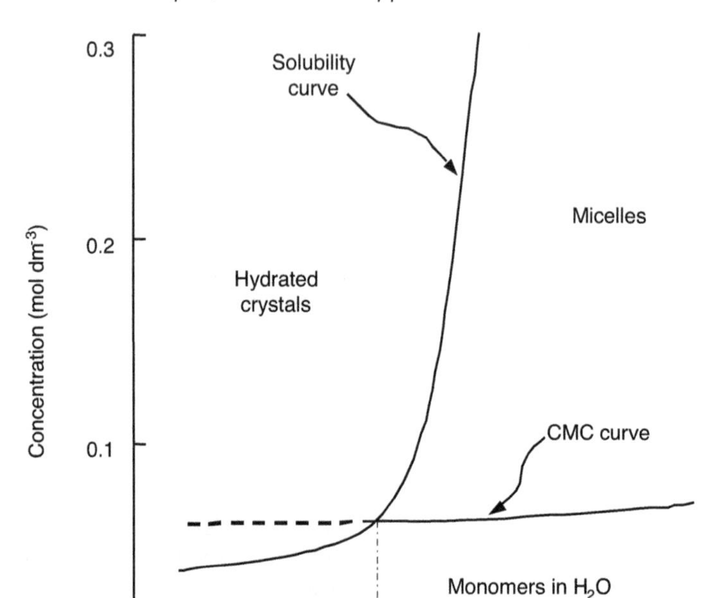

Figure 4.4 *The Krafft temperature T_K is the point at which surfactant solubility equals the critical micelle concentration. Above T_K surfactant molecules form a dispersed phase; below T_K hydrated crystals are formed*

applications since below T_K the surfactant will clearly not perform efficiently; hence typical characteristics such as maximum surface tension lowering and micelle formation cannot be achieved. The development of surfactants with a lower Krafft point but still being very efficient at lowering surface tension (i.e. long-chain compounds) is usually achieved by introducing chain branching, multiple bonds in the alkyl chain or bulkier hydrophilic groups thereby reducing intermolecular interactions that would tend to promote crystallisation.

4.5.2 The Cloud Point

For non-ionic surfactants, a common observation is that micellar solutions tend to become visibly turbid at a well-defined temperature. This is often referred to as the cloud point, above which the surfactant solution phase separates. Above the cloud point, the system consists of an almost micelle-free dilute solution at a concentration equal to its CMC at that temperature, and a surfactant-rich phase. This separation is caused by a sharp increase in aggregation number and a decrease in intermicellar repulsions (34, 35) that produces a difference in density of the surfactant-rich and surfactant-poor phases. Since much larger particles are formed, the solution becomes visibly turbid with large micelles efficiently scattering light. As with Krafft temperatures, the cloud point depends on chemical structure. For poly(ethylene oxide) (PEO) non-ionics, the cloud point increases with increasing EO content for a given hydrophobic group, and at constant EO content it may be lowered by increasing the hydrophobe size, broadening the PEO chain-length distribution and branching in the hydrophobic group (36).

4.6 Micellisation

In addition to forming oriented interfacial monolayers, surfactants can aggregate to form *micelles*, provided their concentration is sufficiently high. Micelles are typically clusters of between 50 to 200 surfactant molecules, whose size and shape are governed by geometric and energetic considerations. Micelle formation occurs over a fairly sharply defined region called the *critical micelle concentration* (CMC). Above the CMC, additional surfactant forms the aggregates, whereas the concentration of the unassociated monomers remains almost constant. As a result, a rather abrupt change in concentration dependence at much the same point can be observed in common equilibrium or transport properties (Figure 4.5).

4.6.1 Thermodynamics of Micellisation

Micelles are dynamic species, in that there is a constant, rapid interchange – typically on a microsecond timescale – of molecules between the aggregate and solution pseudo-phases. This constant formation–dissociation process relies on a subtle balance of interactions. These come from contacts between (i) hydrocarbon chain–water, (ii) hydrocarbon–hydrocarbon chains, (iii) head group–head group, and (iv) from solvation of the head group. Therefore, the net free energy change upon micellisation, ΔG_m, can

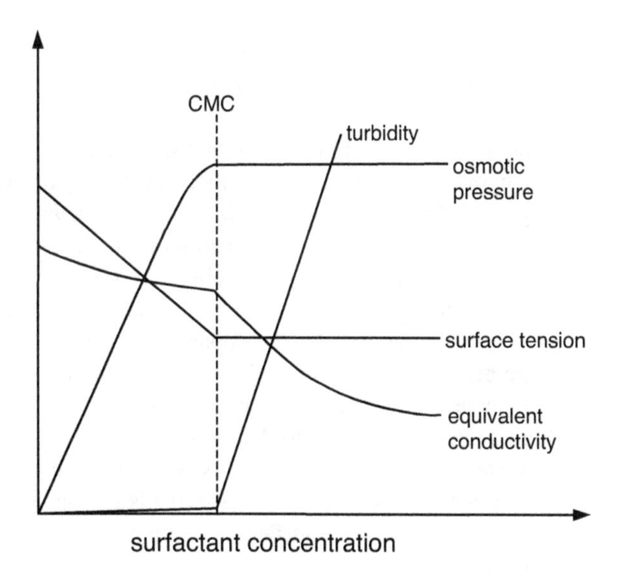

Figure 4.5 Schematic representation of the concentration dependence of some physical properties for solutions of a micelle-forming surfactant

be written as

$$\Delta G_{\rm m} = \Delta G(\text{HC}) + \Delta G(\text{contact}) + \Delta G(\text{packing}) + \Delta G(\text{HG}) \tag{4.21}$$

where:

- ΔG(HC) is the free energy associated with transferring hydrocarbon chains out of water and into the oil-like interior of the micelle
- ΔG(contact) is a surface free energy attributed to solvent–hydrocarbon contacts in the micelle
- ΔG(packing) is a positive contribution associated with confining the hydrocarbon chain to the micelle core
- ΔG(HG) is a positive contribution associated with head group interactions, including electrostatic as well as head group conformation effects.

Aggregation of surfactant molecules partly results from the tendency of the hydrophobic groups to minimise contacts with water by forming oily microdomains within the solvent. There, alkyl–alkyl interactions are maximised, while hydrophilic head groups remain surrounded by water.

The traditional picture of micelle formation thermodynamics is based on the Gibbs–Helmholtz equation ($\Delta G_{\rm m} = \Delta H_{\rm m} - T\Delta S_{\rm m}$). At room temperature the process is characterised by a small, positive enthalpy, $\Delta H_{\rm m}$, and a large, positive entropy of micellisation, $\Delta S_{\rm m}$. The latter is considered as the main contribution to the negative $\Delta G_{\rm m}$ value, and so has led to the (controversial) idea that micellisation is an entropy-driven process. High positive values of $\Delta S_{\rm m}$ are indeed surprising since aggregation, in terms of configurational entropy, should result in a negative contribution (i.e. formation of ordered aggregates from free surfactant monomers). In addition, large values of $\Delta H_{\rm m}$ would have been expected since hydrocarbon groups have very little solubility in water, and consequently a high enthalpy of solution.

One mechanism that accounts for such conflicts is that when alkyl groups of free monomers are surrounded by water, the H_2O molecules form clathrate cavities (i.e. stoichiometric crystalline solids in which water forms cages around solutes), thereby increasing either the strength or number of effective hydrogen bonds (37). Therefore, the predominant effect of the hydrocarbon molecule is to increase the degree of structure in the immediately surrounding water. This is one of the main features of the *hydrophobic effect*, a subject that was explored in detail by Tanford (2) to account for the very slight solubility of hydrocarbons in water. During the formation of micelles, the reverse process occurs: as lyophobic residues aggregate, the highly structured water around each chain collapses back to ordinary bulk water thereby accounting for the apparent large overall gain in entropy, $\Delta S_{\rm m}$. This water–structure effect was also invoked by other researchers (38, 39).

Such an interpretation, however, has been strongly challenged by more recent studies of aqueous systems at high temperatures (up to 166 °C) and micellisation in hydrazine solutions (40). In these systems water loses most of its peculiar structural properties and the formation of structured water around lyophobic species is no longer possible.

The mechanism of micelle formation from surfactant monomers, S, can be described by a series of step-wise equilibria:

$$\text{S} + \text{S} \overset{K_2}{\longleftrightarrow} \text{S}_2 + \text{S} \overset{K_3}{\longleftrightarrow} \text{S}_3 \ldots \overset{K_n}{\longleftrightarrow} \text{S}_n + \text{S} \longleftrightarrow \ldots \tag{4.22}$$

with equilibrium constants K_n for $n = 2 - \infty$, and where the various thermodynamic parameters ($\Delta G^{\ominus}, \Delta H^{\ominus}, \Delta S^{\ominus}$) for the aggregation process can be expressed in terms of K_n. However, each K_n cannot be measured individually, so different approaches have been proposed to model the energetics of the process of self-association. Although not totally accurate, two simple models are generally encountered: the closed-association and the phase separation models. In the closed-association model, with the size range of spherical micelles around the CMC being very limited, it is assumed that only one of the K_n values is dominant, and micelles and monomeric species are considered to be in chemical equilibrium.

$$n\mathrm{S} \longleftrightarrow \mathrm{S}_n \qquad (4.23)$$

n is the number of molecules of surfactant, S, associating to form the micelle (i.e. the aggregation number). In the phase separation model, the micelles are considered to form a new phase within the system at and above the critical micelle concentration, and

$$n\mathrm{S} \longleftrightarrow m\mathrm{S} + \mathrm{S}_n \qquad (4.24)$$

where m is the number of free surfactant molecules in the solution and S_n the new phase. In both cases, equilibrium between monomeric surfactant and micelles is assumed with a corresponding equilibrium constant, K_m, given by

$$K_\mathrm{m} = \frac{[\text{micelles}]}{[\text{monomers}]^n} = \frac{[\mathrm{S}_n]}{[\mathrm{S}]^n} \qquad (4.25)$$

where brackets indicate molar concentrations and n is the number of monomers in the micelle, the aggregation number. Although micellisation is itself a source of non-ideality (41, 42), it is assumed in Equation (4.25) that activities may be replaced by concentrations.

From Equation (4.25), the standard free energy of micellisation per mole of micelles is given by

$$\Delta G^{\ominus}_\mathrm{m} = -RT \ln K_\mathrm{m} = -RT \ln S_\mathrm{n} + nRT \ln S \qquad (4.26)$$

while the standard free energy change per mole of surfactant is

$$\frac{\Delta G^{\ominus}_\mathrm{m}}{n} = -\frac{RT}{n} \ln S_\mathrm{n} + RT \ln S \qquad (4.27)$$

Assuming n is large ($\sim$100) the first term on the right-hand side of Equation 4.27 can be neglected, and an approximate expression for the free energy of micellisation per mole of a neutral surfactant becomes

$$\Delta G^{\ominus}_\mathrm{M,m} \approx RT \ln(\mathrm{CMC}) \qquad (4.28)$$

In the case of ionic surfactants, the presence of the counter-ion and its degree of association with the monomer and micelle must be considered. The mass–action equation becomes

$$nS^x + (n-p)C^y \leftrightarrow S^\alpha_n \qquad (4.29)$$

where C is the concentration of free counter-ions. The degree of dissociation of the surfactant molecules in the micelle, α, the micellar charge, is given by $\alpha = p/n$.

The ionic equivalent to Equation (4.25) is then

$$K_{\mathrm{m}} = \frac{[\mathrm{S}_n]}{[S^x]^n \times [C^y]^{(n-p)}} \tag{4.30}$$

where p is the concentration of free counter-ions associated with, but not bound to the micelle. The standard free energy of micelle formation becomes

$$\Delta G^{\ominus}_{\mathrm{m}} = -RT\{\ln [S_n] - n \ln [S^x] - (n-p) \ln [C^y]\} \tag{4.31}$$

At the CMC $[\mathrm{S}^{-(+)}] = [\mathrm{C}^{+(-)}] = \mathrm{CMC}$ for a fully ionised surfactant, and the standard free energy change per mole of surfactant can be obtained from the approximation

$$\Delta G^{\ominus}_{\mathrm{M,m}} \approx RT\left(2 - \frac{p}{n}\right) \ln (\mathrm{CMC}) \tag{4.32}$$

When the ionic micelle is in a solution of high electrolyte content, the situation described by Equation (4.32) reverts to the simple non-ionic case given by Equation (4.28).

From the Gibbs function and second law of thermodynamics, $\Delta S^{\ominus}$ for non-ionic surfactants is given as

$$\Delta S^{\ominus} = -\frac{\mathrm{d}(\Delta G^{\ominus})}{\mathrm{d}T} = -RT\frac{\mathrm{d} \ln (\mathrm{CMC})}{\mathrm{d}T} - R \ln (\mathrm{CMC}) \tag{4.33}$$

From the Gibbs function and Equations (4.28) and (4.33), the enthalpy of micellisation for non-ionic surfactants, $\Delta H^{\ominus}$, is given by

$$\Delta H^{\ominus} = \Delta G^{\ominus} + T\Delta S^{\ominus} = -RT^2 \frac{\mathrm{d} \ln (\mathrm{CMC})}{\mathrm{d}T} \tag{4.34}$$

and similarly for ionics,

$$\Delta H^{\ominus} = -RT^2\left(2 - \frac{p}{n}\right) \frac{\mathrm{d} \ln (\mathrm{CMC})}{\mathrm{d}T} \tag{4.35}$$

Both the phase separation and closed association models have advantages and disadvantages. One difficulty is activity coefficients: assuming ideality can be erroneous considering the large effective micelle size and charge in comparison to dilute solutions of surfactant monomers. Another disadvantage is the assumption of micellar monodispersity. To counteract this problem, the multiple equilibrium model was proposed, which is an extension of the closed association model. It allows a distribution function of aggregation numbers in micelles to be calculated. A full account of this model and its derivation can be found in elsewhere (43–45).

4.6.2 Factors Affecting the CMC

Many factors are known to affect strongly the CMC. Of major effect is the structure of the surfactant, as will be described below. Also important, but to a lesser extent, are parameters such as counter-ion nature, presence of additives and change in temperature.

4.6.2.1 *The Hydrophobic Group: the Tail*

The length of the hydrocarbon chain is a major factor determining the CMC. For a homologous series of linear single-chain surfactants the CMC decreases logarithmically with carbon number. The relationship usually fits the Klevens equation (46):

$$\log_{10}(\text{CMC}) = A - Bn_{\text{C}} \quad (4.36)$$

where A and B are constants for a particular homologous series and temperature, and n_{c} is the number of carbon atoms in the chain, C_nH_{2n+1}. The constant A varies with the nature and number of hydrophilic groups, while B is constant and is approximately equal to $\log_{10} 2$ ($B \approx 0.29$–0.30) for all paraffin chain salts having a single ionic head group (i.e. reducing the CMC to approximately one-half per each additional $-CH_2-$ group).

Interestingly, for straight-chain dialkyl sulfosuccinates Equation (4.36) is still valid (47) and $B \approx 0.62$, which essentially doubles the value for the single-chain compounds. Alkyl chain branching and double bonds, aromatic groups or some other polar character in the hydrophobic part produce noticeable changes in CMC. In hydrocarbon surfactants, chain branching gives a higher CMC than a comparable straight-chain surfactant (29), and introduction of a six membered benzene ring in the chain is equivalent to only about 3.5 carbon atoms.

4.6.2.2 *The Hydrophilic Group*

For surfactants with the same hydrocarbon chain, varying the hydrophile nature (i.e. from ionic to non-ionic) has an important effect on the CMC values. For instance, for a C_{12} hydrocarbon the CMC with an ionic head group lies in the range of 1×10^{-3} mol dm^{-3}, while a C_{12} non-ionic material exhibits a CMC in the range of 1×10^{-4} mol dm^{-3}.

4.6.2.3 *Counter-ion Effects*

In ionic surfactants micelle formation is related to the interactions of solvent with the ionic head group. Since electrostatic repulsions between ionic groups are greatest for complete ionisation, an increase in the degree of ion binding will decrease the CMC. For a given hydrophobic tail and anionic head group, the CMC decreases as $Li^+ > Na^+ > K^+ > Cs^+ > N(CH_3)_4{}^+ > N(CH_2CH_3)_4{}^+ > Ca^{2+} \approx Mg^{2+}$. For cationic series such as the dodecyltrimethylammonium halides, the CMC decreases in the order $F^- > Cl^- > Br^- > I^-$. In addition, varying counter-ion valency produces a significant effect. Changing from monovalent to divalent or trivalent counter-ions produces a sharp decrease in the CMC.

4.6.2.4 *Effect of Added Salt*

The presence of an indifferent electrolyte causes a decrease in the CMC of most surfactants. The greatest effect is found for ionic materials. The principal effect of the salt is to partially screen the electrostatic repulsion between the head groups and so lower the CMC. For ionics, the effect of adding electrolyte can be empirically quantified viz.

$$\log_{10}(\text{CMC}) = -a \log_{10} C_i + b \quad (4.37)$$

Non-ionic and zwitterionic surfactants display a much smaller effect and Equation (4.37) does not apply.

4.6.2.5 *Effect of Temperature*

The influence of temperature on micellisation is usually weak, reflecting subtle changes in bonding, heat capacity and volume that accompany the transition. This is, however, quite a complex effect. It was shown, for example, that the CMC of most ionic surfactants passes through a minimum as the temperature is varied from 0 °C to 70 °C (48). As already mentioned (Section 4.5), the major effects of temperature are the Krafft and cloud points. For polymeric surfactants strong effects of temperature on CMC are observed and it is common to define a critical micelle temperature (CMT) for this class of surfactants.

4.6.3 Structure of Micelles and Molecular Packing

Early studies (49, 50) showed that, with ionic single alkyl chain compounds spherical micelles form. In particular, in 1936 Hartley (51) described such micelles as spherical aggregates whose alkyl groups form a hydrocarbon liquid-like core, and whose polar groups form a charged surface. Later, with the development of zwitterionic and non-ionic surfactants, micelles of very different shapes were encountered. The different geometries were found to depend mainly on the structure of the surfactant, as well as environmental conditions (e.g. concentration, temperature, pH, electrolyte content).

In the micellisation process, molecular geometry plays an important role and it becomes important to understand how surfactants can pack. The main structures encountered are spherical micelles, vesicles, bilayers or inverted micelles. As described previously, two opposing forces control the self-association process: hydrocarbon–water interactions that favour aggregation (i.e. pulling surfactant molecules out of the aqueous environment), and head group interactions that work in the opposite sense. These two contributions can be considered as an attractive interfacial tension term due to hydrocarbon tails and a repulsion term depending on the nature of the hydrophilic group. More recently, this basic idea was reviewed and quantified by Mitchell and Ninham (52) and Israelachvili (53), resulting in the concept that aggregation of surfactants is controlled by a balanced molecular geometry. In brief, the geometric treatment separates the overall free energy of association to three critical geometric terms (Figure 4.6):

- the minimum interfacial area occupied by the head group, a_0
- the volume of the hydrophobic tail(s), v
- the maximum extended chain length of the tail in the micelle core, l_c.

Formation of a spherical micelle requires l_c to be equal to (or just less than) the micelle core radius, R_{mic}. Then for such a shape, an aggregation number, N, can be expressed either as the ratio of micellar core volume, V_{mic}, and that for the tail, v

$$N = \frac{V_{mic}}{v} = \frac{\frac{4}{3}\pi R_{mic}^3}{v} \tag{4.38}$$

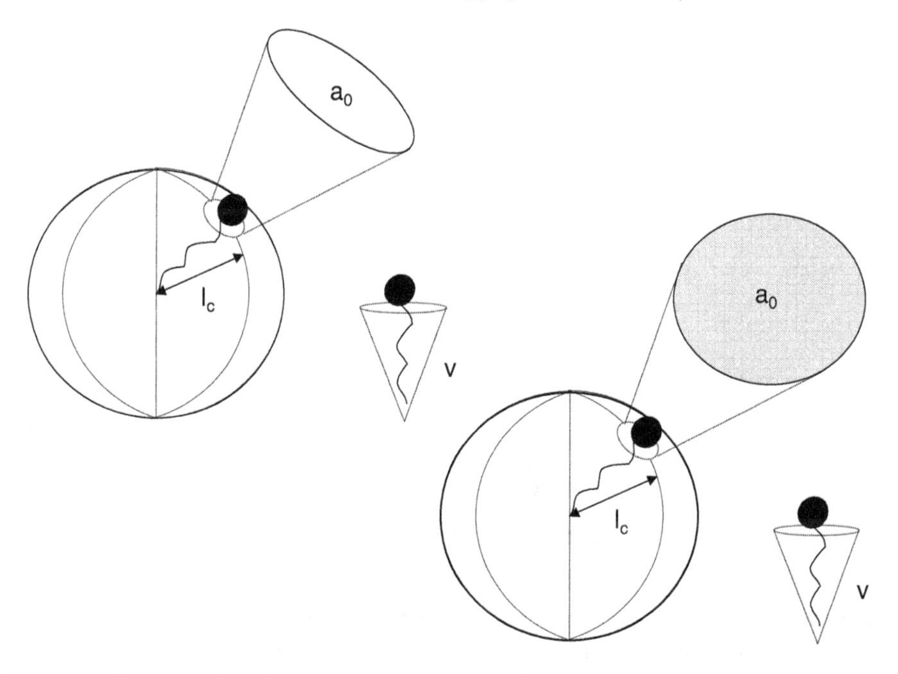

Figure 4.6 *The critical packing parameter P_c (or surfactant number) relates the head group area, the extended length and the volume of the hydrophobic part of a surfactant molecule into a dimensionless number $P_c = v/a_o l_c$*

or as the ratio between the micellar area, A_{mic}, and the cross-sectional area, a_0

$$N = \frac{A_{\text{mic}}}{a_0} = \frac{4\pi R_{\text{mic}}^2}{a_0} \tag{4.39}$$

Equating Equations (4.38) and (4.39)

$$\frac{v}{a_0 R_{\text{mic}}} = 1/3 \tag{4.40}$$

Since l_c cannot exceed R_{mic} for a spherical micelle

$$\frac{v}{a_0 l_c} \leq 1/3 \tag{4.41}$$

More generally, this defines a critical packing parameter, P_c, as the ratio

$$P_c = \frac{v}{a_0 l_c} \tag{4.42}$$

The parameter v varies with the number of hydrophobic groups, chain unsaturation, chain branching and chain penetration by other compatible hydrophobic groups, while a_0 is mainly governed by electrostatic interactions and head group hydration. P_c is a useful quantity since it allows the prediction of aggregate shape and size. The predicted

Table 4.4 *Expected aggregate characteristics in relation to surfactant critical packing parameter, $P_c = v/a_0 l_c$*

P_c	General surfactant type	Expected aggregate structure
<0.33	Single-chain surfactants with large head groups	Spherical or ellipsoidal micelles
0.33–0.5	Single-chain surfactants with small head groups, or ionics in the presence of large amounts of electrolyte	Large cylindrical or rod-shaped micelles
0.5–1.0	Double-chain surfactants with large head groups and flexible chains	Vesicles and flexible bilayer structures
1.0	Double-chain surfactants with small head groups or rigid, immobile chains	Planar extended bilayers
>1.0	Double-chain surfactants with small head groups, very large and bulky hydrophobic groups	Reversed or inverted micelles

aggregation characteristics of surfactants cover a wide range of geometric possibilities, and the main types are presented in Table 4.4 and Figures 4.7 and 4.8.

4.7 Liquid Crystalline Mesophases

Micellar solutions, although the subject of extensive studies and theoretical considerations, are only one of several possible aggregation states. A complete understanding of the aqueous behaviour of surfactants requires knowledge of the entire spectrum of self-assembly. The existence of liquid crystalline phases constitutes an equally important aspect and a detailed description can be found in the literature (e.g. 54, 55). The common features of liquid crystalline phases are summarised below.

4.7.1 Definition

When the volume fraction of surfactant in a micellar solution is increased, typically above a threshold of about 40%, a series of regular geometries is commonly encountered. Interactions between micellar surfaces are repulsive (from electrostatic or hydration forces), so that as the number of aggregates increases and micelles get closer to one another, the only way to maximise separation is to change shape and size. This explains the sequence of surfactant phases observed in the concentrated regime. Such phases are known as mesophases or lyotropic (solvent-induced) liquid crystals.

As the term suggests, liquid crystals are characterised by having physical properties intermediate between crystalline and fluid structures: the degree of molecular ordering is between that of a liquid and a crystal and in terms of rheology the systems are neither simple viscous liquids nor crystalline elastic solids. Certain of these phases have at least one direction that is highly ordered so that liquid crystals exhibit optical birefringence.

Two general classes are encountered, depending on whether one is considering surfactants or other types of material. These are *thermotropic* liquid crystals, in which the

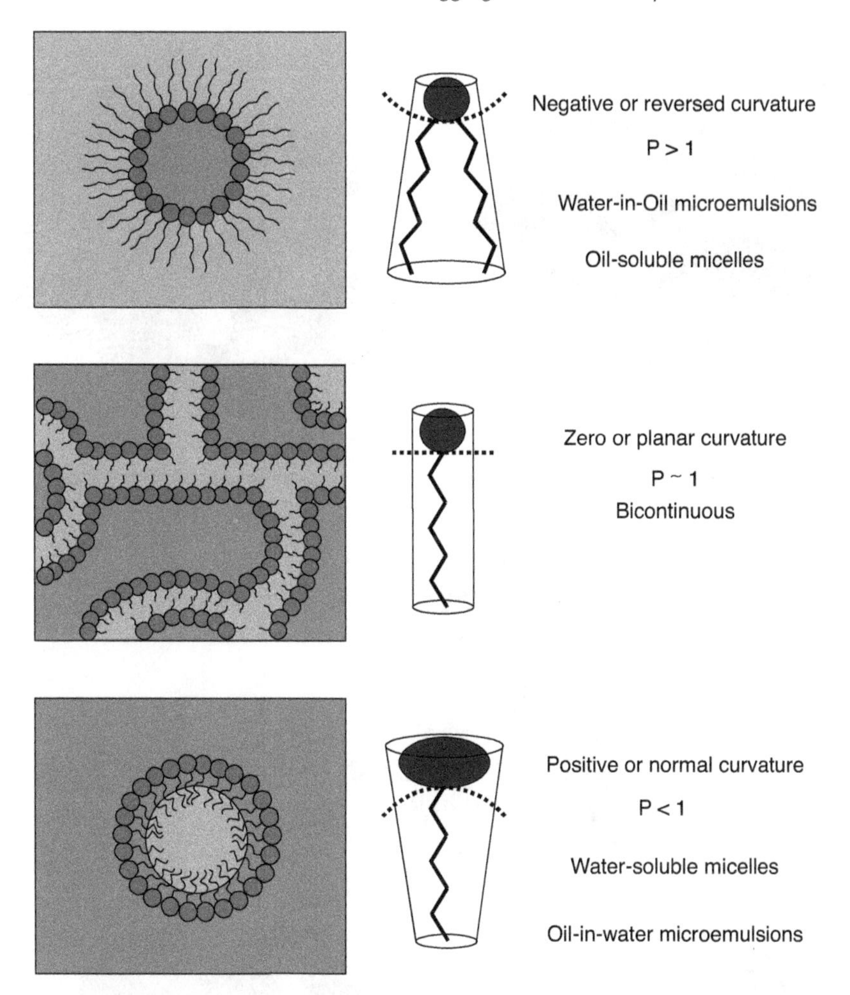

Figure 4.7 *Changes in the critical packing parameters (P_c) of surfactant molecules give rise to different aggregation structures*

structure and properties are determined by temperature (such as employed in LCD cells). For *lyotropic* liquid crystals structure is determined by specific interactions between solute and solvent: surfactant liquid crystals are normally lyotropic.

4.7.2 Structures

The main structures associated with two-component surfactant–water systems are hexagonal (normal), lamellar and several cubic phases. Table 4.5 summarises the notations commonly associated with these phases and their structures are shown in Figure 4.8.

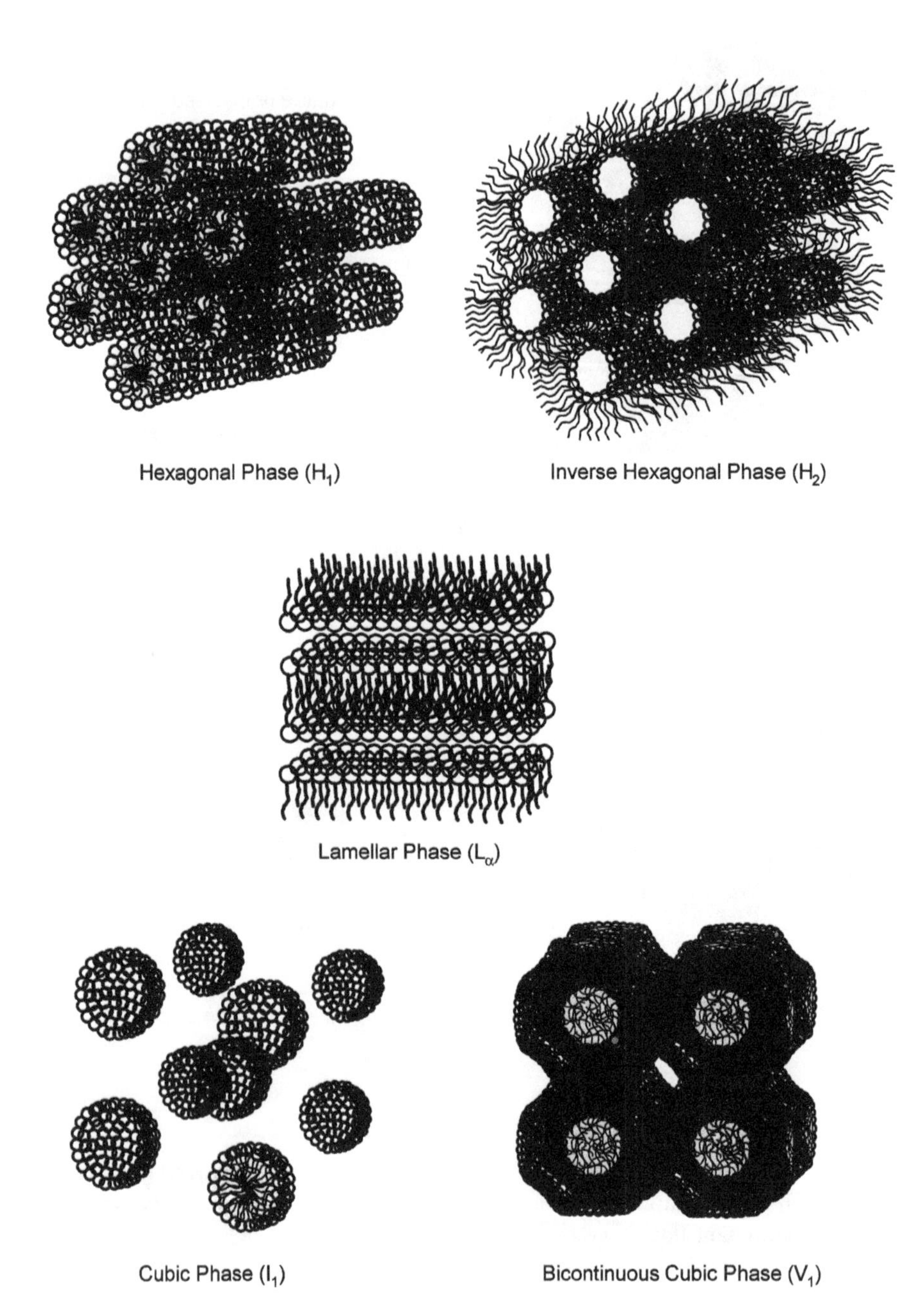

Figure 4.8 *Common surfactant liquid crystalline phases. See Table 4.5 for identification of phase structures*

Table 4.5 *Most common lyotropic liquid crystalline and other phases found in surfactant systems*

Phase structure	Symbol	Other names
Lamellar	L_α	Neat
Hexagonal	H_1	Middle
Reversed hexagonal	H_2	
Cubic (normal micellar)	I_1	Viscous isotropic
Cubic (reversed micellar)	I_2	
Cubic (normal bicontinuous)	V_1	Viscous isotropic
Cubic (reversed bicontinuous)	V_2	
Micellar	L_1	
Reversed micellar	L_2	

The *hexagonal phase* is composed of a close-packed array of long cylindrical micelles, arranged in a hexagonal pattern. The micelles may be 'normal' (in water, H_1) in that the hydrophilic head groups are located on the outer surface of the cylinder, or 'inverted' (H_2), with the hydrophilic groups located internally. Since all the space between adjacent cylinders is filled with hydrophobic groups, the cylindrical micelles are more closely packed than those found in the H_1 phase. As a result, H_2 phases occupy a much smaller region of the phase diagram and are much less common.

The *lamellar phase* (L_α) is built up of alternating water–surfactant bilayers. The hydrophobic chains possess a significant degree of mobility, and the surfactant bilayer can range from being stiff and planar to being very flexible and undulating. The level of disorder may vary smoothly or change abruptly, depending on the specific system, so that it is possible for a surfactant to pass through several distinct lamellar phases.

The *cubic phase* may have a wide variety of structural variations and occurs in different parts of the phase diagram. These are optically isotropic systems and so cannot be characterised by polarising light microscopy. Two main groups of cubic phases have been identified.

- The micellar cubic phases (I_1 and I_2) – built up of regular packing of small micelles (or reversed micelles in the case of I_2). The micelles are short prolates arranged in a body-centred cubic close-packed array (56, 57).
- The bicontinuous cubic phases (V_1 and V_2) – thought to be rather extended, porous, connected structures in three dimensions. They are considered to be formed by either connected rod-like micelles, similar to branched micelles, or bilayer structures. Denoted V_1 and V_2, they can be normal or reverse structures and are positioned between H_1 and L_α and between L_α and H_2 respectively.

In addition to having different structures these common forms also show different viscosities, in the order:

$$\text{cubic} > \text{hexagonal} > \text{lamellar}$$

Cubic phases are generally the more viscous since they have no obvious shear plane and so layers of surfactant aggregates cannot slide easily relative to each other. Hexagonal phases typically contain 30–60% water by weight but are very viscous since cylindrical aggregates can move freely only along their length. Lamellar phases are generally less viscous than the hexagonal phases due to the ease with which each parallel layers can slide over each other during shear.

4.7.3 Phase Diagrams

The sequence of mesophases can be identified simply by using a polarising microscope and the isothermal technique known as a phase cut. Briefly, starting from a small amount of surfactant, a concentration gradient is set up spanning the entire phase diagram, from pure water to pure surfactant. Since crystal hydrates and some of the liquid crystalline phases are birefringent, viewing in the microscope between crossed polars shows up the complete sequence of mesophases.

Transformations between different mesophases are controlled by a balance between molecular packing geometry and inter-aggregate forces. As a result, the system characteristics are highly dependent on the nature and amount of solvent present. Generally, the main types of mesophases tend to occur in the same order and in roughly the same position in the phase diagram. Figure 4.9 shows a classic binary phase diagram of a non-ionic surfactant $C_{16}EO_8$–water. The sequence of phases is common to most non-ionic surfactants of the kind

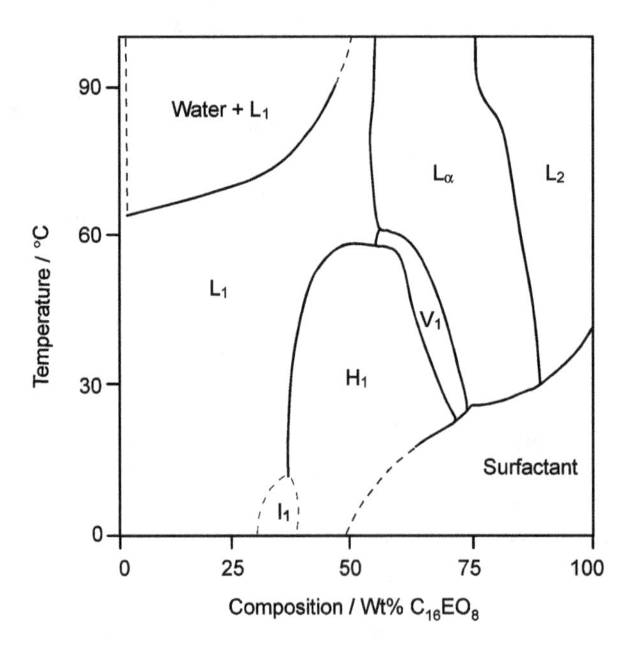

Figure 4.9 *Phase diagram for the non-ionic $C_{16}EO_8$ illustrating the various liquid crystalline phases. L_1 and L_2 are isotropic solutions. See Table 4.5 for details of the other phases. (Reprinted with permission from Mitchell D. J. et al. J. Chem. Soc. Faraday Trans. I 1983, 79, 975). Copyright (1983) Royal Society of Chemistry*

C_iE_j, although the positions of the phase boundaries, in terms of temperature and concentration limits, depend somewhat on the chemical identity of the surfactant.

4.8 Advanced Surfactants

Attention has begun to focus on advanced functionalised surfactants, for example compounds bearing polymerisable (4, 5, 58), pH sensitive (4, 5) and light sensitive (4, 5, 59) groups as the hydrophilic and/or hydrophobic moieties. For example, if a surfactant molecule contains a suitable chromophore, illumination can be used to achieve different physical photo-induced responses. Light can be used to cause *cis–trans* isomerisations, dimerisations, photoscission, polymerisations or polarity changes in these surfactants. Obviously such dramatic changes in hydrophilic head groups and hydrophobic tails could feed through to changes in the surface activity, aggregation structure, viscosity, (micro)emulsion separation and solubilisation. Recent studies have established feasibility of employing photosurfactants to drive molecular changes in interfacial and colloidal systems. The effects are quite general (59), and point to exciting potential applications in light-directed phase, interface and aggregate stability control, delivery of active components and photo-rheology (see Chapter 12). An example is shown in Figure 4.10, where a reduction in surface tension, and a commensurate increase in surface wettability, of water drops containing a custom made photosurfactant can be seen after irradiation with UV light.

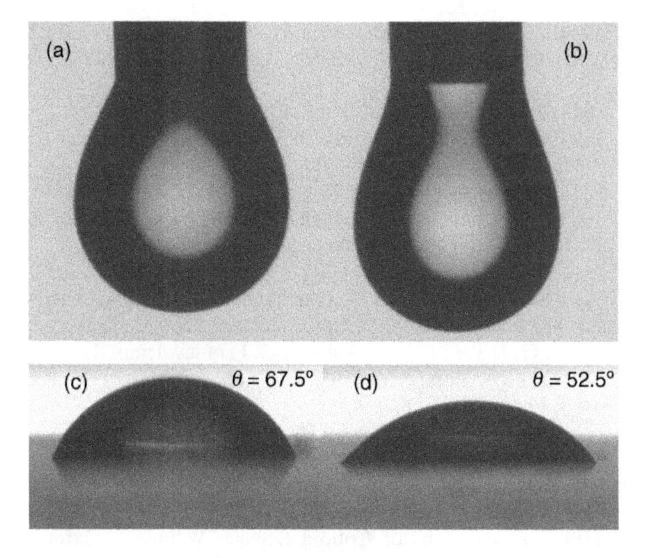

Figure 4.10 *Samples of aqueous photosurfactant solutions studied by surface tensiometry and contact angles (see Chapter 10). Pendant drops 1.0 mmol dm*$^{-3}$ *photosurfactant: non-irradiated and irradiated samples (a) and (b) respectively. Sessile drops 0.24 mmol dm*$^{-3}$ *of non-irradiated and irradiated samples (c) and (d), showing changes in contact angle*

References

(1) Evans, D. F., Wennerström, H. (1999) The Colloidal Domain, Wiley-VCH, New York.
(2) Tanford, C. (1978) The Hydrophobic Effect: Formation of Micelles and Biological Membranes. John Wiley & Sons, Ltd, New York.
(3) Ogino, K., Abe, M., eds. (1993) Mixed Surfactants Systems. Marcel Dekker, New York.
(4) Robb, I. D. (1997) Specialist Surfactants. Blackie Academic & Professional, London.
(5) Holmberg, K. Ed. (1998) Novel Surfactants. Marcel Dekker, New York.
(6) Hollis, G. Ed. (1976) Surfactants UK. Tergo-Data.
(7) The Soap and Detergent Association home page, http://www.sdahq.org/.
(8) Karsa, D. R., Goode, J.M., Donnelly, P.J. eds. (1991) Surfactants Applications Directory. Blackie & Son, London.
(9) Dickinson, E. (1992) An Introduction to Food Colloids. Oxford University Press, Oxford.
(10) Solans, C., Kunieda, H. eds. (1997) Industrial Applications of Microemulsions. Marcel Dekker, New York.
(11) Tanford, C. (1978) The Hydrophobic Effect: Formation of Micelles and Biological Membranes. John Wiley & Sons, Ltd, USA.
(12) Dukhin, S. S., Kretzschmar, G., Miller, R. (1995) Dynamics of Adsorption at Liquid Interfaces. Elsevier, Amsterdam.
(13) Rusanov, A. I., Prokhorov, V. A. (1996) Interfacial Tensiometry. Elsevier, Amsterdam.
(14) Chang, C.-H., Franses, E. I. (1995) Colloid Surf., 100: 1.
(15) Miller, R., Joos, P., Fainermann, V. (1994) Adv. Colloid Interface Sci., 49: 249.
(16) Lin, S. -Y., McKeigue, K., Maldarelli, C. (1991) Langmuir, 7: 1055.
(17) Hsu, C.-H., Chang, C.-H., Lin, S.-Y. (1999) Langmuir, 15: 1952.
(18) Eastoe, J., Dalton, J. S. (2000) Adv. Colloid Interface Sci., 85: 103.
(19) Gibbs, J. W. (1931) The Collected Works of J. W. Gibbs. Longmans, Green, New York. Vol. I: p. 219.
(20) Elworthy, P. H., Mysels, K. J. (1966) J. Colloid Interface Sci., 21: 331.
(21) Lu, J. R., Li, Z. X., Su, T. J., Thomas, R. K., Penfold, J. (1993) Langmuir, 9: 2408.
(22) Bae, S., Haage, K., Wantke, K., Motschmann, H. (1999) J. Phys. Chem. B, 103: 1045.
(23) Downer, A., Eastoe, J., Pitt, A. R., Penfold, J., Heenan, R. K. (1999) Colloids Surf. A, 156: 33.
(24) Eastoe, J., Nave, S., Downer, A., Paul, A., Rankin, A., Tribe, K., Penfold, J. (2000) Langmuir, 16: 4511.
(25) Langmuir, I. (1948) J. Am. Chem. Soc., 39: 1917.
(26) Szyszkowski, B. (1908) Z. Phys. Chem., 64: 385.
(27) Frumkin, A. (1925) Z. Phys. Chem., 116: 466.
(28) Guggenheim, E. A., Adam, N. K. (1933) Proc. Roy. Soc. (London), A139: 218.
(29) Rosen, M. J. (1989) Surfactants And Interfacial Phenomena, John Wiley & Sons, Ltd, USA.
(30) Traube, I. (1891) Justus Liebigs Ann. Chem., 265: 27.
(31) Tamaki, K., Yanagushi, T., Hori, R. (1961) Bull. Chem. Soc. Jpn., 34: 237.
(32) Pitt, A. R., Morley, S. D., Burbidge, N. J., Quickenden, E. L. (1996) Coll. Surf. A, 114: 321.
(33) Hato, M., Tahara, M., Suda, Y. (1979) J. Coll. Interface Sci., 72: 458.
(34) Staples, E. J., Tiddy, G. J. T. (1978) J. Chem. Soc., Faraday Trans. 1, 74: 2530.
(35) Tiddy, G. J. T. (1980) Phys. Rep., 57: 1.
(36) Schott, H. (1969) J. Pharm. Sci., 58: 1443.
(37) Frank, H. S., Evans, M. W. (1945) J. Chem. Phys., 13: 507.
(38) Evans, D. F., Wightman, P. J. (1982) J. Colloid Interface Sci., 86: 515.
(39) Patterson, D., Barbe, M. (1976) J. Phys. Chem., 80: 2435.
(40) Evans, D. F. (1988) Langmuir, 4: 3.
(41) Hunter, R. J. (1987) Foundations of Colloid Science Volume I. Oxford University Press, New York.
(42) Evans, D. F., Ninham, B. W. (1986) J. Phys. Chem., 90: 226.
(43) Corkhill, J. M., Goodman, J. F., Walker, T., Wyer, J. (1969) Proc. Roy. Soc. (London), A, 312: 243.

(44) Mukerjee, P. (1972) J. Phys. Chem., 76: 565.
(45) Aniansson, E. A. G., Wall, S. N. (1974) J. Phys. Chem., 78: 1024.
(46) Klevens, H. (1953) J. Am. Oil Chem. Soc., 30 (7): 4.
(47) Williams, E. F., Woodberry, N. T., Dixon, J. K. (1957) J. Colloid Interface Sci., 12: 452.
(48) Kresheck, G. C. (1975) In Water - a Comprehensive Treatise. Ed. F. Franks, Plenum Press, New York, pp. 95–167.
(49) McBain, J. W. (1913) Trans. Faraday Soc., 9: 99.
(50) Reychler, F. (1913) Kolloid-Z., 12: 283.
(51) Hartley, G. S. (1936) Aqueous Solutions of Paraffin Chain Salts. Hermann & Cie, Paris.
(52) Mitchell, D. J., Ninham, B. W. (1981) J. Chem. Soc. Faraday Trans. 2, 77: 601.
(53) Israelachvili, J. N. (1985) Intermolecular and Surface Forces. Academic Press, London, p. 251.
(54) Laughlin, R. G. (1994) The Aqueous Phase Behaviour of Surfactants. Academic Press, London.
(55) Chandrasekhar, S. (1992) Liquid Crystals. Cambridge University Press, New York.
(56) Fontell, K., Kox, K. K., Hansson, E. (1985) Mol. Cryst. Liquid Cryst. Lett., 1: 9.
(57) Fontell, K. (1990) Coll. Polymer Sci., 268: 264.
(58) Summers, M., Eastoe, J. (2003) Adv. Coll. Int. Sci., 100–102: 137.
(59) Eastoe, J., Vesperinas, A. (2005) Soft Matter, 1: 338.

5

Microemulsions

Julian Eastoe

School of Chemistry, University of Bristol, UK

5.1 Introduction

This chapter is devoted to another important property of surfactants, that of stabilisation of water–oil films and formation of microemulsions. These are a special kind of colloidal dispersion that have attracted a great deal of attention because of their ability to solubilise otherwise insoluble materials. Industrial applications of microemulsions have escalated in the last 40 years following an increased understanding of formation, stability and the role of surfactant molecular architecture. This chapter reviews the main theoretical features relevant to the present work and some common techniques used to characterise micro-emulsion phases.

5.2 Microemulsions: Definition and History

Microemulsions were first identified in the early 1940s and initially they were referred to as hydrophilic oleomicelles or oleophillic hydromicelles. The term microemulsion was coined in the late 1950s, but until the mid-1970s they were viewed as something of a scientific curiosity with little research being conducted on them. Research interest picked up during the 'oil crisis' in the early 1970s because microemulsions can be used in tertiary oil recovery (that is the partial removal of the residual oil remaining in the well rock), but faded again as the oil crisis receded and tertiary oil recovery became commercially unrealistic due to its high cost.

Colloid Science: Principles, methods and applications, Second Edition Edited by Terence Cosgrove

One popular definition of microemulsions is from Danielsson and Lindman (1) '*a microemulsion is a system of water, oil and an amphiphile which is a single optically isotropic and thermodynamically stable liquid solution*'. In some respects, microemulsions can be considered as small-scale versions of emulsions, i.e. droplet type dispersions either of oil-in-water (o/w) or of water-in-oil (w/o), with a size range in the order of 1–50 nm in drop radius. Such a description, however, lacks precision since, as amplified in Section 5.3, there are significant differences between microemulsions and ordinary emulsions (or macroemulsions). In particular, in emulsions the average drop size grows continuously with time so that phase separation ultimately occurs under gravitational force, i.e. they are thermodynamically unstable and their formation requires input of work. The drops of the dispersed phase are generally large (>0.1 μm) so that they often take on a milky appearance.

On the other hand, for microemulsions, once the conditions are right, spontaneous formation occurs. Hence, microemulsions are thermodynamically stable mixtures of two immiscible (or partially immiscible) liquids; the thermodynamic stability is a direct result of strong adsorption of highly effective surfactants at the interface between the two liquid phase domains. As outlined below in Section 5.3, the result of these conditions is to generate large interfacial area systems, and from a structural viewpoint microemulsions generally comprise nanometre-sized domains of one liquid phase, dispersed in another liquid phase, coated by stabilising surfactant monolayers.

As for simple aqueous systems, microemulsion formation is dependent on surfactant type and structure. If the surfactant is ionic and contains a single hydrocarbon chain (e.g. sodium dodecylsulfate, SDS) microemulsions are only formed if a co-surfactant (e.g. a medium size aliphatic alcohol) and/or electrolyte (e.g. 0.2 M NaCl) are also present. With double chain ionics (e.g. Aerosol-OT) and some non-ionic surfactants a co-surfactant is not necessary. This results from one of the most fundamental properties of microemulsions, that is, an ultra-low interfacial tension between the oil and water phases, $\gamma_{o/w}$. The main role of the surfactant is to reduce $\gamma_{o/w}$ sufficiently – i.e. lowering the energy required to increase the surface area – so that spontaneous dispersion of water or oil droplets occurs and the system is thermodynamically stable. As described in Section 5.3.1 ultra-low tensions are crucial for the formation of microemulsions and depend on system composition.

Microemulsions were not really recognised until the work of Hoar and Schulman in 1943, who reported a spontaneous emulsion of water and oil on addition of a strong surface-active agent (2). The term 'microemulsion' was first used even later by Schulman *et al.* (3) in 1959 to describe a multiphase system consisting of water, oil, surfactant and alcohol, which forms a transparent solution. There has been much debate about the word 'microemulsion' to describe such systems (4). Although not systematically used today, some prefer the names 'micellar emulsion' (5) or 'swollen micelles' (6). Microemulsions were probably discovered well before the studies of Schulmann: Australian housewives have used since the beginning of last century water/eucalyptus oil/soap flake/white spirit mixtures to wash wool, and the first commercial microemulsions were probably the liquid waxes discovered by Rodawald in 1928. Interest in microemulsions really stepped up in the late 1970s and early 1980s when it was recognised that such systems could improve oil recovery and when oil prices reached levels where tertiary recovery methods became profit earning (7).

Nowadays this is no longer the case, but other microemulsion applications were discovered, e.g. catalysis, preparation of submicrometre particles, solar energy conversion, liquid–liquid extraction (mineral, proteins, etc.). Recent developments and applications are discussed in Section 5.5. Together with classical applications in detergency and lubrication, the field remains sufficiently important to continue to attract a number of scientists. From the fundamental research point of view, a great deal of progress has been made in the last 20 years in understanding microemulsion properties. In particular, interfacial film stability and microemulsion structures can now be characterised in detail owing to the development of new and powerful techniques such as small-angle neutron scattering (SANS, as described in Chapter 13). The following sections deal with fundamental microemulsion properties, i.e. formation and stability, surfactant films, classification and phase behaviour.

5.3 Theory of Formation and Stability

5.3.1 Interfacial Tension in Microemulsions

A simple picture for describing microemulsion formation is to consider a subdivision of the dispersed phase into very small droplets. Then the configurational entropy change, ΔS_{conf}, can be approximately expressed as (8):

$$\Delta S_{conf} = -nk_B[\ln\phi + \{(1-\phi)/\phi\}\ln(1-\phi)] \tag{5.1}$$

where n is the number of droplets of dispersed phase, k_B is the Boltzmann constant and ϕ is the dispersed phase volume fraction. The associated free energy change can be expressed as a sum of the free energy for creating new area of interface, $\Delta A\gamma_{12}$, and configurational entropy in the form (9):

$$\Delta G_{form} = \Delta A\gamma_{12} - T\Delta S_{conf} \tag{5.2}$$

where ΔA is the change in interfacial area A (equal to $4\pi r^2$ per droplet of radius r) and γ_{12} is the interfacial tension between phases 1 and 2 (e.g. oil and water) at temperature T (in Kelvin). Substituting Equation (5.1) into Equation (5.2) gives an expression for obtaining the maximum interfacial tension between phases 1 and 2. On dispersion, the droplet number increases and ΔS_{conf} is positive. If the surfactant can reduce the interfacial tension to a sufficiently low value, the energy term in Equation (5.2) ($\Delta A\gamma_{12}$) will be relatively small and positive, thus allowing a negative (and hence favourable) free energy change, that is, spontaneous microemulsification.

In surfactant-free oil–water systems, $\gamma_{o/w}$ is of the order of 50 mN m^{-1}, and during microemulsion formation the increase in interfacial area, ΔA, is very large, typically a factor of 10^4 to 10^5. Therefore in the absence of surfactant, the second term in Equation (5.2) is of the order of 1000 k_BT, and in order to fulfil the condition $\Delta A\gamma_{12} \leq T\Delta S_{conf}$, the interfacial tension should be very low (approximately 0.01 mN m^{-1}). Some surfactants (double chain ionics (10, 11) and some non-ionics (12)) can produce extremely low interfacial tensions – typically 10^{-2} to 10^{-4} mN m^{-1} – but in most cases, such low values cannot be achieved by a single surfactant. An effective way to further decrease $\gamma_{o/w}$ is to include a second surface-active species (either a surfactant or medium-chain alcohol), that is a co-surfactant. This can be understood in terms of the Gibbs equation extended to multicomponent systems (13–15). It relates the interfacial tension to the surfactant film composition and the chemical potential, μ, of each component in the system, i.e.

$$d\gamma_{o/w} = -\sum_i (\Gamma_i d\mu_i) \approx -\sum_i (\Gamma_i RT d \ln C_i) \tag{5.3}$$

where C_i is the molar concentration of component i in the mixture, and Γ_i the surface excess (mol m^{-2}). Assuming that surfactants and co-surfactants, with concentration C_s and C_{co} respectively, are the only adsorbed components (i.e. $\Gamma_{water} = \Gamma_{oil} = 0$), Equation (5.3) becomes:

$$d\gamma_{o/w} = -\Gamma_s RT d \ln C_s - \Gamma_{co} RT d \ln C_{co} \tag{5.4}$$

Integration of Equation (5.4) gives:

$$\gamma_{o/w} = \gamma^0_{o/w} - \int_0^{C_s} \Gamma_s RT d \ln C_s - \int_0^{C_{co}} \Gamma_{co} RT d \ln C_{co} \tag{5.5}$$

Equation (5.5) shows that $\gamma^0_{o/w}$ is lowered by two terms, both from the surfactant and co-surfactant (of surface excesses Γ_s and Γ_{co} respectively) so their effects are additive.

Figure 5.1 shows typical low interfacial tensions found in microemulsions, in this case spanning ~1 to 10^{-3} mN m^{-1}. The effect of salt concentration is consistent with

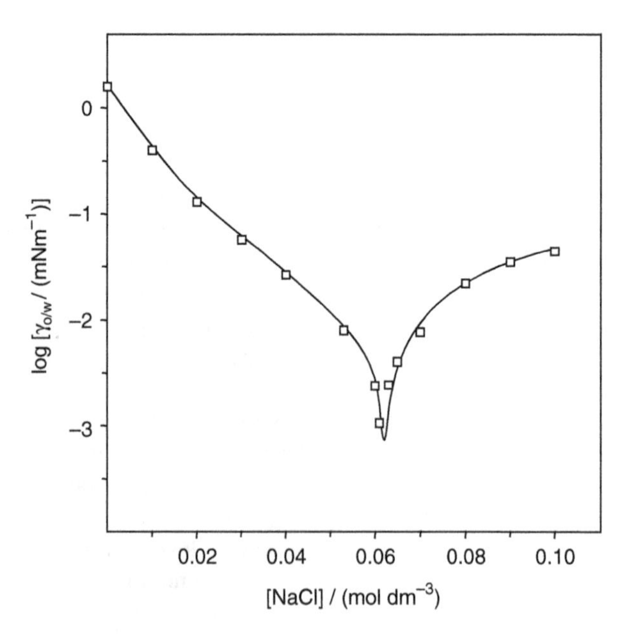

Figure 5.1 *Oil–water interfacial tension between n-heptane and aqueous NaCl solutions as a function of salt concentration in the presence of AOT surfactant. The values were determined by spinning drop tensiometry. The AOT surfactant concentration is 0.050 mol dm^{-3}, temperature 25 °C*

changes in the phase behaviour, which are discussed in more detail in Section 5.4 and Figure 5.2 below.

5.3.2 Kinetic Instability

Internal contents of the microemulsion droplets are known to exchange, typically on the millisecond time scale (16, 17): they diffuse and undergo collisions. If collisions are sufficiently violent, then the surfactant film may rupture thereby facilitating droplet exchange, that is the droplets are kinetically unstable. However, if one disperses emulsions as sufficiently small droplets (<500 Å), the tendency to coalesce will be counteracted by an energy barrier. Then the system will remain dispersed and transparent for a long period of time (months) (18). Such an emulsion is said to be kinetically stable (19). The mechanism of droplet coalescence has been reported for AOT w/o microemulsions (16); the droplet exchange process was characterised by a second-order rate constant k_{ex}, which is believed to be activation controlled (hence the activation energy, E_a, barrier to fusion) and not purely diffusion controlled. Other studies (20) have shown that the dynamic aspects of microemulsions are affected by the flexibility of the interfacial film, that is film rigidity (see Section 5.4.2), through a significant contribution to the energy barrier. Under the same experimental conditions, different microemulsion systems can have different k_{ex} values (16): for AOT w/o system at room temperature, k_{ex} is in the range 10^6–$10^9\ dm^3\ mol^{-1}\ s^{-1}$, and for non-ionics C_iE_j, 10^8–$10^9\ dm^3\ mol^{-1}\ s^{-1}$ (16, 17, 20). In any case, an equilibrium droplet shape and size is always maintained and this can be studied by different techniques (20).

5.4 Physicochemical Properties

This section gives an overview of the main parameters characterising microemulsions. References will be made to related behaviour for planar interfaces presented in Chapter 4.

5.4.1 Predicting Microemulsion Type

A well-known classification of microemulsions is that of Winsor (21) who identified four general types of phase equilibria.

- *Type I:* the surfactant is preferentially soluble in water and oil-in-water (o/w) microemulsions form (Winsor I). The surfactant-rich water phase coexists with the oil phase where surfactant is only present as monomers at small concentration.
- *Type II:* the surfactant is mainly in the oil phase and water-in-oil (w/o) microemulsions form. The surfactant-rich oil phase coexists with the surfactant-poor aqueous phase (Winsor II).
- *Type III:* a three-phase system where a surfactant-rich middle phase coexists with both excess water and oil surfactant-poor phases (Winsor III or middle-phase microemulsion).
- *Type IV:* a single-phase (isotropic) micellar solution forms upon addition of a sufficient quantity of amphiphile (surfactant plus alcohol).

Depending on surfactant type and sample environment, types I, II, III or IV form preferentially, the dominant type being related to the molecular arrangement at the interface

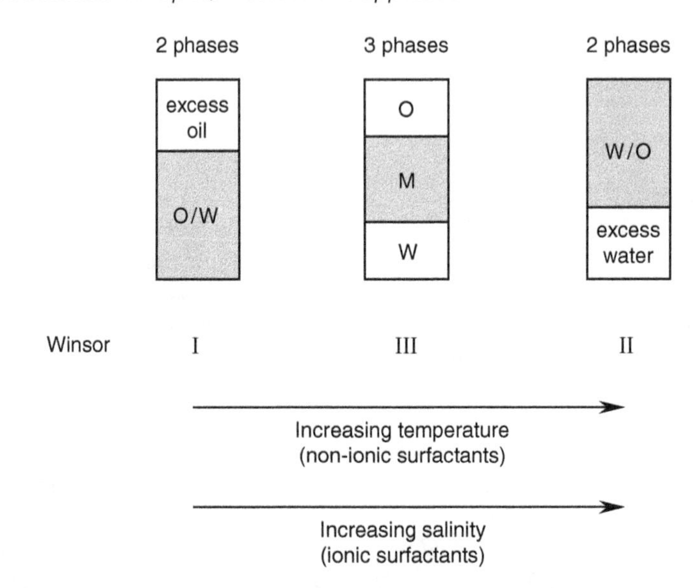

Figure 5.2 *Winsor classification and phase sequence of microemulsions encountered as temperature or salinity is scanned for non-ionic and ionic surfactant respectively. Most of the surfactant resides in the shaded area. In the three-phase system the middle-phase microemulsion (M) is in equilibrium with both excess oil (O) and water (W)*

(see below). As illustrated in Figure 5.2, phase transitions are brought about by increasing either electrolyte concentration (in the case of ionic surfactants) or temperature (for non-ionics). Table 5.1 summarises the qualitative changes in phase behaviour of anionic surfactants when formulation variables are modified (22).

Various investigators have focused on interactions in an adsorbed interfacial film to explain the direction and extent of interfacial curvature. The first concept was that of Bancroft (23) and Clowes (24) who considered the adsorbed film in emulsion systems to be duplex in nature, with an inner and an outer interfacial tension acting independently (25). The interface would then curve such that the inner surface was one of higher tension. Bancroft's rule was stated as '*that phase will be external in which the emulsifier is most*

Table 5.1 *Qualitative effect of several variables on the observed phase behaviour of anionic surfactants. Reprinted with permission from Ref. (22). Copyright (1984) Elsevier Ltd. Concentration effect for low molecular weight alcohols: [a]methanol, ethanol, propanol; [b]higher alkanols*

Scanned variables (increase)	Ternary diagram transition
Salinity	I → III → II
Oil: Alkane carbon number	II → III → I
Alcohol: low M.W.[a]	I → III → II
high M.W.[b]	I → III → II
Surfactant: lipophilic chain length	I → III → II
Temperature	I → III → I

soluble'; i.e. oil-soluble emulsifiers will form w/o emulsions and water-soluble emulsifiers o/w emulsions. This qualitative concept was largely extended and several parameters have been proposed to quantify the nature of the surfactant film. They are briefly presented in this section. Further details concerning the these microemulsion types and their location in the phase diagram will be given in Section 5.4.3.

5.4.1.1 The R Ratio

The R ratio was first proposed by Winsor (21) to account for the influence of amphiphiles and solvents on interfacial curvature. The primary concept is to relate the energies of interaction between the amphiphile layer and the oil and water regions. Therefore, this R ratio compares the tendency for an amphiphile to disperse into oil, to its tendency to dissolve in water. If one phase is favoured, the interfacial region tends to take on a definite curvature. A brief description of the concept is given below, and a full account can be found elsewhere (26).

In micellar or microemulsion solutions, three distinct (single or multicomponent) regions can be recognised: an aqueous region, W, an oil or organic region, O, and an amphiphilic region, C. As shown in Figure 5.3, it is useful to consider the interfacial zone as having a definite composition, separating essentially bulk-phase water from bulk-phase oil. In this simple picture, the interfacial zone has a finite thickness, and will contain, in addition to surfactant molecules, some oil and water.

Cohesive interaction energies therefore exist within the C layer, and these determine interfacial film stability. They are depicted schematically in Figure 5.3: the cohesive energy between molecules x and y is defined as A_{xy}, and is positive whenever interaction between molecules is attractive. A_{xy} is depicted as the cohesive energy per unit area between surfactant, oil and water molecules residing in the anisotropic interfacial C layer. For surfactant–oil and surfactant–water interactions A_{xy} can be considered to be composed of two additive contributions:

$$A_{xy} = A_{Lxy} + A_{Hxy} \tag{5.6}$$

where A_{Lxy} quantifies interaction between non-polar portions of the two molecules (typically London dispersion forces) and A_{Hxy} represents polar interactions, especially hydrogen bonding or Coulombic interactions. Thus, for surfactant–oil and surfactant–water interactions, cohesive energies to be considered are:

$$A_{co} = A_{Lco} + A_{Hco} \tag{5.7}$$

$$A_{cw} = A_{Lcw} + A_{Hcw} \tag{5.8}$$

A_{Hco} and A_{Lcw} are generally very small values and can be ignored.

Other cohesive energies are those arising from the following interactions:

- water–water, A_{ww}
- oil–oil, A_{oo}
- hydrophobic–hydrophobic parts (L) of surfactant molecules, A_{LL}
- hydrophilic–hydrophilic parts (H) of surfactant molecules, A_{HH}.

The cohesive energy A_{co} evidently promotes miscibility of the surfactant molecules with the oil region, and A_{cw} with water. On the other hand, A_{oo} and A_{LL} oppose miscibility with oil,

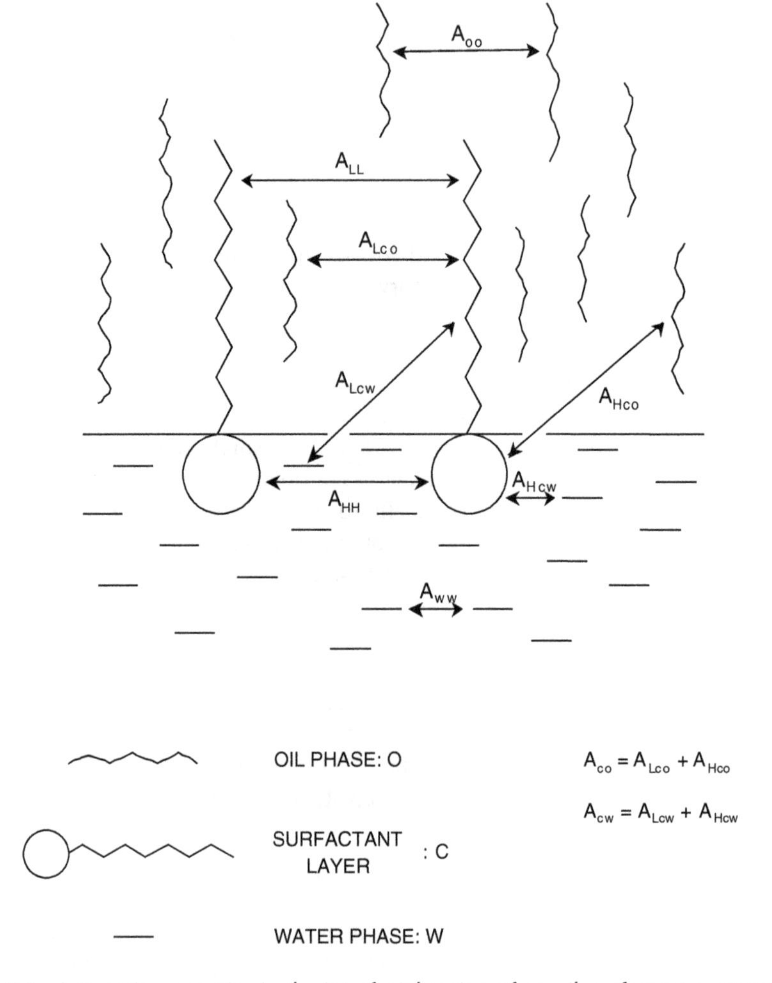

Figure 5.3 *Interaction energies in the interfacial region of an oil–surfactant–water system*

while A_{ww} and A_{HH} oppose miscibility with water. Therefore, interfacial stability is ensured if the difference in solvent interactions in C with oil and water bulk phases is sufficiently small. Too large a difference, i.e. too strong affinity of C for one phase or the other, would drive to a phase separation.

Winsor expressed qualitatively this variation in dispersing tendency by:

$$R = \frac{A_{\mathrm{co}}}{A_{\mathrm{cw}}} \tag{5.9}$$

To account for the structure of the oil, and the interactions between surfactant molecules, an extended version of the original R ratio was proposed (26):

$$R = \frac{(A_{\mathrm{co}} - A_{\mathrm{oo}} - A_{\mathrm{LL}})}{(A_{\mathrm{cw}} - A_{\mathrm{ww}} - A_{\mathrm{HH}})} \tag{5.10}$$

As mentioned before, in many cases, A_{Hco} and A_{Lcw} are negligible, so A_{co} and A_{cw} can be approximated respectively to A_{Lco} and A_{Hcw}.

In brief, Winsor's primary concept is that this R ratio of cohesive energies, stemming from interaction of the interfacial layer with oil, divided by energies resulting from interactions with water, determines the preferred interfacial curvature. Thus, if $R > 1$, the interface tends to increase its area of contact with oil while decreasing its area of contact with water. Thus oil tends to become the continuous phase and the corresponding characteristic system is type II (Winsor II). Similarly, a balanced interfacial layer is represented by $R = 1$.

5.4.1.2 *Packing Parameter and Microemulsion Structures*

Changes in film curvature and microemulsion type can be addressed quantitatively in terms of geometric requirements. This concept was introduced by Israelachivili *et al.* (27) and is widely used to relate surfactant molecular structure to interfacial topology. As described in Section 4.6.3, the preferred curvature is governed by relative areas of the head group, a_o, and the tail group, v/l_c (see Figure 4.7 for the possible aggregate structures). In terms of microemulsion type:

- if $a_o > v/l_c$, then an oil-in-water microemulsion forms
- if $a_o < v/l_c$, then a water-in-oil microemulsion forms
- if $a_o \approx v/l_c$, then a middle-phase microemulsion is the preferred structure.

5.4.1.3 *Hydrophilic–Lipophilic Balance (HLB)*

Another concept relating molecular structure to interfacial packing and film curvature is HLB, the hydrophilic–lipophilic balance. It is generally expressed as an empirical equation based on the relative proportions of hydrophobic and hydrophilic groups within the molecule. The concept was first introduced by Griffin (28) who characterised a number of surfactants, and derived an empirical equation for non-ionic alkyl polyglycol ethers (C_iE_j) based on the surfactant chemical composition (29):

$$\mathrm{HLB} = \frac{\mathrm{E}_j\ \mathrm{wt\%} + \mathrm{OH\ wt\%}}{5} \tag{5.11}$$

where E_j wt% and OH wt% are the weight percent of ethylene oxide and hydroxide groups respectively.

Davies (30) proposed a more general empirical equation that associates a constant to the different hydrophilic and hydrophobic groups:

$$\mathrm{HLB} = [(n_{\mathrm{H}} \times H) - (n_{\mathrm{L}} \times L)] + 7 \tag{5.12}$$

where H and L are constants assigned to hydrophilic and hydrophobic groups respectively, and n_H and n_L the number of these groups per surfactant molecule.

For bicontinuous structures, i.e. zero curvature, it was shown that HLB $\approx$ 10 (31). Then w/o microemulsions form when HLB $<$ 10, and o/w microemulsion when HLB $>$ 10. HLB and packing parameter describe the same basic concept, though the latter is more suitable for

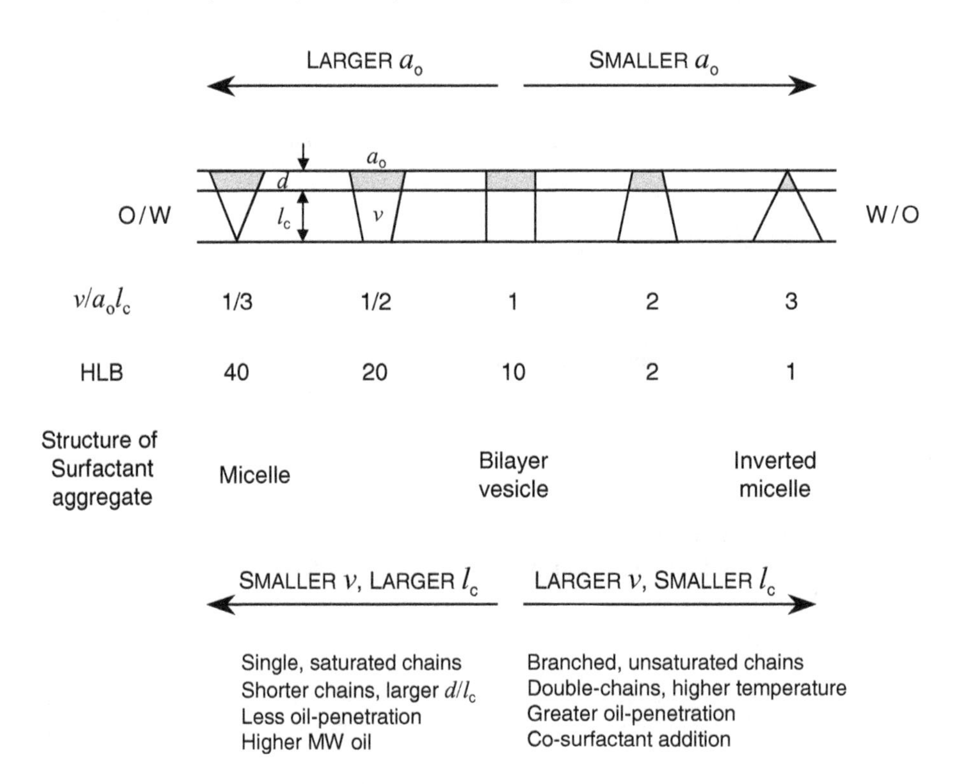

Figure 5.4 *Effect of molecular geometry and system conditions on the packing parameter and HLB number. Reprinted with permission from Ref. (31)*

microemulsions. The influence of surfactant geometry and system conditions on HLB numbers and packing parameter is illustrated in Figure 5.4.

5.4.1.4 Phase Inversion Temperature (PIT)

Non-ionic surfactants form water–oil microemulsions (and emulsions) with a high temperature sensitivity. In particular, there is a specific phase inversion temperature (PIT) and the film curvature changes from positive to negative. This critical point was defined by Shinoda and Saito (32):

- if $T <$ PIT, an oil-in-water microemulsion forms (Winsor I)
- if $T >$ PIT, a water-in-oil microemulsion forms (Winsor II)
- at $T =$ PIT, a middle-phase microemulsion exists (Winsor III) with a spontaneous curvature equal to zero, and a HLB number (Equation 5.11) approximately equal to 10.

The HLB number and PIT are therefore connected; hence the term HLB temperature is sometimes employed (33).

5.4.2 Surfactant Film Properties

An alternative, more physically realistic, approach is to consider mechanical properties of a surfactant film at an oil–water interface. This film can be characterised by three phenomenological constants: tension, bending rigidity and spontaneous curvature. Their relative importance depends on the constraints felt by the film. It is important to understand how these parameters relate to interfacial stability since surfactant films determine the static and dynamic properties of microemulsions (and emulsions). These include phase behaviour and stability, structure and solubilisation capacity.

5.4.2.1 Ultra-low Interfacial Tension

Interfacial (or surface) tensions, γ, were defined in Chapter 4 for planar surfaces, and the same principle applies for curved liquid–liquid interfaces, i.e. it corresponds to the work required to increase interfacial area by unit amount. As mentioned in Section 5.3.1, microemulsion formation is accompanied by ultra-low interfacial oil–water tensions, $\gamma_{o/w}$, typically 10^{-2} to 10^{-4} mN m^{-1}. They are affected by the presence of a co-surfactant, as well as electrolyte and/or temperature, pressure and oil chain length. Several studies have been reported on the effect of such variables on $\gamma_{o/w}$. In particular, Aveyard and co-workers performed several systematic interfacial tension measurements on both ionics (34, 35) and non-ionics (36), varying oil chain length, temperature and electrolyte content. For example, as shown in Figure 5.1 in the system water–AOT–*n*-heptane, at constant surfactant concentration (above its CMC), a plot of $\gamma_{o/w}$ as a function of electrolyte (NaCl) concentration shows a deep minimum that corresponds to the Winsor phase inversion; i.e. upon addition of NaCl, $\gamma_{o/w}$ decreases to a minimum critical value (Winsor III structure), then increases to a limiting value close to 0.2–0.3 mN m^{-1} (Winsor II region). At constant electrolyte concentration, varying temperature (34), oil chain length and co-surfactant content (35) have a similar effect. With non-ionics, a similar tension curve and phase inversion are observed, but instead with increasing temperature (36). In addition, when increasing surfactant chain length, the interfacial tension curves shift to higher temperatures and the minimum in $\gamma_{o/w}$ decreases (37). Ultra-low interfacial tensions cannot be measured with standard techniques such as Du Nouy ring, Wilhelmy plate or drop volume (DVT). Appropriate techniques for this low tension range are spinning drop tensiometry (SDT) and surface light scattering (38).

5.4.2.2 Spontaneous Curvature

Spontaneous (or natural or preferred) curvature C_o is defined as the curvature formed by a surfactant film when a system consists of equal amounts of water and oil. Then, there is no constraint on the film, which is free to adopt the lowest free energy state. Whenever one phase is predominant, there is a deviation from C_o. In principle, every point on a surface possesses two principal radii of curvature, R_1 and R_2 and their associated principal curvatures are $C_1 = 1/R_1$ and $C_2 = 1/R_2$. Mean and Gaussian curvatures are used to define the bending of surfaces (39):

- mean curvature: $C = \frac{1}{2}(1/R_1 + 1/R_2)$
- Gaussian curvature: $\kappa = 1/R_1 \times 1/R_2$

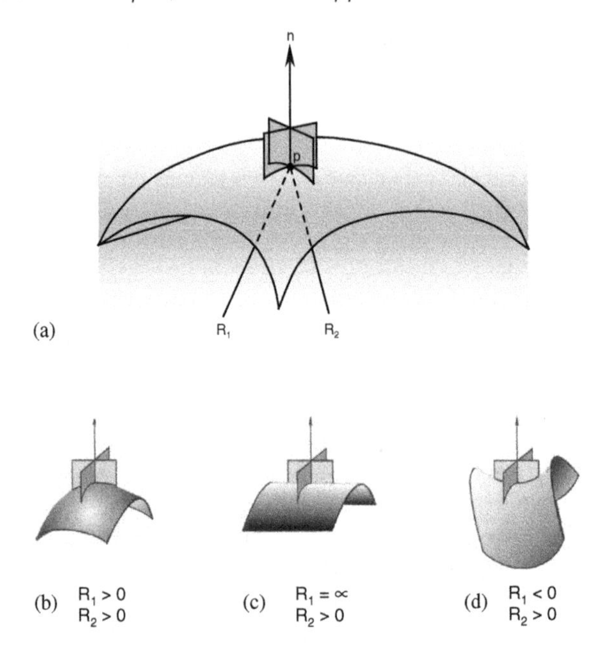

Figure 5.5 *Principal curvatures of different surfaces. (a) Intersection of the surfactant film surface with planes containing the normal vector (n) to the surface at the point p. (b) convex curvature, (c) cylindrical curvature, (d) saddle-shaped curvature. Reprinted with permission from Ref. (39). Copyright (1994) Elsevier Ltd*

C_1 and C_2 are determined as follows: every point on the surface of the surfactant film has two principal radii of curvature, R_1 and R_2 as shown in Figure 5.5. If a circle is placed tangentially to a point p on the surface and if the circle radius is chosen so that its second derivative at the contact point equals that of the surface in the direction of the tangent (of normal vector, n), then the radius of the circle is a radius of curvature of the surface. The curvature of the surface is described by two such circles chosen in orthogonal (principal) directions as shown in Figure 5.5a.

For a sphere, R_1 and R_2 are equal and positive (Figure 5.5b). For a cylinder R_2 is indefinite (Figure 5.5c) and for a plane, both R_1 and R_2 are indefinite. In the special case of a saddle, $R_1 = -R_2$, i.e. at every point the surface is both concave and convex (Figure 5.5d). Both a plane and saddle have the property of zero mean curvature.

The curvature C_o depends both on the composition of the phases it separates and on surfactant type. One argument applied to the apolar side of the interface is that oil can penetrate to some extent between the surfactant hydrocarbon tails. The more extensive the penetration, the more curvature is imposed toward the polar side. This results in a decrease of C_o since, by convention, positive curvature is toward oil (and negative toward water). The longer the oil chains, the less they penetrate the surfactant film and the smaller the effect on C_o. Eastoe *et al.* have studied the extent of solvent penetration in microemulsions stabilised by di-chained surfactants, using SANS and selective deuteration. Results suggested that oil penetration is a subtle effect, which depends on the chemical structures of both surfactant and oil. In particular, unequal surfactant chain length (40–43) or presence of C=C bonds (44) result in a more disordered surfactant–oil interface, thereby providing a region of enhanced oil mixing. For symmetric di-chained surfactants (e.g. DDAB and AOT), however, no evidence for oil mixing was found (42). The effect of alkane structure and molecular volume on the oil penetration was also investigated with *n*-heptane and cyclohexane. The results indicate that heptane is essentially absent from the layers, but the more compact cyclohexane has a greater penetrating effect (43).

Surfactant type, and nature of the polar head group, also influences C_o through different interactions with the polar (aqueous) phase.

- For ionic surfactants electrolyte content and temperature affect the spontaneous curvature in opposite ways. An increase in salt concentration screens electrostatic head group repulsions – i.e. decreases head group area – so the film curves more easily toward water, leading to a decrease in C_o. Raising temperature has two effects: (i) an increase in electrostatic repulsions between head groups due to higher counter-ion dissociation, so C_o increases; (ii) more gauche conformations are induced in the surfactant chains, which become more coiled, resulting in a decrease in C_o. Therefore the combined effects of temperature on the apolar chains and on electrostatic interactions are competitive. The electrostatic term is believed to be slightly dominant, so C_o increases weakly with increasing temperature.
- For non-ionic surfactants, unsurprisingly, electrolytes have very little effect on C_o, whereas temperature is a critical parameter due to the strong dependence of their solubility (in water or oil) on temperature. For surfactants of the C_iE_j type as temperature increases water becomes a less good solvent for the hydrophilic units and penetrates less into the surfactant layer. In addition, on the other side of the film, oil can penetrate further into the hydrocarbon chains, so that increasing temperature for this type of surfactant causes a strong decrease in C_o. This phenomenon explains the strong temperature effects on the phase equilibria of such surfactants as shown in Figure 5.8.

Thus, by changing external parameters such as temperature, nature of the oil or electrolyte concentration, the spontaneous curvature can be tuned to the appropriate value, and so drive transitions between Winsor systems. Other factors affect C_o in a similar fashion; they include varying the polar head group, type and valency of counter-ions, length and number of apolar chains, adding a co-surfactant, or mixing surfactants.

5.4.2.3 *Film Bending Rigidity*

Film rigidity is an important parameter associated with interfacial curvature. The concept of film bending energy was first introduced by Helfrich (45) and is now considered as an essential model for understanding microemulsion properties. It can be described by two elastic moduli (46) that measure the energy required to deform the interfacial film from a preferred mean curvature.

- The mean bending elasticity (or rigidity), K, associated with the mean curvature, that represents the energy required to bend unit area of surface by unit amount. K is positive, i.e. spontaneous curvature is favoured.
- The factor $\bar{K}$ is associated with Gaussian curvature, and hence accounts for film topology. $\bar{K}$ is negative for spherical structures or positive for bicontinuous cubic phases.

Theoretically, it is expected that bending moduli should depend on surfactant chain length (47), area per surfactant molecule in the film (48) and electrostatic head group interactions (49).

The film rigidity theory is based on the interfacial free energy associated with film curvature. The free energy, F, of a surfactant layer at a liquid interface may be given by the sum of an interfacial energy term, F_i, a bending energy term, F_b, and an entropic term, F_{ent}. For a droplet type structure this is written as (50):

$$F = F_i + F_b + F_{ent} = \gamma A + \int \left[\frac{K}{2}(C_1 + C_2 - 2C_o)^2 + \bar{K} C_1 C_2\right] dA + nk_B T f(\phi) \quad (5.13)$$

where γ is the interfacial tension, A is the total surface area of the film, K is the mean elastic bending modulus, $\bar{K}$ is the Gaussian bending modulus, C_1 and C_2 are the two principal curvatures, C_o the spontaneous curvature, n is the number of droplets, k_B is the Boltzmann constant, and $f(\phi)$ is a function accounting for the entropy of mixing of the microemulsion droplets, where ϕ is the droplet core volume fraction. For dilute systems where $\phi < 0.1$, it was shown that $f(\phi) = [\ln(\phi) - 1]$ (50). Microemulsion formation is associated with ultra-low interfacial tension, γ, so the γA term is small compared to F_b and F_{ent} and can be ignored as an approximation.

As mentioned previously, the curvatures C_1, C_2 and C_o can be expressed in terms of radii as $1/R_1$, $1/R_2$ and $1/R_o$ respectively. For spherical droplets, $R_1 = R_2 = R$, and the interfacial area is $A = n4\pi R^2$. Note that R and R_o are core radii rather than droplet radii (50). Solving Equation (5.13) and dividing by area A, the total free energy, F, for spherical droplets (of radius R) is expressed as:

$$\frac{F}{A} = 2K\left(\frac{1}{R} - \frac{1}{R_o}\right)^2 + \frac{\bar{K}}{R^2} + \left[\frac{k_B T}{4\pi R^2} f(\phi)\right] \quad (5.14)$$

For systems where the solubilisation boundary is reached (WI or WII region), a microemulsion is in equilibrium with an excess phase of the solubilisate and the droplets have achieved their maximum size, i.e. the maximum core radius, R^{av}_{max}. Under this condition the minimisation of the total free energy leads to a relation between the spontaneous radius, R_o, and the elastic constants K and $\bar{K}$ (51):

$$\frac{R^{av}_{max}}{R_o} = \frac{2K + \bar{K}}{2K} + \frac{k_B T}{8\pi K} f(\phi) \quad (5.15)$$

A number of techniques have been used to determine K and $\bar{K}$ separately, in particular, ellipsometry, X-ray reflectivity and small-angle X-ray scattering (SAXS) techniques (52–54). De Gennes and Taupin (55) have developed a model for bicontinuous microemulsions. For $C_o = 0$ the layer is supposed to be flat in the absence of thermal fluctuations. They introduced the term ξ_K, the persistence length of the surfactant layer that relates to K via:

$$\xi_K = a \exp(2\pi K / k_B T) \quad (5.16)$$

where a is a molecular length and ξ_K is the correlation length, i.e. the distance over which this layer remains flat in the presence of thermal fluctuations. ξ_K is extremely sensitive to the magnitude of K. When $K \gg k_B T$, ξ_K is macroscopic, i.e. the surfactant layer is flat over large distances and ordered structures such as lamellar phases may form. If K is reduced to $\sim k_B T$ then ξ_K is microscopic, ordered structures are unstable and disordered phases such as microemulsions may form. Experiments reveal that K is typically between $100 k_B T$ for condensed insoluble monolayers (56) and about $10 k_B T$ for lipid bilayers (57–59) but can decrease below $k_B T$ in microemulsion systems (60). The role of $\bar{K}$ is also important; however, there are few measurements of this quantity in the literature (e.g. 53, 61). Its importance in determining the structure of surfactant–oil–water mixtures is still far from clear.

An alternative, more accessible, method to quantify film rigidities is to calculate the composite parameter $(2K + \bar{K})$ using tensiometry and SANS techniques. This parameter can be derived for droplet microemulsion at the solubilisation boundary, WI or WII system, by combining the radius of the droplet with interfacial tensions or droplet polydispersity. Two expressions can be derived from Equations (5.14) and (5.15).

1 Using the Interfacial Tension $\gamma_{o/w}$ and the Maximum Mean Core Radius R^{av}_{max}

The interfacial tension $\gamma_{o/w}$ at the interface between microemulsion and excess phases at the solubilisation boundary can be measured by surface light scattering (SLS), or spinning-drop tensiometery (SDT). Taken together with the mean droplet size R^{av}_{max} measured by SANS, this tension can be used to estimate these elastic moduli (52). Any new area created must be covered by a monolayer of surfactant, and so this energy may be calculated in the case of WI or WII systems since the surfactant monolayer is taken from around the curved microemulsion droplets (56). To do this it is necessary to unbend the surfactant film, introducing a contribution from K, of $2K/(R^{av}_{max})^2$. The resulting change in the number of microemulsion droplets introduces an entropic contribution and a contribution due to the change in topology involving $\bar{K}$, of $\bar{K}/(R^{av}_{max})^2$. So the interfacial tension between the microemulsion and excess phase is given by:

$$\gamma_{o/w} = \frac{2K + \bar{K}}{(R^{av}_{max})^2} + \frac{k_B T}{4\pi (R^{av}_{max})^2} f(\phi) \tag{5.17}$$

which gives for the bending moduli:

$$2K + \bar{K} = \gamma_{o/w} (R^{av}_{max})^2 - \frac{k_B T}{4\pi} f(\phi) \tag{5.18}$$

2 Using the Schultz Polydispersity Width $p = \sigma / R^{av}_{max}$ Obtained from SANS Analysis (See Chapter 13)

Droplet polydispersity relates to the bending moduli through thermal fluctuations of the microemulsion droplets. Safran (62) and Milner (63) described the thermal fluctuations by an expansion of the droplet deformation in terms of spherical harmonics. The principal contribution to these fluctuations was found to arise from the deformation mode $l = 0$ only (50); and $l = 0$ deformations are fluctuations in droplet size, i.e. changes of the mean droplet radius and hence the droplet polydispersity. In the case of the two phase equilibria at maximum solubilisation (WI or WII), this polydispersity, p, may be expressed as a function of K and $\bar{K}$:

$$p^2 = \frac{u_o^2}{4\pi} = \frac{k_B T}{8\pi (2K + \bar{K}) + 2 k_B T f(\phi)} \tag{5.19}$$

where u_o is the fluctuation amplitude for the $l = 0$ mode. This polydispersity is given by the SANS Schultz polydispersity parameter σ/R_{max}^{av} (64), and Equation (5.19) can be written:

$$2K + \bar{K} = \frac{k_B T}{8\pi(\sigma/R_{max}^{av})^2} - \frac{k_B T}{4\pi} f(\phi) \qquad (5.20)$$

Therefore Equations (5.17) and (5.20) give two accessible expressions for the sum $(2K + \bar{K})$ using data from SANS and tensiometry. This approach has been shown to work well with non-ionic films in WI systems (50, 65), and also cationic (64) and zwitterionic (66) layers in WII microemulsions. Figure 5.6 shows results for these latter two classes of system, as a function of surfactant alkyl carbon number n-C. The good agreement between

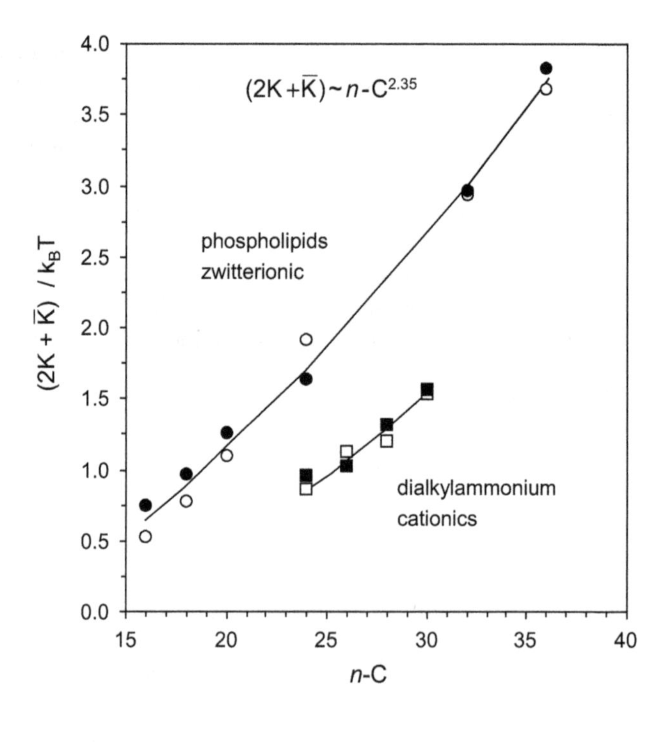

SANS $\qquad (2K+\bar{K}) = \frac{k_B T}{8\pi(\sigma/R_{max}^{av})^2} - \frac{k_B T}{4\pi} f(\varphi)$ ● ■

SANS & tensiometry $\qquad (2K+\bar{K}) = \gamma_{o/w}(R_{max}^{av})^2 - \frac{k_B T}{4\pi} f(\varphi)$ ○ □

Figure 5.6 *Film rigidities $(2K + \bar{K})$ as a function of the surfactant chain total alkyl carbon number n-C from Winsor II microemulsions. The lines are guides to the eye*

Equations (5.18) and (5.20) suggests they can be used with confidence. These values are in line with current statistical mechanical theories (48), which suggest that K should vary as n-$C^{2.5}$ to n-C^3, whereas there is only a small effect on $\bar{K}$.

5.4.3 Phase Behaviour

Solubilisation and interfacial properties of microemulsions depend upon pressure, temperature and also on the nature and concentration of the components. The determination of phase stability diagrams (or phase maps), and location of the different structures formed within these water (salt)–oil–surfactant–alcohol systems in terms of variables are, therefore, very important. Several types of phase diagram can be identified depending on the number of variables involved. In using an adequate mode of representation, it is possible to describe not only the limits of existence of the single and multiphase regions, but also to characterise equilibria between phases (tie-lines, tie-triangles, critical points, etc.). Below is a brief description of ternary and binary phase maps, as well as the phase rule that dictates their construction.

5.4.3.1 Phase Rule

The phase rule enables the identification of the number of variables (or degrees of freedom) depending on the system composition and conditions. It is generally written as (67):

$$F = C - P + 2 \tag{5.21}$$

where F is the number of possible independent changes of state or degrees of freedom, C the number of independent chemical constituents, and P the number of phases present in the system. A system is called invariant, monovariant, bivariant, and so on, according to whether F is zero, 1, 2, and so on. For example, in the simplest case of a system composed of three components and two phases, F is univariant at a fixed temperature and pressure. This means that the mole or weight fraction of one component in one of the phases can be specified but all other compositions in both phases are fixed. In general, microemulsions contain at least three components: oil (O), water (W) and amphiphile (S), and as mentioned previously a co-surfactant (alcohol) and/or an electrolyte are usually added to tune the system stability. These can be considered as simple O–W–S systems: whenever a co-surfactant is used, the ratio oil:alcohol is kept constant and it is assumed that the alcohol does not interact with any other component so that the mixture can be treated (to a first approximation) as a three-component system. At constant pressure, the composition–temperature phase behaviour can be presented in terms of a phase prism, as illustrated in Figure 5.7. However, the construction of such a phase map is rather complex and time consuming so it is often convenient to simplify the system by studying specific phase-cuts. The number of variables can be reduced either by keeping one term constant and/or by combining two or more variables. Then, ternary and binary phase diagrams are produced.

5.4.3.2 Ternary Phase Diagrams

At constant temperature and pressure, the ternary phase diagram of a simple three-component microemulsion is divided into two or four regions as shown in Figure 5.8. In

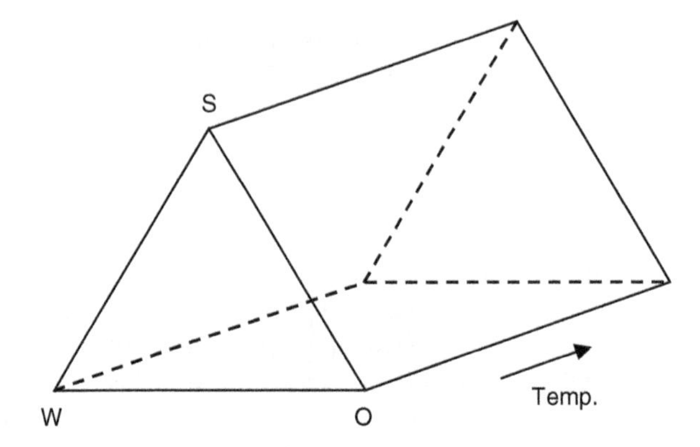

Figure 5.7 The phase prism, describing the phase behaviour of a ternary system at constant pressure

each case, every composition point within the single-phase region above the demixing line corresponds to a microemulsion. Composition points below this line correspond to multiphase regions comprising in general microemulsions in equilibrium with either an aqueous or an organic phase or both, i.e. Winsor type systems (see Section 5.4.1).

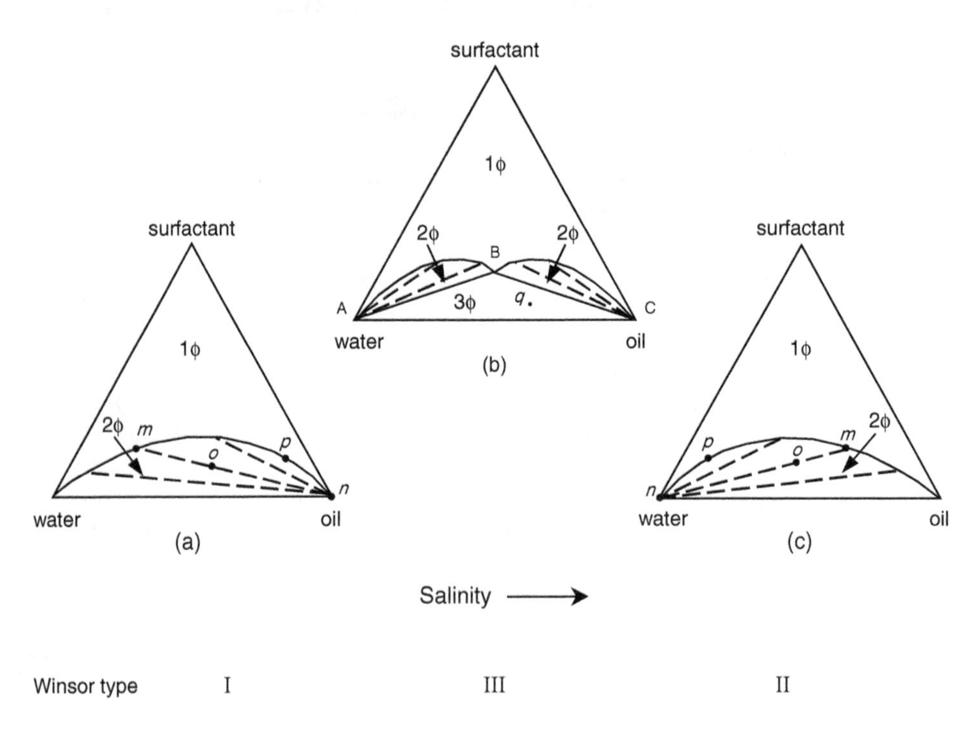

Figure 5.8 Ternary diagram representations of two- and three-phase regions formed by simple water–oil–surfactant systems at constant temperature and pressure. (a) Winsor I type, (b) Winsor II type, (c) Winsor III type systems. For ionic surfactants the changes will be driven by increases in salinity

Any system whose overall composition lies within the two-phase region (e.g. point o in Figures 5.8a and 5.8c) will exist as two phases whose compositions are represented by the ends of the 'tie-line', i.e. a segment formed by phases m and n. Therefore, every point on a particular tie-line has identical coexisting phases (m and n) but of different relative volumes. When the two conjugate phases have the same composition ($m = n$), this corresponds to the plait (or critical) point, p.

If three phases coexist (Figure 5.8b), i.e. corresponding to WIII, the system at constant temperature and pressure is, according to the phase rule, invariant. Then, there is a region of the ternary diagram that consists of three-phase systems having invariant compositions and whose boundaries are tie-lines in the adjacent two-phase regions that surround it. This region of three-phase invariant compositions is therefore triangular in form and called 'tie-triangle' (26). Any overall composition, such as point q (Figure 5.8b) lying within the tie-triangle will divide into three phases having compositions corresponding to the vertices a, b and c of the triangle. The compositions a, b and c are invariant in the sense that varying the position q, the overall composition, throughout the triangle will result in variations in the amounts of the phases a, b and c but not in their composition.

5.4.3.3 Binary Phase Diagrams

As mentioned previously, ternary diagrams can be further simplified by fixing some parameters and/or combining two variables together (e.g. water and electrolyte into brine, or water and oil into water-to-oil ratio), i.e. reducing the degrees of freedom. Then, determining the phase diagram of such systems reduces to a study of a planar section through the phase prism. Examples of such pseudo-binary diagrams are given in Figures 5.9–5.11 for non-ionic and anionic surfactants.

Figure 5.9 shows the schematic phase diagram for a non-ionic surfactant–water–oil ternary system. Since temperature is a crucial variable in the case of non-ionics, the pseudo-binary diagram is represented by the planar section defined by $\phi_w = \phi_o$, where ϕ_w and ϕ_o are the volume fractions of water and oil respectively. Then, at constant pressure, defining the system in a single-phase region requires the identification of two independent variables ($F = 2$), i.e. temperature and surfactant concentration. The section shown in Figure 5.9b can be used to determine T_L and T_U, the lower and upper temperatures, respectively, of the phase equilibrium W + M + O (with M, the microemulsion phase), and the minimum amount of surfactant necessary to solubilise equal amounts of water and oil, denoted C_s^* (68). The lower C_s^* the more efficient the surfactant. Figure 5.10 illustrates the determination of a second possible section for a non-ionic surfactant–water–oil ternary system: pressure and surfactant concentration are kept constant, leaving the two variables, temperature and water-to-oil ratio ($\phi_{w\text{-}o}$). This diagram shows the various surfactant phases obtained as a function of temperature and water-to-oil ratio (68). The third example (Figure 5.11) concerns an anionic surfactant, Aerosol-OT. In order to obtain $F = 2$ when defining the ternary W–O–S system in a single-phase region at constant pressure, the surfactant concentration parameter is fixed. Then, the two variables are temperature and w, the water-to-surfactant molar ratio defined as $w = [\text{water}]/[\text{surfactant}]$. w represents the number of water molecules solubilised per surfactant molecule, so that this phase diagram characterises the surfactant efficiency, as a microemulsifier.

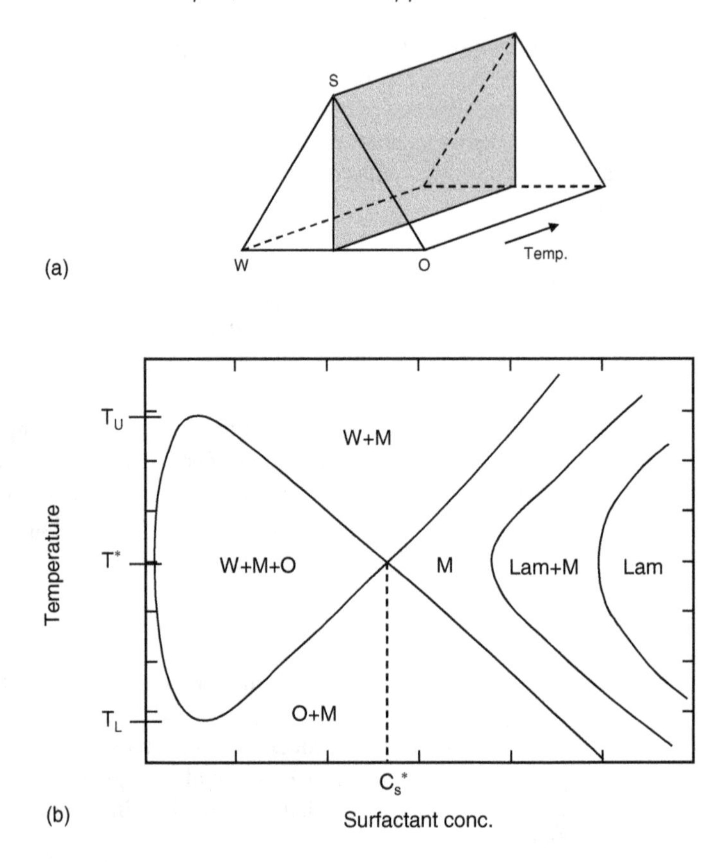

Figure 5.9 *Binary phase behaviour in ternary microemulsion systems formed with non-ionic surfactants. (a) Illustration of the section through the phase prism at equal water and oil content. (b) Schematic phase diagram plotted as temperature versus surfactant concentration C_s. T_L and T_U are the lower and upper temperatures, respectively, of the phase equilibrium $W + M + O$. T^* is the temperature at which the three-phase triangle is an isosceles, i.e. when the middle-phase microemulsion contains equal amounts of water and oil. This condition is also termed 'balanced'. C_s^* is the surfactant concentration in the middle-phase microemulsion at balanced conditions. 'Lam' denotes a lamellar liquid crystalline phase. Reprinted with permission from Ref. (68). Copyright (1997) Elsevier B.V.*

5.5 Developments and Applications

5.5.1 Microemulsions with Green and Novel Solvents

There is a need to reduce volatile organic solvents and compounds (VOCs) that are employed in many industrial processes since they pose major environmental threats. This has led to a drive to find suitable green non-VOC solvents: since they can be considered 'universal solvents' microemulsions offer attractive prospects in this area. Two main approaches have been explored: (a) super critical carbon dioxide (sc-CO_2) (69) and (b) room temperature ionic liquids (RTILs) (70) as such replacements. In potential sc-CO_2 represents an excellent green solvent due to the ease of solvent removal, tuneability and

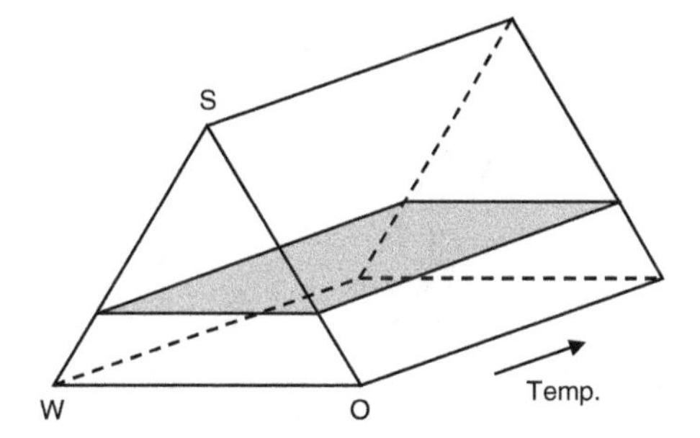

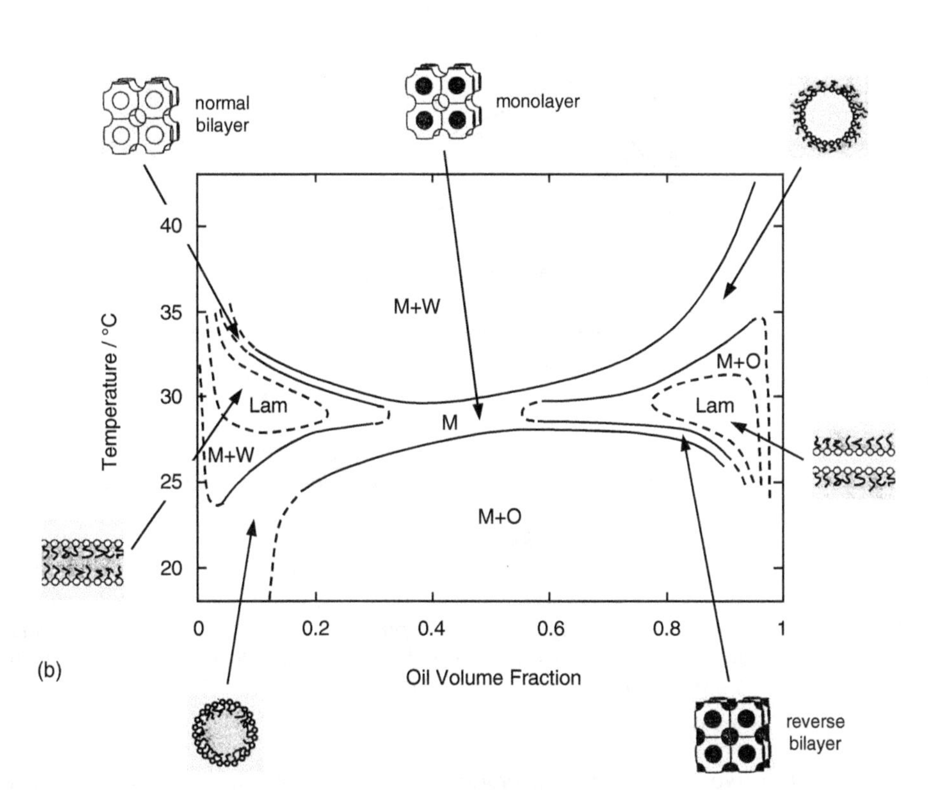

Figure 5.10 *Binary phase behaviour in ternary microemulsion systems formed with non-ionic surfactants. (a) Illustration of a section at constant surfactant concentration through the phase prism. (b) Schematic phase diagram, plotted as temperature versus volume fraction of oil, ϕ_o, at constant surfactant concentration. Also shown are various microstructures found in different regions of the microemulsion phase, M. At higher temperatures the liquid phase is in equilibrium with excess water (M + W), and at lower temperatures with excess oil (M + O). At intermediate temperatures a lamellar phase is stable at higher water contents and higher oil contents, respectively. Reprinted with permission from Ref. (68)*

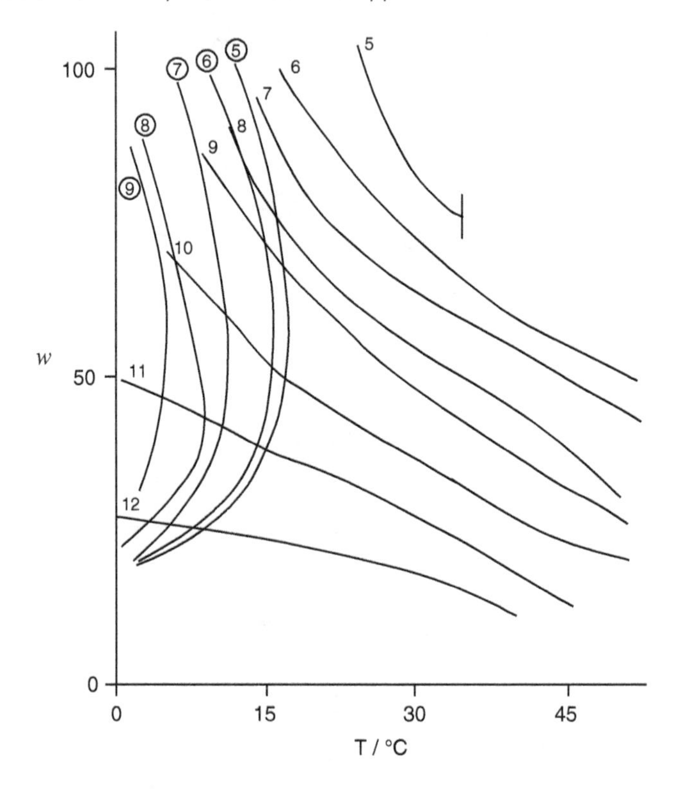

Figure 5.11 *Pseudo-binary phase diagram in ternary microemulsion systems formed with the anionic surfactant Aerosol-OT (AOT) in various straight-chain alkane solvents. The water-to-surfactant molar ratio, w, is plotted versus temperature at constant surfactant concentration and pressure. Alkane carbon numbers are indicated; ringed numbers correspond to the lower temperature (solubilisation) boundary,* T_L*, and un-ringed numbers to the upper temperature (haze) boundary,* T_U*. The single phase microemulsion region is located between* T_L *and* T_U*. Below* T_L *the system consists of a microemulsion phase in equilibrium with excess water (WII type), and above* T_U *the single microemulsion phase separates into a surfactant-rich phase and an oil phase. Reprinted with permission from Ref. (16). Copyright (1994) Elsevier Ltd*

recyclability of solvent quality by temperature and pressure, the easily accessible critical point ($P_c = 72.8$ bar, $T_c = 31.1$ °C), it is non-flammable, non-toxic, environmentally benign, biocompatible, cheap and abundant. On the other hand RTILs are salts made of sterically mismatched ions, which hinder crystallisation, they are 'trapped' in the liquid state. As such RTILs are also green solvent candidates due to tuneability, polar solvation properties and zero volatility. By forming microemulsions with these unusual liquids their properties can be enhanced: water-in-CO_2 systems (69) serve to extend the capability with polar solutes, and oil-in-RTIL microemulsions provide compatibility of the highly polar RTIL for organic and hydrophobic components. The advent of custom-made CO_2-philic, and RTIL-philic surfactants (which are not necessarily also very hydrophilic or hydrophobic) has greatly stimulated research in this area. As an example small-angle neutron scattering (see Chapter 13) results proving micelles of a custom made CO_2-philic surfactant (TC14)

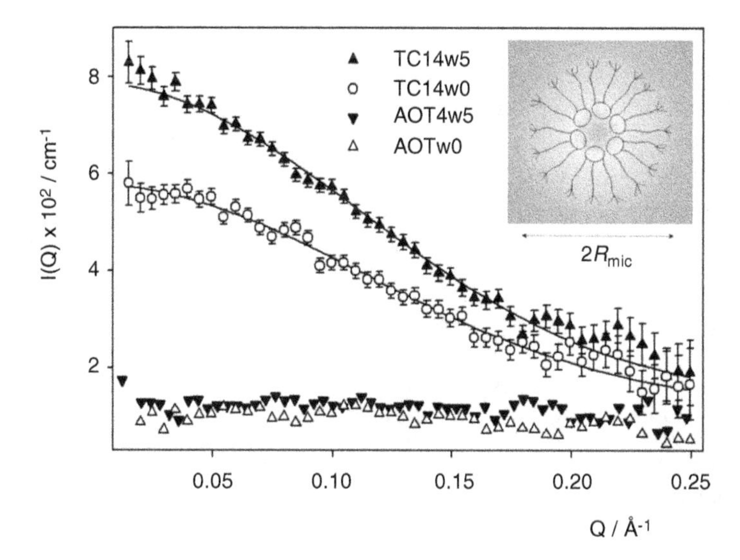

Figure 5.12 *SANS profiles of TC14-stabilised dry (w0) and hydrated (w5) micelles obtained in liquid CO_2 at 360 bar, 25 °C. The scattering obtained from formulated AOT4 w5 and dry AOT w0 systems is also shown for comparison. Smooth lines represent model fits to a spherical form factor scattering model (see Chapter 13), consistent with 11 Å radii (±10%) for both TC14-stabilised micelles*

are formed in dense CO_2 are shown in Figure 5.12. On the other hand, the normal AOT-like compounds (AOT and AOT4) do not show any evidence for aggregation in this challenging solvent environment.

Hydrofluorocarbon (HFC) solvents are recognised as attractive alternatives to fully fluorinated solvents, owing to both low toxicity and flammability. A significant development is replacement of CFCs by HFCs as refrigerants, as propellants in metered-dose inhalers and drug delivery devices for respiratory tract infections. These HFC carrier solvents are hydrophobic, creating solubility problems for the pharmaceutically active components: however, application of suitable HFC-compatible surfactants could allow for the dispersion of aqueous drug solutions as microemulsions. Progress has been made (71) towards designing surfactants for HFC microemulsions for potential drug delivery applications.

With regard to drug delivery and encapsulation, the first commercialised oil-in-water microemulsion formation is for the immunosuppressant drug cyclosporin, which is marketed as Neoral (72). The success of Neoral points to a future role for microemulsions in the pharma and allied medical sciences, an area popularly termed 'nano-medicine'.

5.5.2 Microemulsions as Reaction Media for Nanoparticles

Another application of microemulsions in nanotechnology is for nanoparticle synthesis (73), which has been a hot research topic since the early 1980s, after the first colloidal solutions of platinum, palladium and rhodium metal nanoparticles were prepared. Since this ground-breaking work, a huge variety of nanoparticles have been synthesised, in both water in oil and water in supercritical fluid microemulsions (73). The particle growth, size and shape have shown to be strongly dependent on the nature and composition of the host microemulsions, especially intermicellar exchange rates. Of particular interest is the ability to use microemulsions for facile generation of nanoparticles with catalytic, semi-conductor and super-conductor, magnetic, luminescent and buffering properties.

Therefore, since the discovery of microemulsions in the second half of the 20th century, they have begun to find various applications in scientific research and practical commercial products and processes.

References

(1) Danielsson, I., Lindman, B. (1981) Colloids Surf. A, 3: 391.
(2) Sjöblom, J., Lindberg, R., Friberg, S. E. (1996) Adv. Colloid Interface Sci., 65: 125.
(3) Schulman, J. H., Stoeckenius, W., Prince, M. (1959) J. Phys. Chem., 63: 1677.
(4) Shinoda, K., Friberg, S. (1975) Adv. Colloid Interface Sci., 4: 281.
(5) Adamson, A. W. (1969) J. Colloid Interface Sci., 29: 261.3.
(6) Friberg, S. E., Mandell, L., Larsson, M. (1969) J. Colloid Interface Sci., 29: 155.
(7) Shah, D. O., ed. (1981) Surface Phenomena in Enhanced Recovery. Plenum Press, New York.
(8) Overbeek, J. Th. G. (1978) Faraday Discuss. Chem. Soc., 65: 7.
(9) Tadros, Th. F., Vincent, B. (1980) in Encyclopaedia of Emulsion Technology, Becher, P., ed., Vol. 1. Marcel Dekker, New York.
(10) Kunieda, H., Shinoda, K. (1980) J. Colloid Interface Sci., 75: 601.
(11) Chen, S. J., Evans, F. D., Ninham, B. W. (1984) J. Phys. Chem., 88: 1631.
(12) Kahlweit, M., Strey, R., Busse, G. (1990) J. Phys. Chem., 94: 3881.
(13) Hunter, R. J. (1994) Introduction to Modern Colloid Science. Oxford University Press, Oxford.
(14) Lekkerkerker, H. N. W., Kegel, W. K., Overbeek, J. Th. G. (1996) Ber. Bunsenges Phys. Chem., 100: 206.
(15) Ruckenstein, E., Chi, J. C. (1975) J. Chem. Soc. Faraday Trans., 71: 1690.
(16) Fletcher, P. D. I., Howe, A. M., Robinson, B. H. (1987) J. Chem. Soc. Faraday Trans. 1, 83: 985.
(17) Fletcher, P. D. I., Clarke, S., Ye, X. (1990) Langmuir, 6: 1301.
(18) Biais, J., Bothorel, P., Clin, B., Lalanne, P. (1981) J. Colloid Interface Sci., 80: 136.
(19) Friberg, S., Mandell, L., Larson, M. (1969) J. Colloid Interface Sci., 29: 155.
(20) Fletcher, P. D. I., Horsup, D. I. (1992) J. Chem. Soc. Faraday Trans. 1, 88: 855.
(21) Winsor, P. A. (1948) Trans. Faraday Soc., 44: 376.
(22) Bellocq, A. M., Biais, J., Bothorel, P., Clin, B., Fourche, G., Lalanne, P., Lemaire, B., Lemanceau, B., Roux, D. (1984) Adv. Colloid Interface Sci., 20: 167.
(23) Bancroft, W. D. (1913) J. Phys. Chem., 17: 501.
(24) Clowes, G. H. A. (1916) J. Phys. Chem., 20: 407.
(25) Adamson, A. W. (1960) Physical Chemistry of Surfaces, Interscience, p 393.
(26) Bourrel, M., Schechter, R. S. (1988) Microemulsions and Related Systems. Marcel Dekker, New York.
(27) Israelachvili, J. N., Mitchell, D. J., Ninham, B. W. (1976) J. Chem. Soc. Faraday Trans. 2, 72: 1525.
(28) Griffin, W. C. (1949) J. Cosmetics Chemists, 1: 311.
(29) Griffin, W. C. (1954) J. Cosmetics Chemists, 5: 249.
(30) Davies, J. T. (1959) Proc. 2nd Int. Congr. Surface Act. Vol. 1. Butterworths, London.
(31) Israelachvili, J. N. (1994) Colloids Surf. A, 91: 1.
(32) Shinoda, K., Saito, H. (1969) J. Colloid Interface Sci., 34: 238.
(33) Shinoda, K., Kunieda, H. (1983) in Encyclopaedia of Emulsion Technology, Becher, P., ed., Vol. 1. Marcel Dekker, New York.
(34) Aveyard, R., Binks, B. P., Clarke, S., Mead, J. (1986) J. Chem. Soc. Faraday Trans. 1, 82: 125.
(35) Aveyard, R., Binks, B. P., Mead, J. (1986) J. Chem. Soc. Faraday Trans. 1, 82: 1755.
(36) Aveyard, R., Binks, B. P., Fletcher, P. D. I. (1989) Langmuir, 5: 1210.

(37) Sottmann, T., Strey, R. (1996) Ber. Bunsenges Phys. Chem., 100: 237.
(38) Langevin, D. ed. (1992) Light Scattering by Liquid Surfaces and Complementary Techniques. Marcel Dekker, New York.
(39) Hyde, S., Andersson, K., Larsson, K., Blum, Z., Landh, S., Ninham, B. W. (1997) The Language of Shape. Elsevier, Amsterdam.
(40) Eastoe, J., Dong, J., Hetherington, K. J., Steytler, D. C., Heenan, R. K. (1996) J. Chem. Soc. Faraday Trans., 92: 65.
(41) Eastoe, J., Hetherington, K. J., Sharpe, D., Dong, J., Heenan, R. K., Steytler, D. C. (1996) Langmuir, 12: 3876.
(42) Eastoe, J., Hetherington, K. J., Sharpe, D., Dong, J., Heenan, R. K., Steytler, D. C. (1997) Colloids Surf. A, 128: 209.
(43) Eastoe, J., Hetherington, K. J., Sharpe, D., Steytler, D. C., Egelhaaf, S., Heenan, R. K. (1997) Langmuir, 13: 2490.
(44) Bumajdad, A., Eastoe, J., Heenan, R. K., Lu, J. R., Steytler, D. C., Egelhaaf, S. (1998) J. Chem. Soc. Faraday Trans., 94: 2143.
(45) Helfrich, W. (1973) Z. Naturforsch., 28c: 693.
(46) Kellay, H., Binks, B. P., Hendrikx, Y., Lee, L. T., Meunier, J. (1994) Adv. Colloid Interface Sci., 9: 85.
(47) Safran, S. A., Tlusty, T. (1996) Ber. Bunsenges. Phys. Chem., 100: 252.
(48) Szleifer, I., Kramer, D., Ben-Shaul, A., Gelbart, W. M., Safran, S. (1990) J. Chem. Phys., 92: 6800.
(49) Winterhalter, M., Helfrich, W. (1992) J. Phys. Chem., 96: 327.
(50) Gradzielski, M., Langevin, D., Farago, B. (1996) Phys. Rev. E, 53: 3900.
(51) Safran, S. A. (1992) in Structure and Dynamics of Strongly Interacting Colloids and Supramolecular Aggregates in Solution, Vol. 369 of NATO Advanced Study Institute, Series C: Mathematical and Physical Sciences, Chen, S. H., Huang, J. S., Tartaglia, P. Ed. Kluwer, Dortrecht.
(52) Meunier, J., Lee, L. T. (1991) Langmuir, 46: 1855.
(53) Kegel, W. K., Bodnar, I., Lekkerkerker, H. N. W. (1995) J. Phys. Chem., 99: 3272.
(54) Sicoli, F., Langevin, D., Lee, L. T. (1993) J. Chem. Phys., 99: 4759.
(55) De Gennes, P. G., Taupin, C. (1982) J. Phys. Chem., 86: 2294.
(56) Daillant, J., Bosio, L., Benattar, J. J., Meunier, J. (1989) Europhys. Lett., 8: 453.
(57) Shneider, M. B., Jenkins, J. T., Webb, W. W. (1984) Biophys. J., 45: 891.
(58) Engelhardt, H., Duwe, H. P., Sackmann, E. (1985) J. Phys. Lett., 46: 395.
(59) Bivas, I., Hanusse, P., Botherel, P., Lalanne, J., Aguerre-Chariol, O. (1987) J. Phys., 48: 855.
(60) Di Meglio, J. M., Dvolaitzky, M., Taupin, C. (1985) J. Phys. Chem., 89: 871.
(61) Farago, B., Huang, J. S., Richter, D., Safran, S. A., Milner, S. T. (1990) Progr. Colloid Polym. Sci., 81: 60.
(62) Safran, S. A. (1983) J. Chem. Phys., 78: 2073.
(63) Milner, S. T., Safran, S. A. (1987) Phys. Rev. A, 36: 4371.
(64) Eastoe, J., Sharpe, D., Heenan, R. K., Egelhaaf, S. (1997) J. Phys. Chem. B, 101: 944.
(65) Gradzielski, M., Langevin, D. (1996) J. Mol. Struct., 383: 145.
(66) Eastoe, J., Sharpe, D. (1997) Langmuir, 13: 3289.
(67) Rock, P. A. (1969) Chemical Thermodynamics, MacMillan, London.
(68) Olsson, U., Wennerström, H. (1994) Adv. Colloid Interface Sci., 49: 113.
(69) Eastoe, J., Gold, S., Steytler, D. C. (2006) Langmuir, 22: 9832.
(70) Eastoe, J., Gold, S., Rogers, S. E., Paul, A., Welton, T., Heenan, R. K., Grillo, I. (2005) J. Am. Chem. Soc., 217: 7302.
(71) Patel, N., Marlow, M., Lawrence, M. J. (2003) J. Coll. Int. Sci., 258: 345.
(72) UK Patent No. 2 222 770 'Pharmaceutical compositions containing cyclosporine'.
(73) Eastoe, J., Hollamby, M. J., Hudson, L. K. (2006) Adv. Colloid Interface Sci., 128–130: 5.

6

Emulsions

Brian Vincent

School of Chemistry, University of Bristol, UK

6.1 Introduction

6.1.1 Definitions of Emulsion Type

An emulsion is a dispersion of one liquid in a second liquid continuous phase, where the two liquids concerned are essentially immiscible (or at least have limited mutual miscibility). In principle, one may classify emulsions according to their type, e.g. oil/water (OW) or water/oil (WO), and by their size:

- microemulsions (see Chapter 5): <100 nm
- mini-emulsions: 100 nm to 1 μm
- macro-emulsions: >1 μm.

However, this size classification is only a guide and not rigorous. Microemulsions are normally considered to be thermodynamically stable systems (see Section 5.2), whereas mini- and macro-emulsions are, at best, metastable (i.e. kinetically stable). A more rigorous definition of thermodynamic stability, with regard to emulsions, is not in terms of droplet size, but rather in terms of Equation (5.2) in Chapter 5:

$$\Delta G_{\mathrm{form}} = \Delta A \gamma_{12} - T \Delta S_{\mathrm{conf}} \tag{6.1}$$

In considering ΔG_{form} for mini- or macro-emulsions, Figure 6.1 is a good basis for discussion.

For an emulsion system to be thermodynamically stable ΔG_{form} must be *negative*; that is, the emulsion must form spontaneously. This requires, in general, a sufficiently low

Colloid Science: Principles, methods and applications, Second Edition Edited by Terence Cosgrove

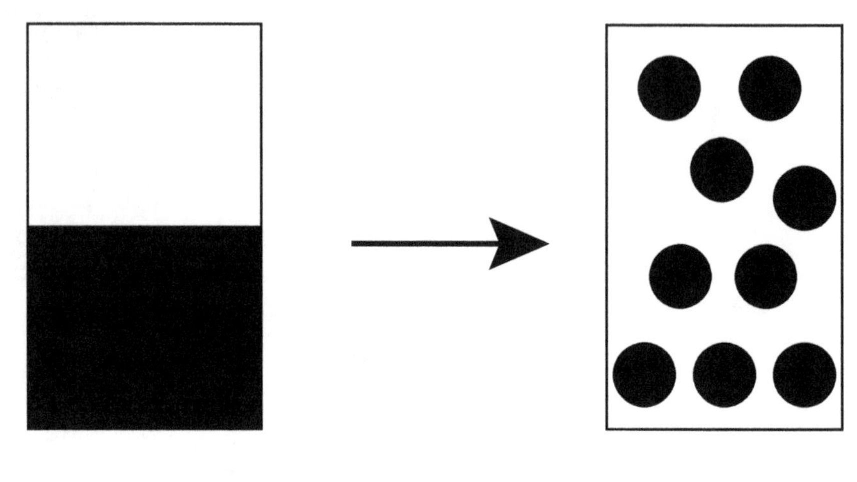

$$\Delta G_{form} = \Delta A.\gamma_{12} - T\Delta S_{conf}$$

Figure 6.1 *The free energy of formation of emulsion droplets*

interfacial tension (γ_{12}) between the two liquid phases ($\sim 10^{-4}$ to 10^{-2} mN m^{-1}), so that the ΔS_{conf} term in Equation (6.1) is greater in magnitude than the $\Delta A\gamma_{12}$ term. This is the basis of formation of microemulsions; they are essentially micelles, swollen with solubilised oil or water, depending on their nature.

If an emulsion is formed by comminution then ΔG_{form} is positive, i.e. work has to be done on the two bulk liquid phases to form the emulsion (see Sections 6.2.1 and 6.2.2). If the emulsion is formed by nucleation and growth (see Section 6.2.3) then ΔG_{form} is negative. However, if steps are not taken to (kinetically) stabilise the droplets, once formed, the system would simply revert to two bulk liquids.

As stated above, for classical macro-emulsions formed by comminution, ΔG_{form} is positive, and the ΔS_{conf} term in Equation (6.1) is now, in general, much lower in magnitude (and indeed is usually negligible) compared to the $\Delta A\gamma_{12}$ term. This is because γ_{12} is now normally around a few tens of mN m^{-1}. As Figure 4.3 in Chapter 4 illustrates, addition of a surfactant to aid formation of the liquid/liquid interface is most efficient in terms of lowering γ_{12} at concentrations around the CMC (in the liquid which is to form the continuous phase). Addition of further surfactant cannot lower the value of γ_{12} any further. The 'tricks' used to obtain the very low values of γ_{12} required for microemulsions in an oil plus water system, are explained in Chapter 5.

Mini-emulsions are interesting in that, in this case, the ΔS_{conf} term in Equation (6.1) is still smaller than the $\Delta A\gamma_{12}$ term, but now much closer in magnitude. Indeed, some systems which have been labelled as 'microemulsions' are in fact strictly mini-emulsions, because the sign of ΔG_{form} is actually net positive, rather than net negative. An example occurs in the early studies of Schulman *et al.* (1) on microemulsions. They studied the system: benzene + water + potassium oleate, which, on stirring, formed a normal WO macro-emulsion. On adding *n*-hexanol, the system became optically translucent. Schulman showed that the continuous oil phase contained very small water droplets of diameter 10–50 nm. It was he who in fact coined the term 'microemulsion' for such systems. Later studies were

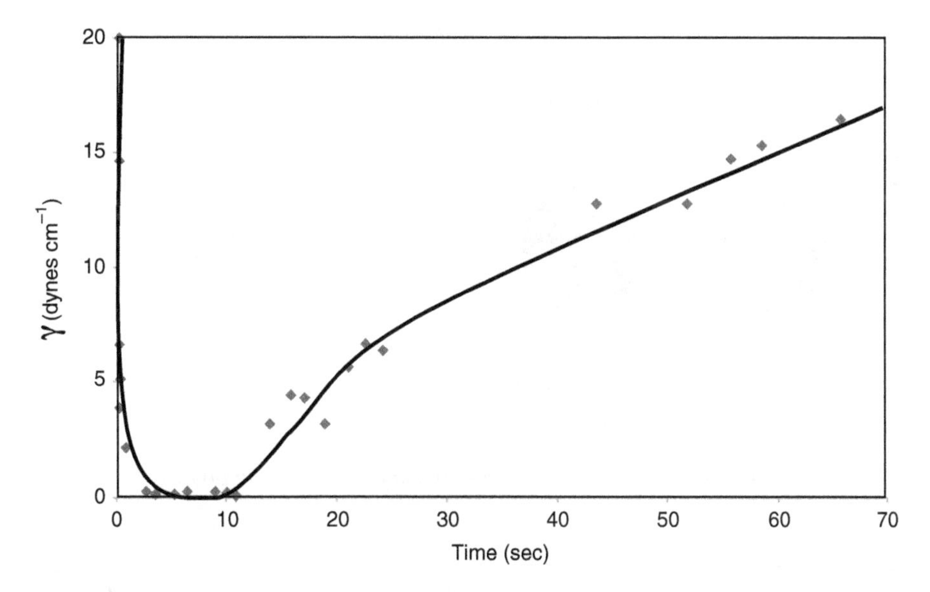

Figure 6.2 *The change in interfacial tension of the benzene–water interface (with potassium oleate present) with time, on adding n-hexanol (Reprinted with permission from Ref. (2). Copyright (1973) Elsevier Ltd)*

made by Gerbacia and Rosano (2) on very similar systems (*n*-hexadecane + water + sodium dodecyl sulfate) but in this case the system was left as two contacting bulk phases. They titrated *n*-pentanol (called a 'co-surfactant') into the *n*-hexadecane phase, and monitored the OW interfacial tension (γ), with time (t), using a Wilhelmy plate. The result is shown in Figure 6.2.

It can be seen that, for a period of ~1 min, γ falls transiently near to zero, as the alcohol molecules transfer across the interface, but then increases again to just less than the original value (~ 20 mN m^{-1}) when the system has re-equilibrated. During the time interval when γ is close to zero, spontaneous droplet formation was observed near the OW interface. These droplets *appear* to be microemulsion droplets, but clearly they are not, since the equilibrium value of γ is too high, and ΔG_{form} must be positive! The conclusion is that droplets formed using the *dynamic*, Schulman route are not microemulsion droplets in terms of the thermodynamic definition, even if they fall into that class in terms of their size. The only true microemulsion droplets are those which are formed by swelling micelles (by a solubilisation route), under *equilibrium* conditions. These days we would label Schulman's droplets as mini-emulsion droplets. Indeed, since about 1980, mini-emulsions have become much more studied in their own right, largely due to their application in making latex particles by emulsion polymerisation routes (3).

Further types of emulsion systems are those containing *three* (or more) bulk phases. Some examples of three-phase systems are given in Figure 6.3.

Type A is perhaps the most familiar and the most readily formed (e.g. by a double emulsification process); here phases α and γ are similar (e.g. both water) and phase β is say oil. This would be labelled a WOW double emulsion. Type B is a variation of A, where the separate, internal α droplets have coalesced to form a bulk α phase. In fact if two immiscible liquids, α and β are stirred together in a third mutually immiscible liquid, γ (or if

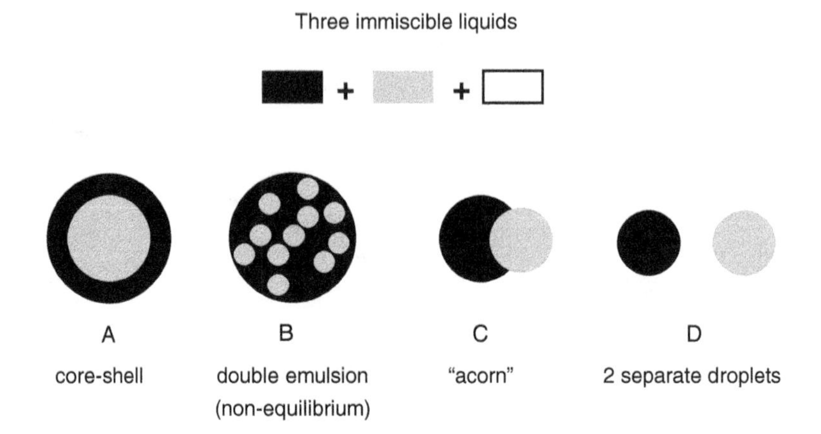

Figure 6.3 *Emulsion systems containing three liquid phases*

an $\alpha\gamma$ emulsion is mixed with a $\beta\gamma$ emulsion), then whether system B, C or D forms depends on the relative magnitudes of the three interfacial tensions: $\gamma_{\alpha\beta}$, $\gamma_{\alpha\gamma}$ and $\gamma_{\beta\gamma}$. The exact equations governing which type of system is formed have been given by Torza and Mason (4), but intuitively one can see, for example, that if $\gamma_{\alpha\gamma}$ is significantly greater than both $\gamma_{\alpha\beta}$ and $\gamma_{\beta\gamma}$, then system B will be preferred as no $\alpha\gamma$ interface is present. Systems of type A and core-shell systems of type B have found applications in the controlled delivery of active molecules (dissolved initially in the α phase and delivered to the γ phase). For type B systems, in particular, if phase β is a polymerisable liquid monomer, then the shell may be turned into a solid polymer. Type C systems, where α and β are immiscible monomers, could be used, in principle, to form *Janus*-type polymer particles.

6.1.2 Novel Features of Emulsion Systems, Compared to Solid/Liquid Dispersions

One major difference between dispersions of liquid droplets, compared to dispersions of solid particles, in a liquid continuous phase, has to do with the deformability of the liquid/liquid interface. Spherical droplets, especially in macro-emulsions, may change their *shape*. This is particularly important at high droplet concentrations and in creamed emulsions, where the dispersed phase volume fractions (ϕ) may exceed the value for hexagonal close-packing of spheres (0.74). Droplets, unlike solid particles, may also coalesce.

A second important difference, as will become apparent later, is that surfactants or polymers (or even nanoparticles), which adsorb at the liquid/liquid interface, play a much more important role in stabilising emulsion droplets, compared to the corresponding solid/liquid interfaces in particulate dispersions.

6.2 Preparation

6.2.1 Comminution – Batch

Comminution (for which, it has already been stated, ΔG_{form} is positive) involves starting with two bulk liquids (most frequently an oil phase and an aqueous phase), and supplying

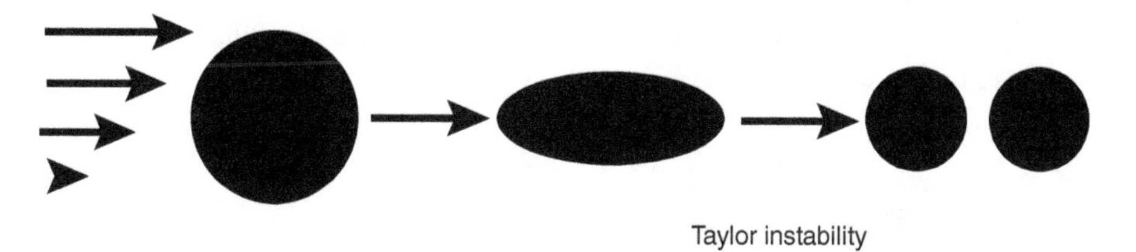

Figure 6.4 *The break-up of an emulsion droplet in laminar flow*

sufficient energy to cause one phase to break up into droplets dispersed in the second phase. In general, the energy supplied far exceeds the actual value of ΔG_{form} and the excess energy appears in the system as heat; thermostatting, if possible, is therefore a good idea. The energy input may take different forms: stirring and ultrasonics are two classical methods. However, these days commercial equipment (so-called 'homogenisers') often makes use of repeated forced flow through narrow orifices. Mostly in comminution methods turbulent flow is involved and the associated hydrodynamics are complex to analyse. However, as illustrated in Figure 6.4, in simple laminar flow gradients larger droplets break down by being 'stretched', and when the axial ratio exceeds a certain value, these elongated droplets break up into smaller droplets (a so-called 'Taylor instability').

It should be remembered that, in all comminution methods, as well as break-up, droplet re-coalescence will also be occurring. In this way, a steady-state droplet size distribution will be achieved, with a mean size that depends primarily on the power input: the larger the power input, the smaller the mean droplet size. However, the presence of so-called 'emulsifiers' will also have a strong effect on the mean droplet size achieved. Such emulsifiers may be surfactants, polymers or even nanoparticles, which adsorb at the liquid/liquid interface involved. In general, surfactants are the most efficient emulsifiers. Their primary role is to lower the interfacial tension (γ_{12}), which reduces ΔG_{form} (see Equation 6.1); they diffuse to, and adsorb more quickly at, newly forming liquid/liquid interfaces during droplet breakdown, and also reduce γ_{12} more effectively than most polymers in general. (For nanoparticles the situation is more complex, as will be discussed later in this section.) An important, secondary role, for surfactants (or polymers and nanoparticles) is that they help to reduce the re-coalescence of droplets (see Section 6.3.3). This is the main factor responsible for the much smaller average droplet size achieved in emulsions, produced with the same power input, when surfactants are present rather than absent. The smallest droplets are generally produced when the surfactant concentration is close to the CMC.

It is possible to relate the expected, average droplet diameter (d_{av}) to the volume (V^{d}) of the liquid to be dispersed, in a given volume of continuous phase (V^{c}), and the surfactant concentration (c_{s}), provided a number of assumptions are made: (i) all the surfactant is in the continuous phase prior to emulsification; (ii) after emulsification there is no free surfactant left in solution, i.e. it is all adsorbed; (iii) the interface is fully packed with a monolayer of surfactant, such that the cross-sectional area occupied by each adsorbed surfactant molecule takes its close-packed, monolayer value (a_{m}). The value of a_{m}, for a given oil–water–surfactant system, may be obtained (see Equation 6.2) from Γ_{m}, the adsorbed excess of surfactant in the monolayer by applying the Gibbs equation to the experimental γ_{12} versus ln c_s plot (see Chapter 4).

$$a_m = \frac{1}{N_A \Gamma_m} \tag{6.2}$$

d_{av} is then given by Equation (6.3):

$$d_{av} = \frac{6V^d}{V^c c_s a_m N_A} \tag{6.3}$$

By way of illustration, an average droplet diameter of 3.5 μm is predicted by Equation (6.3) for the following system: $V^d/V^c = 0.1$, $c_s = 10^{-3}$ M and $a_m = 40$ Å^2.

An important question to consider when emulsifying a given oil and water mix is which type of emulsion (i.e. O/W or W/O) will be formed? In the absence of any added emulsifier, this depends on the volume ratio of oil to water used and also on the relative viscosities of the two bulk liquids. Simple stirring of oil and water, say, will produce both types of droplets, i.e. both nascent oil droplets in water and vice-versa. However, if the volume ratio is high, then the liquid present with the smaller volume will tend, statistically, to form the dispersed phase. If the volume ratio is ~1, then the liquid having the smaller viscosity will tend to form the dispersed phase. This is because the more viscous the liquid, the more difficult it is to perform the deformation process illustrated in Figure 6.4, and also, once droplets are formed, re-coalescence is slower the more viscous the continuous phase. However, if an emulsifier is present, both these factors tend to be overridden. The dominating factor is now the natural curvature that the emulsifier imparts to the oil/water interface. This will be discussed first for nanoparticle emulsifiers, then for surfactants and polymers.

The situation for nanoparticle emulsifiers is illustrated in Figure 6.5a which shows a droplet of oil, carrying a monolayer of adsorbed nanoparticles, dispersed in water.

The key parameter here is the oil/water contact angle (θ) at the solid surface. For the case shown, $90° < \theta < 180°$ (i.e. the solid is partially *water-wetted* - see Chapter 10), and so the particles protrude more into the continuous aqueous phase, forcing the curvature of the interface to be as shown, i.e. an OW emulsion is formed. Clearly, if $0° < \theta < 90°$ (i.e. the solid surface is partially *oil-wetted*) then the opposite situation would prevail and a WO emulsion would form. If full wetting occurs, i.e. $\theta < 0°$ or $\theta > 180°$, then the particles would not adsorb at the interface, but remain dispersed in either the water or the oil phase, respectively; hence, no emulsion will be formed.

Emulsion droplets, stabilised by adsorbed nanoparticles, are called 'Pickering emulsions' after an early investigator of such systems (5); a large revival of interest in emulsions of this type has been generated in recent years, largely through the work of Binks *et al.* (6).

Adsorbed surfactants and polymers also tend to 'force' the curvature of the liquid/liquid interface concerned, but in this case the preferred curvature depends on the structure (or 'shape') of the adsorbed molecules at the interface. We will only consider surfactants here, but polymers behave similarly, although the structure (conformation) of the adsorbed polymer molecules cannot be 'guessed' *a priori*; detailed experimental analysis is required

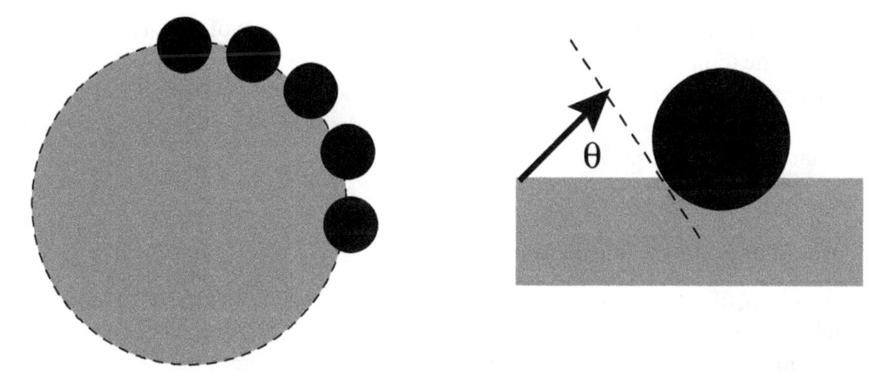

Figure 6.5 *The adsorption of nanoparticles at a liquid/liquid interface*

(e.g. neutron scattering or reflection studies – see Chapters 8 and 13). Surfactant molecules, being less flexible than polymer molecules tend, in general, to retain a similar conformation at the oil/water interface to the one they have in solution. Hence, it is possible to make an intelligent prediction as to what the preferred curvature will be. This problem has already been discussed in some detail in Chapter 5 (Section 5.4.1) in regard to microemulsions. The basic parameter now is the packing parameter (p), introduced by Israelachvili (7); it is defined as the ratio of the area that the hydrophilic head group, subtended at the oil/water interface, has to that of the hydrophobic tail group (see Figure 6.6).

If $p > 1$ then an OW microemulsion is preferred, but if $p < 1$ then a WO microemulsion is preferred. It so happens that the same rule seems to apply to macro-emulsions as well. This is

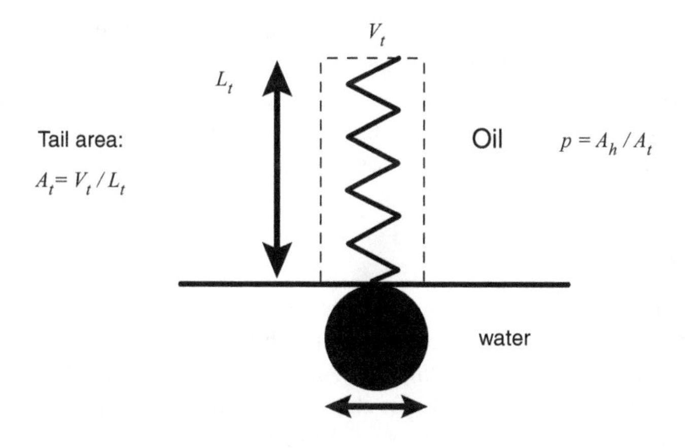

Figure 6.6 *Areas subtended by the head-group and tail of a surfactant molecule adsorbed at an oil/water interface*

most likely associated with the high curvature induced at the ends of the stretched droplets during emulsification (see Figure 6.4).

The packing parameter concept has now largely replaced older concepts, used previously to predict macro-emulsion type, such as the HLB (hydrophilic–lipophilic balance) value of a surfactant (see Section 5.4.1).

6.2.2 Comminution – Continuous

Continuous emulsification is best achieved by passing one liquid phase through an orifice (or a parallel series of orifices, as in a frit or membrane), such that droplets emerge and are dispersed into the second liquid phase. A successful computer-controlled, cross-flow membrane device has been described, for example, by Peng and Williams (8). The tubular membrane, which is cylindrical in shape, is made of steel or a ceramic material and has an array of similar-sized, perforated holes (typically $\sim$10 μm or greater in size). The liquid to be dispersed flows up the centre of the tube and droplets are forced out, under pressure, into the continuous phase liquid (containing emulsifier), flowing in a second, outer concentric cylinder. Both the inner and outer chambers usually have a continuous loop configuration, so that droplet formation is continuous; flow is stopped when the droplet concentration has reached the desired value. Pressure differences of several atmospheres are normally required for efficient droplet production. The pores must be sufficiently separated so that neighbouring droplets do not touch and possibly coalesce. Also the pores need to be wetted by the dispersed phase liquid. The smallest droplets that may be produced in this manner are typically about three times the pore diameter, so droplets less than $\sim$50 μm are difficult to produce by this method. However, emulsions produced by continuous flow methods have the advantage that they are generally more monodisperse than similar-sized droplets produced by more conventional batch emulsification methods. Dowding *et al.* (9) have described a simpler, cross-flow membrane device which is operated manually and uses a flat-disc membrane.

A very interesting, alternative approach for continuous emulsification, which is capable of producing highly monodisperse droplets, over a wide size-range, is based on microfluidics (10). Here droplets are formed at the tip of a narrow capillary tube (usually elongated and narrowed by stretching under heating), by flowing the dispersed phase along the capillary and out into the continuous phase.

6.2.3 Nucleation and Growth

Nucleation and growth is well-established as the classical method for producing monodisperse colloidal *particles*, but has been applied much less in the area of droplet formation. Vincent *et al.* (11) have reviewed this field. In nucleation and growth processes phase separation is induced either physically or chemically in an initially homogeneous system.

Physically induced phase separation is most conveniently achieved either by a change in temperature in a two-component system or by a change in concentration in a three-component system. In either case it is necessary to cross a binodal boundary line, from a one-phase region into a two-phase region, of the corresponding phase diagram. Vincent *et al.* (11) have described both types of phase diagram based on mixtures of two low

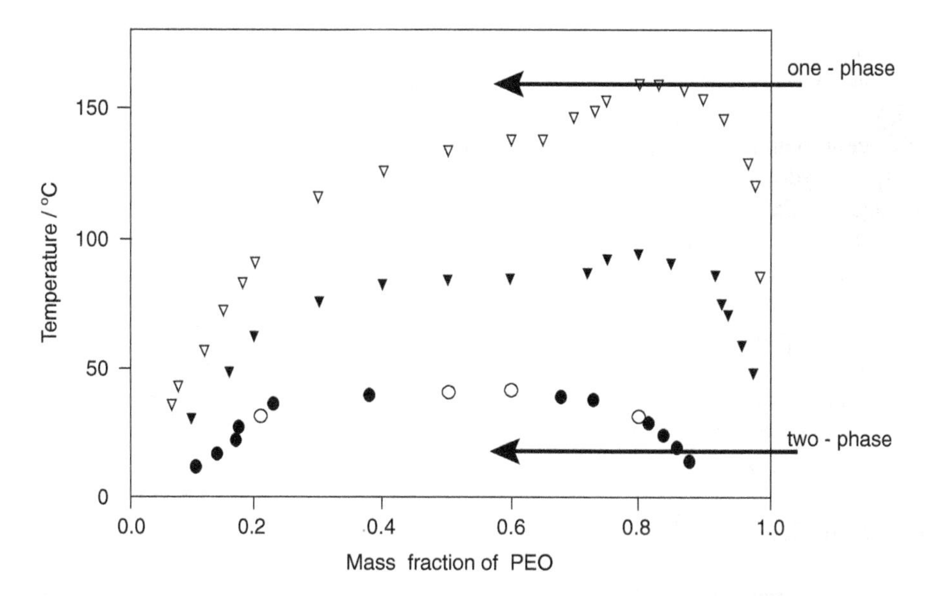

Figure 6.7 *Temperature–composition phase diagram for binary mixtures of liquid polymers: PDMS and PEO. Molar mass of PDMS = 222 in all cases; molar mass of PEO = 311 (● ○), 770 (▼), and 2000 (▽). (Reprinted with permission from Ref. (11). Copyright (1998) Royal Society of Chemistry)*

molecular-weight polymers: poly(ethylene oxide) (PEO) and poly(dimethylsiloxane) (PDMS). Binary mixtures of these two polymers exhibit an upper consolute temperature (see Figure 6.7), which is strongly molecular-weight dependent.

By cooling from the one-phase region across the binodal boundary line into the two-phase region, droplet formation of one phase within the second phase occurs by a nucleation and growth mechanism. Eventually complete phase separation into two co-existing phases would occur if a stabiliser of some sort, to prevent droplet coalescence, were not present (see Section 6.3.4). In the work of Vincent *et al.* (11), this was a PEO-PDMS block copolymer, dissolved in the homopolymer mixture at high temperatures. For relatively dilute mixtures of PEO and PDMS, i.e. less than 10% PEO in PDMS, or vice-versa, reasonably monodisperse droplets of PEO in PDMS, or PDMS in PEO, respectively, could be produced by this cooling route, with average droplet diameters less than ~1 μm. An alternative procedure, carried out at a *fixed* temperature (e.g. 30 °C), was to dissolve both homopolymers (plus the block copolymer) in a common solvent (toluene) and then to evaporate the toluene (under vacuum). Reasonably monodisperse emulsion droplets, in the size range 1 to 2 μm, could be produced by this route, even at fairly high droplet phase:continuous phase volume ratios.

A related temperature-jump procedure for producing macro-emulsions from microemulsions (12) involves crossing the phase (solubilisation) boundary line separating the one-phase (microemulsion region) into the two-phase (microemulsion plus excess oil or water) region (see Section 5.4.3, in particular Figure 5.9, which illustrates these one- and two-phase regions).

Although commonly used for producing colloidal *particles*, chemical methods for producing macro-emulsion *droplets*, based on nucleation and growth, are rare in the literature. The basic concept is to somehow produce the dispersed phase by a chemical reaction, but the reaction product must form liquid droplets rather than solid particles. Obey and Vincent (13) made μm in diameter oligomeric PDMS ('silicone oil', a liquid-state polymer at room temperature) in water using this concept. The basic chemistry is the base-catalysed hydrolysis of dimethyldiethoxysilane ethanol–water mixtures. This reaction produces a mixture of short-chain linear and cyclic PDMS, the ratio depending on the concentration of ethanol used (12). The ethanol may be removed by dialysis of the emulsion at the end of the reaction. An interesting feature of this method is that no added emulsifier or stabiliser is required; the droplets are *electrostatically* stabilised, due to the negative charge developed at the PDMS/water interface from dissociation of –OH groups at the end of the PDMS chains. In this regard, these monodisperse silicone oil droplets resemble the monodisperse silica particles produced using the Stöber method (14) from the base-hydrolysis of tetraethoxysilane in ethanol–water mixtures.

6.3 Stability

6.3.1 Introduction

'Stability' is a ubiquitous word when used in reference to colloid systems, including emulsions. In general, it refers to the ability of the colloidal system concerned to withstand some breakdown process, induced by the presence of forces acting externally on, or internally within, the system. Examples of external forces are gravity, centrifugal, electrostatic or magnetic fields. Sedimentation (or creaming) is the response to gravity, or an applied centrifugal force, acting on the droplets. Internal forces occur at both the inter-droplet and the intermolecular level. Inter-droplet forces, if net attractive, may give rise to droplet aggregation. An imbalance of intermolecular forces at the liquid/liquid interface of droplets is the origin of the interfacial tension and the associated interfacial free energy; the emulsion will try to minimise this interfacial free energy by coalescence and/or Ostwald ripening. If subjected to temperature or concentration changes, minimisation of the interfacial free energy may also drive phase inversion of the emulsion. Each of these breakdown processes is described in turn below, along with methods adopted to prevent them. An important feature of emulsion systems to bear in mind, however, is that some of these breakdown processes may occur concurrently, e.g. sedimentation, aggregation and coalescence.

6.3.2 Sedimentation and Creaming

The steady-state velocity (v_s) of (isolated) droplets, undergoing sedimentation or creaming, in a dilute emulsion, is given by Equation (6.4).

$$v_s = \frac{2\pi\Delta\rho g a^2}{9\eta} \tag{6.4}$$

where $\Delta\rho$ is the density difference between the dispersed and the continuous phase (positive for sedimentation, negative for creaming), g is the acceleration due to gravity, a is the droplet radius and η is the viscosity of the continuous phase.

For concentrated emulsions undergoing sedimentation (or creaming), the situation is more complex. Droplets no longer sediment independently. Various theories have been developed to attempt to account for the many-body hydrodynamic interactions now occurring. A simple, semi-empirical approach is to express the sedimentation velocity as a virial expansion in the droplet volume fraction (ϕ):

$$v_s = v_{s,o}(1 + a\phi + b\phi^2 + \cdots) \quad (6.5)$$

Here $v_{s,o}$ is the limiting value of v_s, at infinite dilution, given by Equation (6.4). For emulsions at reasonably moderate values of ϕ, the empirical virial coefficients, a and b (etc.), may be found by fitting the experimental data for v_s as a function of ϕ. A major challenge is to devise theories which allow for the calculation of these virial coefficients from first principles.

For a given emulsion system, the only parameter which can be readily adjusted to reduce v_s is η (see Equation 6.4). This may be achieved, for example, by adding polymer to increase the continuous phase viscosity (but avoiding any induced depletion aggregation – see Section 6.3.2). If sufficient polymer is added (i.e. well in excess of the so-called 'critical polymer overlap concentration') then the continuous phase will form a weak physical gel, with a yield stress (and this should also prevent any depletion aggregation occurring). If this yield stress is greater than the sedimentation force acting on the individual droplets, then v_s will effectively be reduced to zero.

6.3.3 Aggregation

Droplet aggregation, in almost all respects, resembles the aggregation of colloidal *particles*, as described in chapters 3 and 9. Chapter 3 is concerned with aggregation in charge-stabilised dispersions, and chapter 9 with the effect of added polymers on the aggregation behaviour of dispersions. To summarise briefly here, for *charge*-stabilised particles or droplets, van der Waals attraction between the colloidal entities leads to aggregation, when the electrostatic repulsion between them is sufficiently reduced. This may occur if their surface charge is sufficiently reduced (e.g. by a change in pH, so that their iso-electric point is approached) or the electrostatic repulsion between them is sufficiently screened (e.g. by the addition of electrolyte). Charge-stabilised *emulsions* are encountered most often where the constituent droplets carry an adsorbed (mono)layer of an *ionic* surfactant.

There has been some interesting discussion in the literature recently (15) concerning the question, if a really pure oil and pure water are emulsified, i.e. without any traces of surface active material being present, then why should emulsions stable to aggregation not be formed? The basic argument is that pure oil/water interfaces do actually have an intrinsic charge (the exact origin is still open to debate, but corresponding zeta potentials of several tens of mV have been measured), so why do the droplets not repel each other electrostatically? The answer is that pure oil droplets are hydrophobic, and small (nano)bubbles, normally present in the water, will adsorb at the oil/water interface; this leads to (bubble) bridging aggregation of the droplets, rapidly followed by droplet coalescence. Oil droplets carrying an adsorbed layer of an ionic surfactant are hydrophilic, and resist such bubble adsorption.

If the droplets carry an adsorbed layer of a *non-ionic* polymer or surfactant (in effect an oligomeric block copolymer) then other factors may come into play. If the emulsion is formed in the presence of the non-ionic polymer or surfactant then it is reasonable to assume that this will be present at monolayer coverage. In that case the droplets will be *sterically*-stabilised against van-der-Waals induced aggregation (see chapter 8 for further details). If a polyelectrolyte is used to form the emulsion, then the stabilisation mechanism is a mixture of electrostatic and steric (i.e. so-called 'electro-steric' stabilisation).

In addition to the ubiquitous van der Waals forces of attraction between droplets in an emulsion, another type of attractive force may occur when *non-adsorbed* polymer (or polyelectrolyte) molecules are present in the continuous phase. This is the so-called 'depletion interaction', which is described in detail in Chapter 9. It basically has to do with the depletion (in concentration) of polymer molecules in the thin film formed between two neighbouring droplets, when they approach to a separation less than the dimensions of the polymer molecules in solution. This sets up an osmotic force pushing the two droplets together. It should be remembered that small nanoparticles, or indeed nanodroplets, present in the continuous phase can also give rise to an (albeit weaker, in general) depletion attraction between droplets. A manifestation of this is the aggregation that has been observed between macro-emulsion oil droplets, when the concentration of excess free surfactant in the continuous aqueous phase is greater than the CMC. In that case the micelles will solubilise oil (from the oil droplets, through the aqueous phase) to form microemulsion droplets co-existing with the macro-emulsion droplets. These microemulsion droplets may lead to depletion aggregation of the larger droplets (16).

6.3.4 Coalescence

When a soft, spherical rubber ball is pressed against a flat surface, the rubber in contact with the surface tends to flatten. Similarly, when two liquid droplets approach each other to within a certain separation, the inertial energy of the two droplets tends to cause some flattening of the thin film of continuous phase liquid between them, forming a parallel-sided film, as illustrated schematically in Figure 6.8.

The degree of flattening clearly depends on the inertial energy of the droplets. If the net interaction energy between the two droplets is repulsive at all separations, then the film between the two droplets will thin to a certain separation and then the two droplets will separate again (just as a squashed rubber ball would bounce off a flat surface). A related but somewhat different situation arises when the droplets are already held in a 'near contact' situation, and the flattened thin film between any two droplets is at some *equilibrium* thickness (h, see Figure 6.8). This situation would occur in a high volume fraction emulsion ($\phi > 0.74$, the value for hexagonal close-packing of monodisperse spheres). Such high values of ϕ occur frequently in creamed or sedimented emulsions; here the (net) repulsive force between the droplets is balanced by the gravitational or centrifugal force acting on

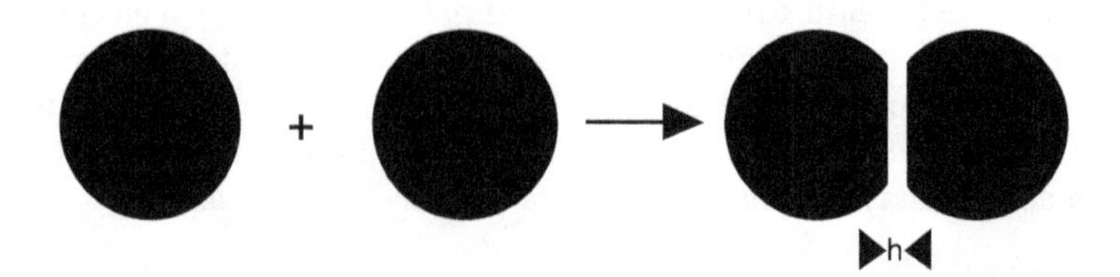

Figure 6.8 *The formation of a thin film between two approaching liquid droplets*

the droplets. A similar situation would also arise in an aggregate of droplets, where the droplets are 'sitting' in a potential energy minimum, at separation h, in the inter-droplet pair potential. This would occur, for example, with droplets carrying an adsorbed, stabilising polymer or surfactant, but where the range of the inter-droplet attractive force (e.g. van der Waals or depletion) is longer than the steric repulsion force.

For coalescence of two droplets to occur, the thin film between them has to somehow rupture. The most likely mechanism for rupture is related to the presence of oscillatory waves in the thin film. Such waves may be generated thermally (naturally occurring waves have been detected in liquid/liquid films by reflective dynamic light scattering), or mechanically (due to external vibrations). As two nodes in the two oscillating interfaces approach, so a 'hole' may form in the film at that point, leading to rupture.

Clearly, in a film with no stabilising moieties adsorbed at the surface, the thin film will drain continuously under the influence of the inter-droplet van der Waals attractive forces and rupture, by the above mechanism, is virtually spontaneous; droplet coalescence will continue until complete phase separation of the two constituent liquids has occurred. If there is adsorbed stabiliser present, on the other hand, then the rate of droplet coalescence will be retarded; in some cases the rate may be reduced effectively to zero. This is the case, for example, if the stabilising species are nanoparticles (i.e. in a Pickering emulsion). Firstly, the nanoparticle monolayer will make the interfacial film effectively 'rigid'. Secondly, the energy required to displace any one nanoparticle from the interface, into either bulk phase, is very high (this displacement energy is a maximum when the contact angle – see Figure 6.5 – is 90°).

For emulsions stabilised by adsorbed surfactant (or polymer) the situation is slightly more complex. Stabilisation against coalescence now has to do more with dampening of the oscillatory waves rather than actual desorption of the stabilising molecules from the interface. The formation of oscillatory waves is opposed by an accompanying increase in interfacial area (ΔA) of the two interfaces comprising the film. It is not so much the associated increase in interfacial free energy ($2\gamma_{12}\Delta A$) which is important, but rather the so-called 'Gibbs-Marangoni' effect. Expansion of the two interfaces causes a depletion of surfactant, and a corresponding local increase in the interfacial tension, at the nodes of the waves. The ease with which this expansion of the interfaces occurs is related to the dilation (or Gibbs elasticity) modulus (ε) of the interface. This is defined in Equation (6.6):

$$\varepsilon = \frac{d\gamma_{12}}{d \ln A} \tag{6.6}$$

The greater ε, the more stable the emulsion droplets will be to coalescence. Plots of experimental values of ε versus surfactant concentration (c_s) in (either) bulk phase (determined by interfacial rheometry) often show a maximum at a given value of c_s, with ε reducing to zero at a sufficiently high value of c_s. This is because the displacement of surfactant within the interface, causing depletion at the wave nodes, is compensated by a flow of surfactant (and associated solvent) within the adjacent solution into the depleted region (the Marangoni effect); these surfactant molecules in the adjacent solution can then adsorb to fill the gaps left by the displaced surfactant molecules. The higher the value of c_s, the greater is this effect. At sufficiently high c_s the rate of displacement of surfactant molecules along the interface is matched by the rate of adsorption from solution, such that

$d\gamma_{12}$, and, hence, ε (Equation 6.6) are both reduced to zero. This implies that there is an optimum value of c_s for stabilising emulsions against coalescence.

6.3.5 Ostwald Ripening

It is frequently observed that emulsions may coarsen in size with time, even though it is clear that no coalescence as such is occurring. This is called 'Ostwald ripening'. The mechanism is associated with the existence of a pressure difference (Δp) across a curved liquid/liquid interface (radius of curvature r and interfacial tension γ), as given by the Laplace equation (Equation 6.7):

$$\Delta p = \frac{2\gamma}{r} \tag{6.7}$$

This means that Δp is greater for the smaller droplets than for the larger ones. In turn, this implies that the chemical potential (μ) of the molecules comprising the droplets is greater in the smaller droplets. It is this difference in chemical potential ($\Delta\mu_L$), arising from the Laplace pressure difference between droplets of different sizes, which drives the migration of molecules from the smaller droplets, across the aqueous continuous phase, into the larger droplets, thereby reducing the total free energy of the system.

Ostwald ripening can occur for both OW and WO emulsions, but it is enhanced by the following factors.

(i) The size distribution of the droplets becomes wider (strictly monodisperse emulsions will not exhibit Ostwald ripening).

(ii) The solubility, in the continuous phase, of the molecules comprising the droplets increases; this will enhance the *rate* of transfer between droplets. Indeed, it is sometimes observed that the presence of surfactant (above the CMC) enhances the rate further. This is because molecules from the droplets will be solubilised in the micelles, increasing their overall effective concentration in the continuous phase. This may happen, inadvertently, if too much surfactant is added to stabilise the emulsion against coalescence.

A method that is sometimes employed to reduce Ostwald ripening is to add a component (X) to the droplet phase which is *insoluble* in the continuous phase. Then, as a droplet decreases in size, the concentration of X will increase; conversely, as droplets grow in size, the concentration of X will decrease (see Figure 6.9).

Let us denote the primary constituent molecules of the droplets as d. The chemical potential (μ_d) of the d molecules in the droplets is given by Equation (6.8) (assuming for present purposes that X and d form an ideal solution).

$$\mu_d = \mu_d^0 + RT \ln x_d = \mu_d^0 + RT \ln(1 - x_X) \sim \mu_d^0 - x_X \tag{6.8}$$

where x_d and x_X are the mole fractions of d and X in the droplets, respectively (and $x_X \ll x_d$); μ_d^0 is the chemical potential of *pure* d, R is the gas constant and T is the absolute temperature. Thus, according to Equation (6.8), if x_X becomes greater in the smaller droplets (as in Figure 6.9), then the component of the chemical potential of the d molecules, associated with the presence of X in the droplets, will be less in the smaller droplets than in the larger droplets. Hence, the chemical potential difference ($\Delta\mu_X$) between the small

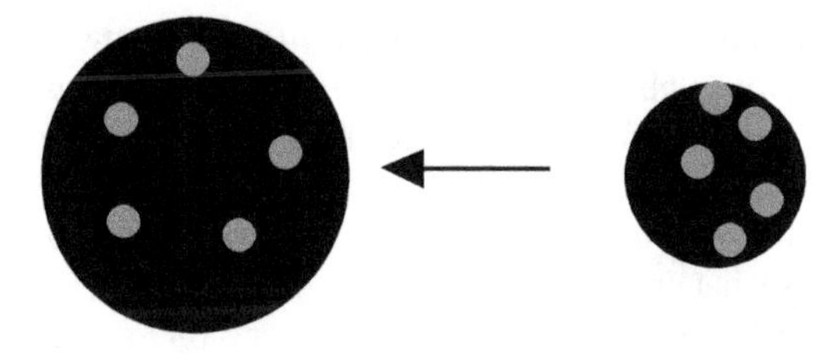

Figure 6.9 *The use of an added component in emulsion droplets, which is insoluble in the continuous phase, to oppose Ostwald ripening*

droplets and the large droplets, associated with the difference in concentration of X, will be opposite in sign to $\Delta\mu_L$. In theory, therefore, an equilibrium situation should be reached, whereby $\Delta\mu_L + \Delta\mu_X = 0$, and Ostwald ripening should cease.

6.3.6 Phase Inversion

Phase inversion is what the name implies: causing an OW emulsion to become a WO emulsion, and vice-versa. It is in this sense that this phenomenon is included here in the section on emulsion stability. There are two primary ways of inducing phase inversion.

(i) By a large change in the relative volume fraction of the two components. If, for example, one were to start with an OW emulsion and then add a sufficiently large amount of oil, it might happen that phase inversion to a WO system would result. Such a change could be conveniently monitored by monitoring the electrical conductivity of the emulsion: an oil continuous phase emulsion will have a much lower conductivity than a water continuous phase emulsion. Such inversion has been termed 'catastrophic' emulsion phase inversion. Clearly, from the discussion in Section 6.2.1 concerning the effect of surfactant structure (through the surfactant packing parameter, p, at the oil/water interface) on the type of emulsion formed initially, one might expect inversion to occur more readily if the value of p is not too different from 1. In that case neither sign of curvature of the interface (i.e. convex to water or oil) is strongly preferred. So inversion of the emulsion, and the resulting change in the sign of the interfacial curvature, is not too costly in terms of the associated free energy change. For emulsions where the value of p is either significantly >1, or <1, catastrophic phase inversion is unlikely to occur. Indeed, adding more oil to an OW emulsion (where $p > 1$) in this case would probably just lead to a higher volume fraction (ϕ) OW emulsion, where it may reach very high values (>0.9), so-called high internal phase emulsions (HIPEs).

(ii) The alternative, and much more common, route to emulsion phase inversion, involves changing the value of p *in situ* in the emulsion. So, to change from an OW emulsion to a WO emulsion, one would need to induce a change from $p > 1$ to $p < 1$. This process is called 'transitional' emulsion phase inversion. Clearly, to achieve such a change in p, one requires systems where the value and sign of p are sensitive to changes in local thermodynamic conditions, such as a change in temperature, salt concentration, or a change in surfactant composition (i.e. using a *mixed* surfactant system). In general, in this regard, emulsions stabilised with *non-ionic* surfactants are ones where transitional phase inversion can be achieved most readily. The strong correlation between the observed phase behaviour (i.e. OW or WO) of macro-emulsions and the phase behaviour of the corresponding microemulsions was discussed in Section 6.2.1; it was shown that the p parameter is what governs the type of system observed in both cases. Figure 6.10 is a simplified version of Figure 5.10 from Chapter 5 on microemulsions. Figure 6.10 shows the *equilibrium* (i.e. prior to any emulsification process), temperature (T)–oil volume fraction phase diagram, at a fixed surfactant concentration, for some unspecified oil + water + non-ionic surfactant system. At a given oil volume fraction (around the middle of the range), and at low temperatures, one sees an OW microemulsion, co-existing with excess oil (here $p > 1$). Conversely, at high temperatures, one sees a WO microemulsion, co-existing with excess water ($p < 1$). Over a small intermediate temperature range a single microemulsion (bicontinuous) phase exists ($p \sim 1$). The changes in p are associated with the dehydration of the PEO tails on raising the temperature. One would predict that if this system (say at an oil volume fraction of 0.5) was emulsified at low temperatures an OW macro-emulsion would form, and whereas at high temperatures a WO macro-emulsion would form. Hence, if one were to take the OW macro-emulsion at a low temperature and heat it sufficiently, the emulsion would invert to a WO emulsion. The intermediate temperature region, where only the single microemulsion phase forms, is known as the 'phase-inversion region' (PIR). Commonly, in practice, for the corresponding macro-emulsion, phase inversion is seen to occur most rapidly at a certain temperature, the so-called 'phase inversion temperature' (PIT), which must lie within the PIR.

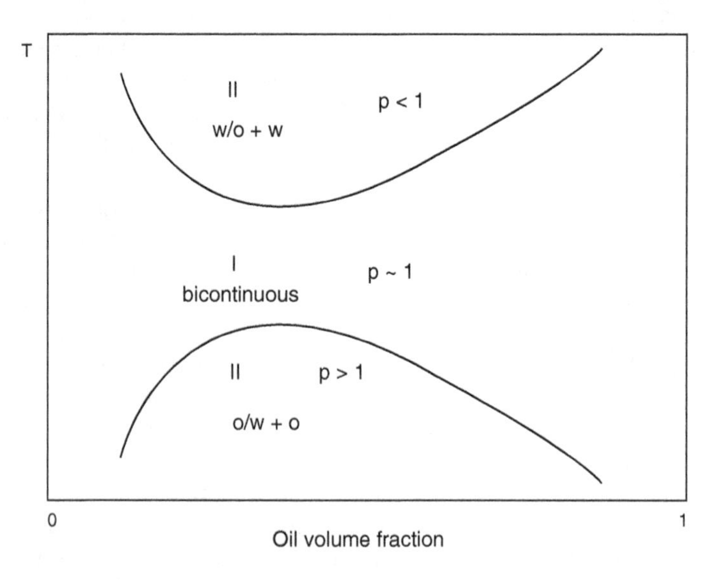

Figure 6.10 *Simplified, schematic temperature versus oil volume fraction phase diagram, for an oil (O) + water (W) + non-ionic surfactant system, at fixed surfactant concentration. Roman numerals indicate the number of coexisting phases in that part of the phase diagram. W/O = water in oil microemulsion; O/W = oil in water microemulsion. p is defined in the text*

Finally, it is of interest to note that Shinoda and Saito (17) were able to produce very stable OW macro-emulsions, with relatively small droplets, in an oil/water/non-ionic surfactant system, by actually stirring the system in the bicontinuous, single phase region, and then cooling rapidly below the PIR.

References

(1) Schulman, J. H., Stoeckenius, W., Prince, C. (1959) J. Phys. Chem., 63: 1672.
(2) Gerbacia, W., Rosano J. (1973) J. Colloid Interface Sci., 44: 242.
(3) Ugelstat, J., El-Aasser, M. S., Vanderhoff, J. W. (1973) Polym. Lett., 11: 503.
(4) Torza, S., Mason, S. G. (1970) J. Colloid Interface Sci., 33: 67.
(5) Pickering, S. (1907) J. Chem. Soc., 91: 200.
(6) Binks, B. P., Lumbsden, S. O. (1999) Phys. Chem. Chem. Phys., 1: 3007.
(7) Israelachvili, J. N., Mitchell, D. J., Ninham, B. W. (1976) J. Chem. Soc. Faraday Trans., 2: 72,: 1525.
(8) Peng, S. J., Williams, R. A. (1998) Chem. Eng. Res. Design, 76: 894.
(9) Dowding, P. J., Goodwin, J. W., Vincent, B. (2001) Colloids and Surfaces A, 180: 301.
(10) Weitz, D. A. *et al.* (2008) Materials Today, 11: 18.
(11) Vincent, B., Kiraly, Z., Obey, T. M. (1998) in Binks, B. P. (ed.) Modern Aspects of Emulsion Science, Royal Society Chemistry, London, p. 100.
(12) Shinoda, K., Friberg, S. E. (1986) Emulsions and Solubilisation. Wiley-Interscience, New York.
(13) Obey, T. M., Vincent, B. (1994) J. Colloid Interface Sci., 163: 454.
(14) Stöber, W., Fink, A., Bohn, E. (1968) J. Colloid Interface Sci., 26: 62.
(15) Pashley, R. M. (2003) J. Phys. Chem. B, 107: 1714.
(16) Aronson, M. P. (1989) Langmuir, 5: 494.
(17) Shinoda, K., Saito, H. (1969) J. Colloid Interface Sci., 30: 258.

Textbooks and General Reading

Shinoda, K., Friberg, S. E. (1986) Emulsions and Solubilisation, Wiley-Interscience, New York.
Dickinson, E. (ed) (1987) Food Emulsions and Foams, Royal Society of Chemistry, London.
Larsson, K., Friberg, S. E. (1990) Food Emulsions, 2nd edn. Dekker, New York.
Sjöblom, J. (ed.) (1992) Emulsions – A Fundamental and Practical Approach. Kluwer, Dordrecht.
Schramm, L. L. (ed.) (1992) Emulsions – Fundamentals and Applications in the Petroleum Industry. Am. Chem. Soc. Symp Series, 231.
Becher, P. (1996) Encyclopedia of Emulsion Technology. Dekker, New York.
Sjöblom, J. (ed.) (1996) Emulsions and Emulsion Stability. Surfactant Science Series 61, Dekker, New York.
Binks, B. P. (1998) Modern Aspects of Emulsion Science. Royal Society of Chemistry, London.
Dickinson, E., Rodriguez-Patino, J. M. (1999) Food Emulsions and Foams. Royal Society of Chemistry, London.
Becher, P. (2001) Emulsions: Theory and Practice, 3rd edn. Oxford University Press, New York.

7

Polymers and Polymer Solutions

Terence Cosgrove

School of Chemistry, University of Bristol, UK

7.1 Introduction

Polymers are long-chain molecules that are made by assembling a series of monomers, which may be of the same type (homopolymer) or a mixture (copolymer). Many polymers exist in nature such as natural rubber (*cis*-polyisoprene) and DNA (a polymer with a helical double chain whose backbone is made of an alternating sequence of sugar and phosphates). The number of monomers we can join together is unlimited but stresses in long chains can cause them to break up. Natural polymers can have over 10^6 monomers in a chain and all the chains can have the same length. Synthetic polymers commonly have up to 10^5 monomers but there is always a distribution of chain lengths. If a typical monomer length is approximately 1 nm then 10^6 monomers would have a contour length of 1 mm. Polymers, however, are rarely fully stretched.

In this chapter we review briefly the basic polymerisation schemes for making polymers, the statistics of chain molecules, their thermodynamics and their physical properties. More detailed texts in this area include the classic texts by Flory (1, 2) and de Gennes (3) as well as several more recent books by Sun (4), Doi (5), Grosberg and Khokhlov (6), Rubinstein and Colby (7) and, for an introduction to polymer synthesis, Stevens (8).

7.2 Polymerisation

There are many methods that can be used to make polymers synthetically and these include condensation, free radical and ionic and emulsion methods.

Colloid Science: Principles, methods and applications, Second Edition Edited by Terence Cosgrove

In practice, the final use of a polymer will dictate the most efficient method to choose. Essentially, each of these methods has the same basic chemical steps, but the differences are due to the control of the rate constants associated with them. For example, a fast termination will give only low molecular weight products. The basic steps are initiation (I), propagation (P) and termination (T), each of which can be characterised by a rate constant as in Equation (7.1). In each of the four methods above these rate constants are controlled in different ways and lead to very different molecular weight distributions:

$$\begin{aligned} A &\rightarrow (k_\mathrm{I}) \rightarrow A^* \\ A + A^* &\rightarrow (k_\mathrm{P}) \rightarrow A_2^* \\ A_n^* &\rightarrow (k_\mathrm{T}) \rightarrow A_n \end{aligned} \tag{7.1}$$

7.2.1 Condensation

Conceptually (and historically), the method of condensation is the easiest to visualise. In this method two bifunctional monomers are mixed together. The reaction often proceeds spontaneously, as in the famous nylon rope trick, when the ingredients meet at an interface and the product can be pulled out as a thread. Typically, the base-catalysed reaction involves an acid chloride and an amine and, after the first step, both reactive groups are still present in the dimer (Equation 7.2). As the polymerisation progresses both monomers are used and the reaction can proceed until the supply of one or the other is exhausted. The reaction kinetics can be approximated by a second-order rate equation and lead to a polydisperse molecular weight distribution:

$$\begin{aligned} &\mathrm{ClOC(CH_2)}_n\mathrm{COCl} + \mathrm{NH_2(CH_2)}_m\mathrm{NH_2} \rightarrow (k_\mathrm{P}) \\ &\rightarrow \mathrm{ClOC(CH_2)}_n\mathrm{CONH(CH_2)}_m\mathrm{NH_2} + \mathrm{HCl} \end{aligned} \tag{7.2}$$

7.2.2 Free Radical

Probably the most common method for polymerisation is the free radical method and an early example was making poly(vinyl chloride). The initiation step can be carried out using benzoyl peroxide, which attacks the double bond in a vinyl monomer giving a polymeric radical, which can then propagate. The termination can come about by several mechanisms and in the example given in Equation (7.3) a combination route is shown.

The polymers produced by this route are also polydisperse because of the competing reactions (propagation, initiation and termination). Recently, advances in free radical methods have meant that the radical species can be protected and this means that much closer control of polydispersity can be achieved:

$$\begin{aligned} &\mathrm{C_6H_5}^\bullet + \mathrm{CH_2} = \mathrm{CHX} \rightarrow (k_\mathrm{I}) \rightarrow \mathrm{C_6H_5CH_2CHX}^\bullet \\ &\mathrm{C_6H_5CH_2CHX}^\bullet + \mathrm{CH_2} = \mathrm{CHX} \rightarrow (k_\mathrm{P}) \rightarrow \mathrm{C_6H_5(CH_2CHX)}_n\mathrm{CH_2CHX}^\bullet \\ &\mathrm{C_6H_5()}_n\mathrm{CH_2CHX}^\bullet + {}^\bullet\mathrm{XHCH_2C()}_n\mathrm{H_5C_6} \rightarrow (k_\mathrm{T}) \\ &\rightarrow \mathrm{C_6H_5()}_n\mathrm{CH_2CHXXHCH_2C()}_n\mathrm{H_5C_6} \end{aligned} \tag{7.3}$$

7.2.3 Ionic Methods

Ionic polymerisation is more technically challenging but is a method for making highly monodisperse polymers, by effectively removing the termination process. So if the initiation is very rapid compared to the propagation then living ionic polymers can be made which can be terminated in a controlled way. One example is the polymerisation of styrene (S):

$$\begin{aligned}
&Na^{+}C_{10}H_{8}^{-} + C_{6}H_{5}CH = CH_{2} \rightarrow (k_{I}) \rightarrow Na^{+}C_{6}H_{5}CH^{-}CH_{2}^{\bullet} + C_{10}H_{8}\\
&2Na^{+}C_{6}H_{5}CH^{-}CH_{2}^{\bullet} \rightarrow Na^{+}C_{6}H_{5}CH^{-}CH_{2}CH_{2}CH^{-}C_{6}H_{5}Na^{+} + \{S\}\\
&\rightarrow (k_{P}) \rightarrow Na^{+}C_{6}H_{5}CH^{-}CH_{2}\{S\}_{n}\{S\}_{m}CH_{2}CH^{-}C_{6}H_{5}Na^{+}
\end{aligned} \tag{7.4}$$

The initiator can be made from sodium and naphthalene in an inert solvent (THF). This gives a carbanion, which combines with styrene to give a radical ion. The dianion can propagate from both ends. The charges stay close because of the low dielectric constant of the medium and termination is precluded (provided contamination is kept to a minimum) but can be controlled by quenching. The resulting polymers can be very monodisperse.

7.3 Copolymers

Copolymers come in many different guises: random, alternating and block as shown in Figure 7.1. Random copolymers can be made by mixing monomers with similar reactivity ratios to give a product, which has the same ratio of monomers as the reactants. They often have properties intermediate between the two monomers used. Block copolymers can be made, in principle, by sequential polymerisation by any of the methods above. Higher order structures such as combs, bottle brushes and ladders can also be made by sequential addition of monomers or macromonomers.

Figure 7.1 The structures of copolymers: the top figure is an alternating copolymer, the middle a block copolymer and the bottom random

Table 7.1 *Properties of high and low molecular weight polymers*

Property	Low molecular weight	High molecular weight
Solution volume	Little/no change	Small change
Viscosity	Newtonian	Non-Newtonian
Film forming	No/brittle	Yes
Dialysis	No	Yes
Stress	Fractures	Viscoelastic
Ultrasound	No effect	Degradation

7.4 Polymer Physical Properties

Polymers have very different physical properties to their respective monomers. In Table 7.1 we explore some of these.

7.4.1 Entanglements

A fascinating topological property of polymers is that they can physically entangle. This property is strongly molecular weight dependent and gives rise to complex rheological and diffusional properties. For viscosity (η) the dependence on molecular weight switches from one which is linear in molecular weight (M) below a critical value, M_c, to a 3.4 power law above it, as given in Equations (7.5) and (7.6):

$$\eta \sim M(M < M_c) \tag{7.5}$$

$$\eta \sim M^{3.4}(M > M_c) \tag{7.6}$$

This switchover is seen in many experimental studies and Figure 7.2 shows how viscosity depends on molecular weight for a series of melt polymers of polydimethylsiloxane. Similarly, films formed by polystyrene are very sensitive to M_c as shown in Figure 7.3; below M_c no film is formed.

7.5 Polymer Uses

Polymers find many uses in a wide range of industries from pharmaceutical to heavy engineering (Table 7.2).

7.6 Theoretical Models of Polymer Structure

Although the contour length of polymer chains may be substantial, because of bond rotation and elasticity, they are rather compact. Flory first suggested that a polymer chain should undergo a random walk through space as shown in Figure 7.4. Ignoring the volume of the chain, this model predicts that the end to end distance, R, is proportional to the square root of the number of bonds, n.

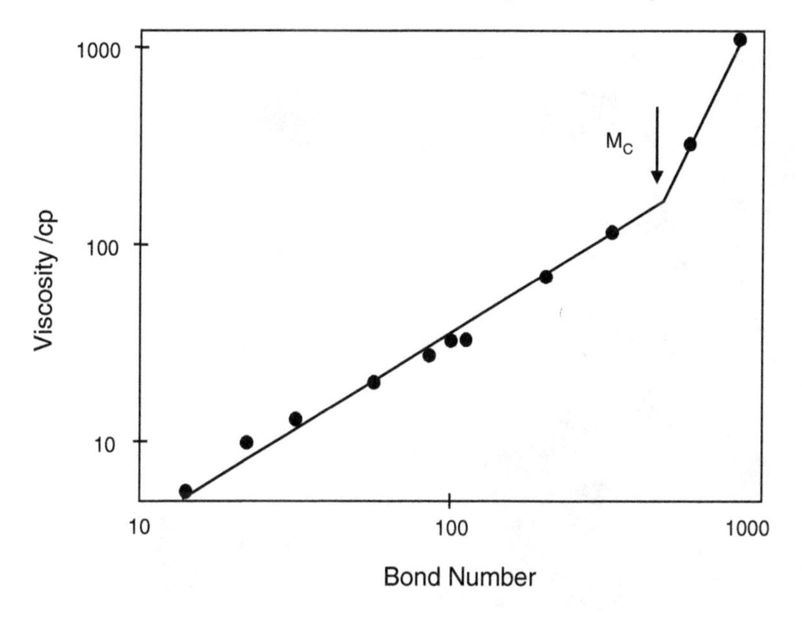

Figure 7.2 *Melt viscosities of polydimethylsiloxane*

The proof of the random walk for n steps is quite straightforward using the vector model. Each step (i) is defined by a vector $\vec{\ell}_i$. As the walk has an equal probability of going in any direction the sum of these vectors, the end to end distance, is zero. We can calculate, however, the mean square end to end distance:

$$\langle R^2 \rangle = \sum_{i=1}^{n} \vec{\ell}_i \sum_{j=1}^{n} \vec{\ell}_i$$

$$\langle R^2 \rangle = \sum_{i \neq j}^{n} \vec{\ell}_i \cdot \vec{\ell}_j + \sum_{i}^{n} \vec{\ell}_i^{\,2} \tag{7.7}$$

The first term in Equation (7.7) is effectively zero as any vector $\vec{\ell}_i$ will be at a random orientation to any vector $\vec{\ell}_j$ and hence the resultant products will be equally positive or negative and cancel in the summation. The second term is just a scalar so

$$\langle R^2 \rangle = n\ell^2 \tag{7.8}$$

which shows that end to end distance is proportional to the square root of the number of steps. This is the same result that one gets for a random walk diffusion process.

7.6.1 Radius of Gyration

It is not straightforward to measure R directly but several experimental methods allow us to measure the radius of gyration, R_G, of a polymer chain by, for example, viscosity or scattering. The value of R_G depends not only on the chain length but also on the shape of the molecule. Equation (7.9) defines R_G, and r_i is the distance of monomer i from the centre of

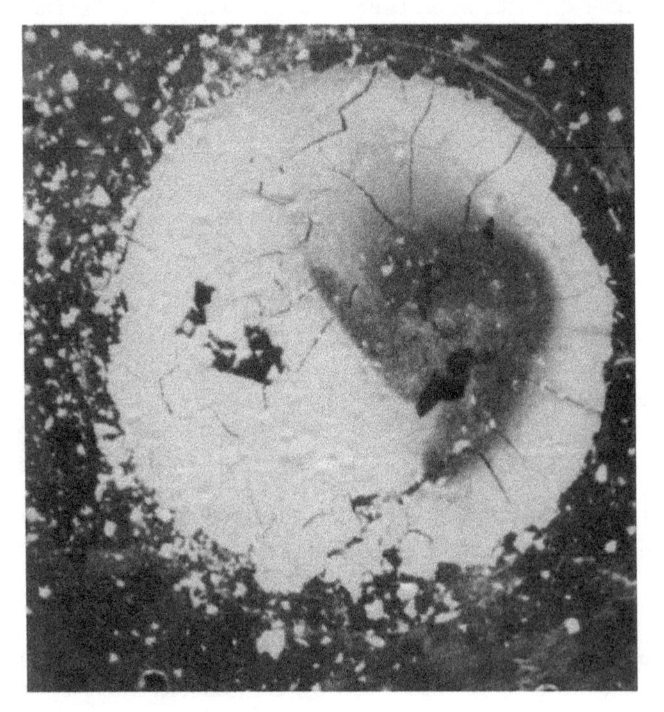

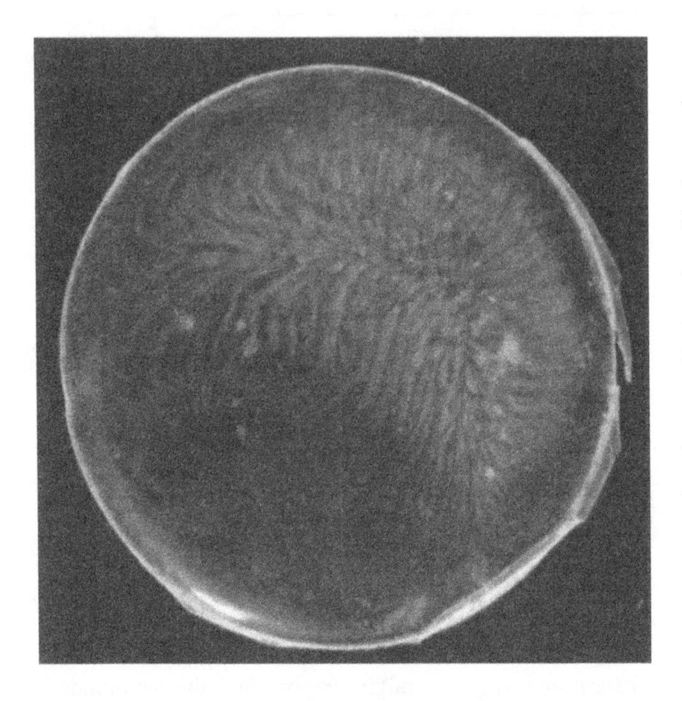

Figure 7.3 Films formed by evaporating polystyrene solutions of different molecular weights above and below the critical entanglement weight. The figure on the left is a sample of $20\,kg\,mol^{-1}$ and on the right that of $51\,kg\,mol^{-1}$

Table 7.2 *Polymer uses*

Property	Usage
Impermeability	Protective coatings, beer glasses
Inertness	Artificial joints, prosthetics
Adsorption	Crystallisation modifiers, colloidal stabilisation, adhesives
Strength	Building materials
Electrical	Conducting polymers and insulators
Fluidity	Lubricants, viscosity modifiers

mass. Hence long rods have a very large radius of gyration. Figure 7.5 illustrates three different shapes a polymer coil could adopt, a random coil, a solid sphere and a rod. For a polyethylene chain of 5000 segments these three shapes would have values of R_G of 8.2, 2.3 and 130 nm respectively. This means that if we measure R_G then we must be careful in equating this value to the molecular size if we do not have some indication of the molecular shape.

$$\langle R_G^2 \rangle = \frac{\sum_{i=1}^{n} m_i r_i^2}{\sum_{i=1}^{n} m_i} \tag{7.9}$$

For a random walk, Debye showed that $\langle R^2{}_G \rangle$ and $\langle R^2 \rangle$ are closely related:

$$\langle R_G^2 \rangle = \frac{\langle R^2 \rangle}{6} \tag{7.10}$$

7.6.2 Worm-like Chains

Real chains have fixed valence angles, and rotations about bonds are not entirely free so the simplistic result above needs to be modified for real chains. We introduce the characteristic ratio C_∞; which can vary between 4 and 20 to correct for chain flexibility. Some examples are given in Table 7.3 which are average values. In a solution, the solvent also plays an

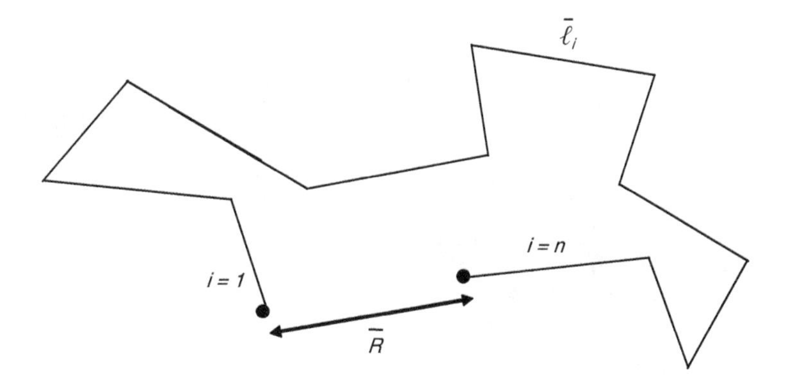

Figure 7.4 *An idealised polymer walk*

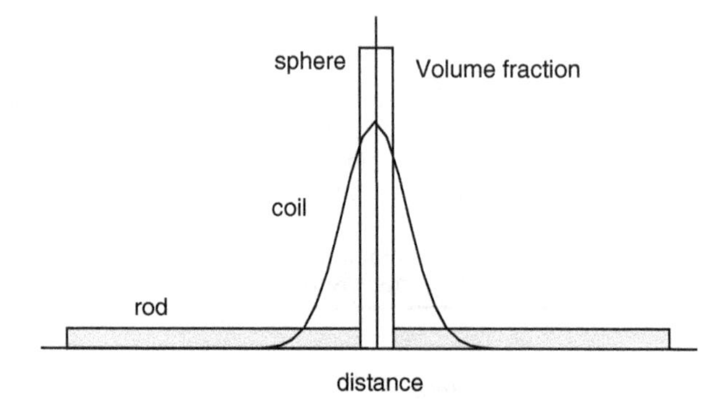

Figure 7.5 Average volume fractions for three different shapes of a polymer chain, a random coil, a rod and a sphere

important role (see Section 7.8). We can rewrite Equation (7.10) as

$$\langle R_G^2 \rangle = \frac{C_\infty n \ell^2}{6} \tag{7.11}$$

Kuhn showed that any chain could be scaled so that it could be described as completely flexible. This is done by dividing the chain into Kuhn lengths which are multiple numbers of monomers depending on C_∞. Another related dimensionless parameter is the persistence, $p = C_\infty/6$.

7.6.3 Radius of Gyration in Ideal Solution

Rewriting Equation (7.11) in a simpler form allows us to use tabulated values to find the radius of gyration of polymers in ideal solvents (defined as $R_G \sim n^{0.5}$). We use the expression

$$\langle R_G \rangle = \alpha M^{0.5} \tag{7.12}$$

and use values of α in Table 7.4 and the required molecular weight.

7.6.4 Excluded Volume

For the ideal chains we have discussed above we have used the random walk model, which allows chains to overlap. This is, of course, not very realistic as the chains must have a finite volume to exist. Theoretically this is a tricky problem but computer simulations and a

Table 7.3 Values of the characteristic ratio

Polymer	C_∞
Ideal chain	1.0
Poly(ethylene oxide)	5.6
Polydimethylsiloxane	5.2
Polystyrene	9.5
Polyethylene	5.3
DNA	600

Table 7.4 Values of α to calculate the radius of gyration from Equation (7.12)

Polymer	$\alpha/10^{-4}$ nm
Polyethylene	435
Poly(ethylene oxide)	330
Polydimethylsiloxane	250
Polystyrene	282

speculation by Flory has shown that we can use Equation (7.13) to account for excluded volume:

$$\langle R_G^2 \rangle \sim M^{\nu} \tag{7.13}$$

The exponent ν is given by $\nu = 6/(D+2)$, where D is the spatial dimension. So in 3D, $\nu = 6/5$ and the chain expands beyond its ideal dimensions ($\nu = 1$). The effect does not seem very significant but for high molecular weight chains it is appreciable. Consider a chain with 10^4 segments. Then R_G is predicted by Equation (7.13) to increase by a factor ≈ 2.5. The ideal chain and the excluded volume chain are two possible models we can use to estimate the polymer chain dimensions. In practice, the ideal chain model works in two real situations, in a poor solvent which causes the chain to collapse due to a net chain–chain attraction (Section 7.8) and in a polymer melt where the intra-chain–chain repulsion is balanced by repulsions from neighbouring chains.

7.6.5 Scaling Theory: Blobs

A rather clever model which encapsulates these two models of polymer chains has been proposed by de Gennes. His idea was to break the chain up into blobs, as shown in Figure 7.6. Inside the blob the chain is self-avoiding but the blobs themselves can overlap and are essentially ideal.

If we have g monomers per blob and the blob size is ξ, then $\xi \sim g^{\nu} \sim g^{3/5}$. Now if the blobs can overlap then $R \sim (n/g)^{0.5}\xi$ and hence

$$R \sim n^{0.5} g^{0.1} \tag{7.14}$$

Figure 7.6 The Blob model

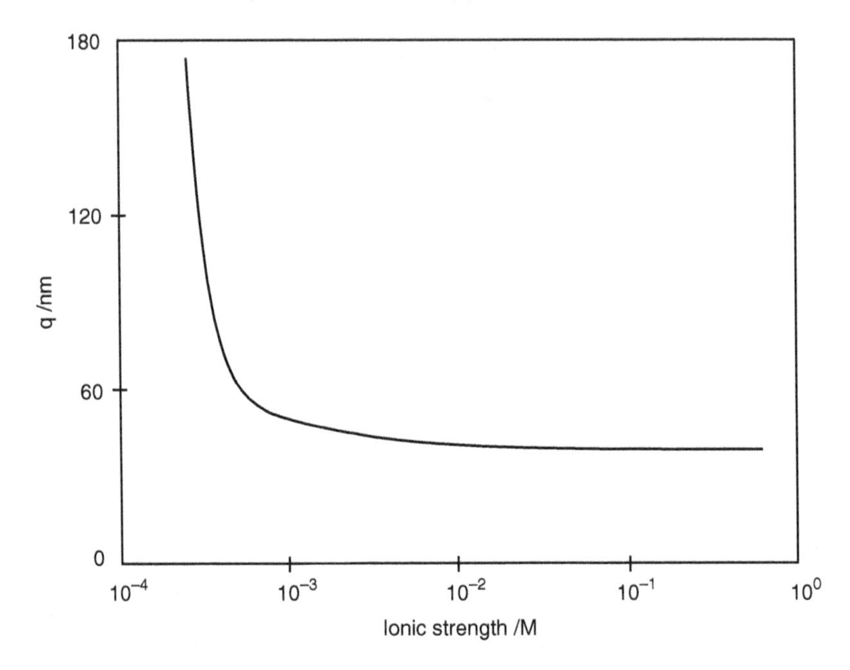

Figure 7.7 *The change in persistence length of DNA in aqueous salt solutions*

For an ideal chain $g = 1$ and for a chain with full excluded volume $n = g$ which recovers the two extreme cases of an ideal and swollen chain above.

7.6.6 Polyelectrolytes

Polyelectrolytes are a very common class of polymer in which each monomer carries a charge. For weak polyelectrolytes this is pH dependent (e.g. poly(acrylic acid)) but is pH independent for strong polyacids (e.g. poly(styrene sulfonate)). As the effect of charge is repulsive these molecules can become highly extended and their size will depend strongly on the solution environment (e.g. by adding salt the charges on the chain will be screened and the chain will contract). Because of this strong stretching we can use the following expression for the end to end distance:

$$\langle R^2 \rangle = L^2(1 - L/3q) \tag{7.15}$$

where L is the contour length and q is the electrostatic contribution to the persistence length. Figure 7.7 shows how q varies with ionic strength for DNA solutions with change in ionic strength.

7.7 Measuring Polymer Molecular Weight

There are many methods to establish the molecular weight of a polymer. However, the problem is not straightforward as polymer chains are not all of the same length and methods which can distinguish between chain lengths are particularly important. Firstly we shall define what we mean by molecular weight.

The number average molecular weight (M_N) is calculated by finding the total weight and dividing by the number of molecules:

$$M_N = \frac{\sum_{i=1}^{N} N_i M_i}{\sum_{i=1}^{N} N_i} \tag{7.16}$$

N_i is the number of molecules with mass M_i.

The weight average (M_W) is defined as

$$M_W = \frac{\sum_{i=1}^{N} N_i M_i^2}{\sum_{i=1}^{N} N_i M_i} \tag{7.17}$$

From these definitions higher molecular weight molecules contribute more to M_W than to M_N. For a monodisperse polymer M_W equals M_N and the ratio of M_W/M_N can be used to indicate the degree of polydispersity. Figure 7.8 illustrates the molecular weight distributions for different values of this ratio and it can be seen that even when M_W/M_N is less than 1.02 there is still appreciable polydispersity.

More generally higher moments can be defined as

$$M_n = \frac{\sum_{i=1}^{N} N_i M_i^n}{\sum_{i=1}^{N} N_i M_i^{n-1}} \tag{7.18}$$

Many techniques allow us to measure the molecular weight of polymers and, as they depend on different physical properties, they give rise to different moments of the distribution.

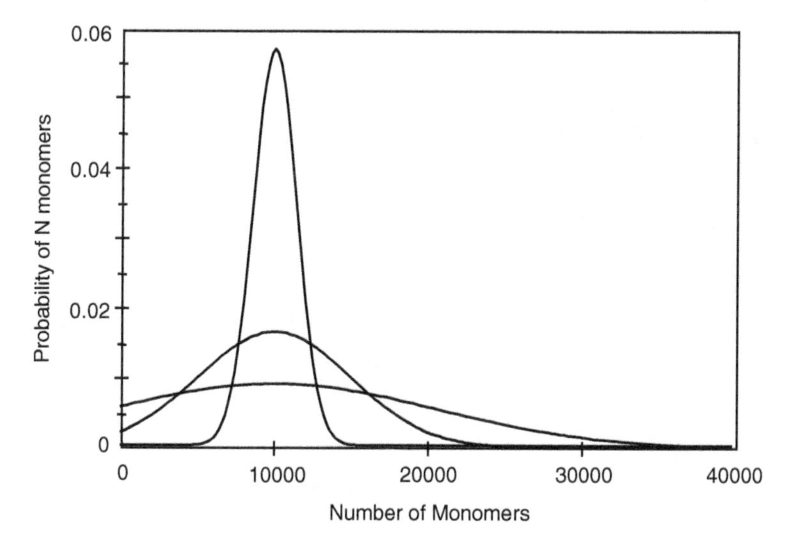

Figure 7.8 *Molecular weight distribution functions for three different values of M_W/M_N: 1.38 (broad peak) 1.20 (medium peak) and 1.02 (narrow peak). The value of M_N was 10 000*

Table 7.5 *A summary of different polymer molecular weight methods and the moments of the distribution that can be measured*

Method	Moment	Absolute/Polydispersity
Osmotic pressure	M_N	No/limited MW range
End group analysis	M_N	No/limited MW range
Chromatography	M_W	Yes/requires calibration
Scattering	M_W	Yes (model required/absolute)
Viscosity	M_V	No/requires calibration
MALDI TOF	M_N	Absolute

Table 7.5 summarises some common methods and the molecular weight average they measure.

7.7.1 Viscosity

A detailed description of polymer viscosity is beyond the scope of this chapter. However, the relationship of viscosity to molecular weight is very important. We can define specific viscosity as a dimensionless quantity with first-order solvent effects removed. First defining the specific viscosity η_{sp} as

$$\eta_{sp} = (\eta_{solution} - \eta_{solvent})/\eta_{solvent} \tag{7.19}$$

Einstein first noted that the specific viscosity was proportional to volume fraction $-\eta_{sp} \sim \phi \sim cR^3/M$ as the volume of a polymer coil is $\sim R^3$ and c is the molar concentration.

The intrinsic viscosity extrapolates this value to infinite dilution to remove intermolecular effects: $[\eta] = \eta_{sp}/c$ as the concentration c tends to zero.

As the end to end distance for the polymer coil $R \sim M^{0.5}$, we can use this to estimate the coil volume and hence

$$[\eta] \sim R^3/M \sim M^{3/2}/M = kM^a \tag{7.20}$$

k and a are known as the Mark–Houwink parameters and the equation gives us a simple semi-empirical formula for the viscosity which we can use to determine molecular weight.

7.8 Flory Huggins Theory

7.8.1 Polymer Solutions

The conformation of polymer chains in solution depends strongly on their architecture and the solvent quality. Flory and Huggins developed a theory for the solubility of polymers based on a net solvent–polymer interaction energy. It is defined as

$$\chi\left[(u_{12} - \frac{1}{2}(u_{11} + u_{22})\right]\frac{z}{k_B T} = \frac{\Delta u}{k_B T} \tag{7.21}$$

where 1 refers to solvent and 2 to a polymer segment. z is a coordination number, k_B is the Boltzmann constant and T is the absolute temperature. The final and initial states that lead to this definition are shown in Figure 7.9 and u are the pairwise energy interactions.

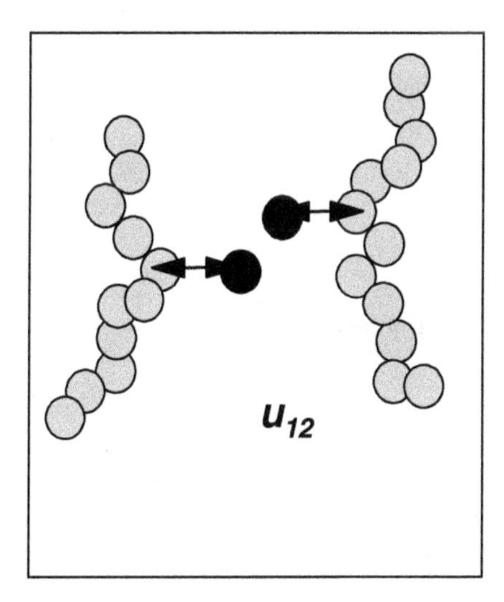

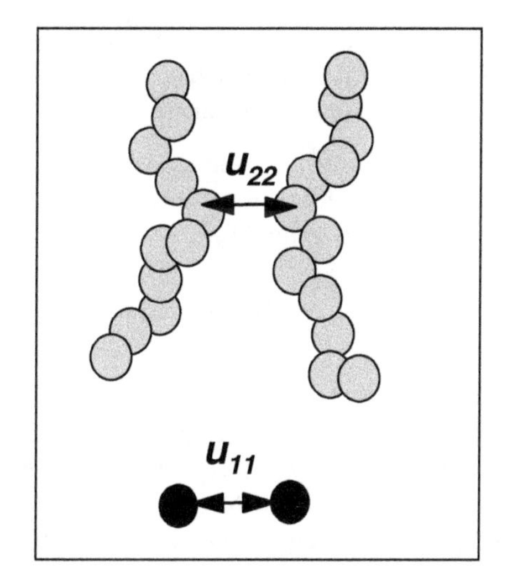

Figure 7.9 *The final (left-hand figure) and initial (right-hand figure) states in mixing pure polymer with pure solvent*

The Flory–Huggins lattice model (Figure 7.10) can be used to find the free energy for mixing ΔA_{m} of a polymer with a solvent and thus to construct a polymer solution phase diagram. The basic model makes the following assumptions.

- The lattice is full with either polymer segments or solvent molecules.
- The segment size and solvent size are the same.

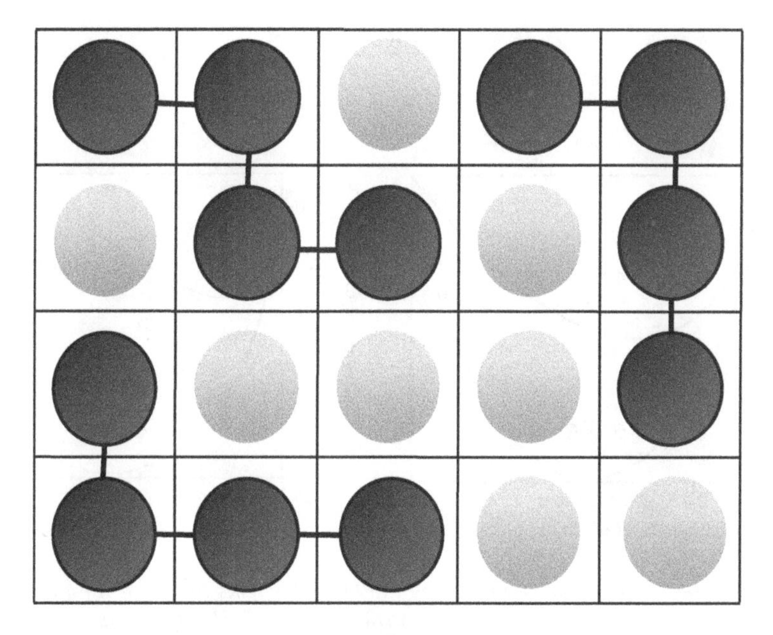

Figure 7.10 *The Flory–Huggins lattice*

- There is random mixing in the lattice.
- The system is homogeneous and the mean field assumption can be used.
- The Flory–Huggins parameter is purely enthalpic.

The assumptions make the model reasonably simple to use but they also impose several limitations, but most of these can be overcome. The final equation for the free energy of mixing, ΔA_{m}, is given by

$$\Delta A_{\mathrm{n}} = k_{\mathrm{B}}TN\left[\frac{\phi_1}{r_1}\ln\phi_1 + \frac{\phi_2}{r_2}\ln\phi_2 + \phi_1\phi_2\chi\right] \tag{7.22}$$

where r is the chain length (1 refers to the solvent and 2 to the polymer), ϕ is the volume fraction and N is the number density of molecules.

From Equation (7.22) it can be seen that the two entropy terms are always negative and so dissolution depends on the value of the χ parameter. Figure 7.11 shows how the free energy changes with volume fraction of polymer for different values of χ. When $\chi = 0$ we have complete mixing for a monomer solution ($r_1 = r_2 = 1$, lower curve) and for a polymer solution ($r_1 = 1$ and $r_2 = 1000$, upper curve) solutions. When $\chi = 1.2$, the monomer solution (lower curve, right-hand figure) is still miscible over the entire volume fraction range, but for the polymer solution, a miscibility gap appears and the solution breaks down into two immiscible solutions: one dilute and one concentrated.

In terms of a phase diagram, the theory predicts an upper critical solution temperature (UCST) and this has been found experimentally for many non-aqueous polymer solutions. However, many aqueous polymer solutions also show a lower critical point as well (LCST). This can be rationalised as a breakdown in the assumption that the system has no free volume. The χ parameter can be modified to take this into account:

$$\chi = \frac{\Delta u}{kT_{\mathrm{B}}} + AT^n \tag{7.23}$$

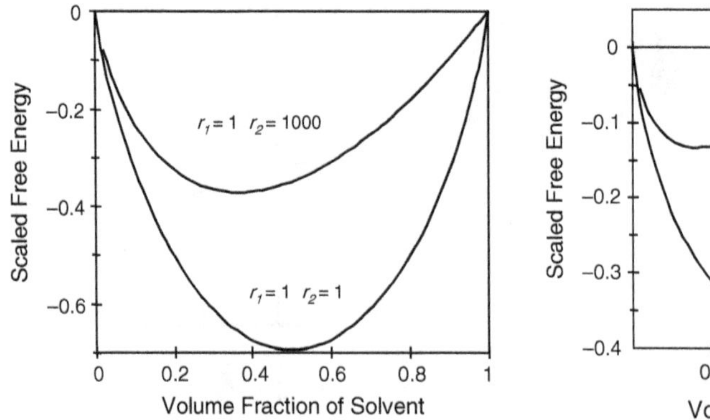

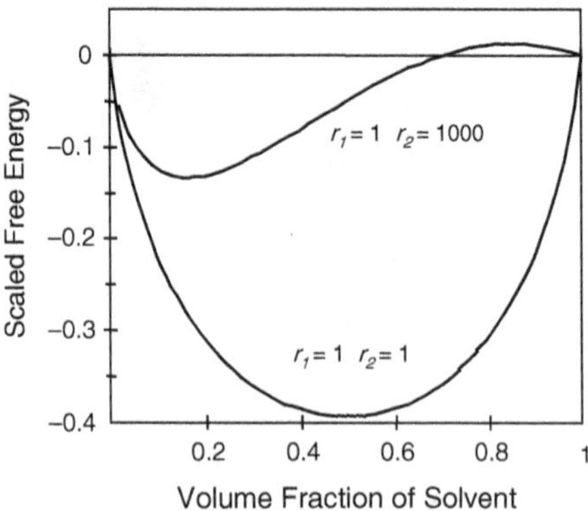

Figure 7.11 *Free energy calculations using the Flory–Huggins model as a function of the χ parameter: left-hand figure $\chi = 0$ and right-hand figure $\chi = 1:5$. The upper curves correspond to $r_1 = 1$ and $r_2 = 1000$ and the bottom curves correspond to $r_1 = 1$ and $r_2 = 1$*

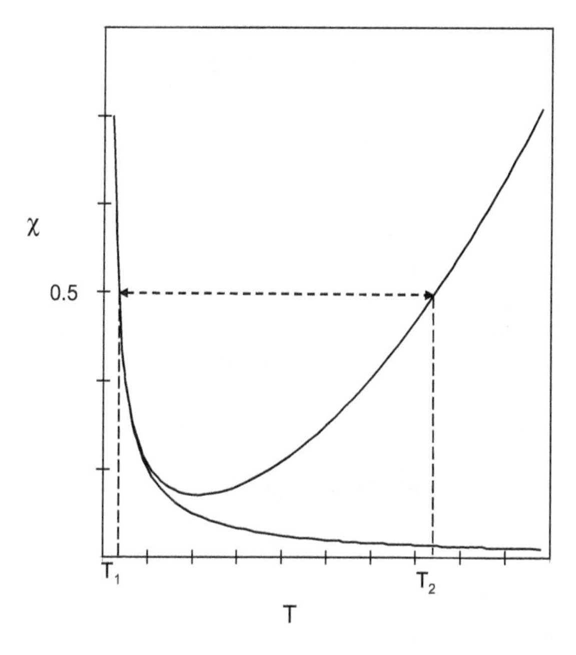

Figure 7.12 *Variation in χ with temperature using Equation (7.22)*

where T is temperature and A and n are constants. The form of Equation (7.23) is shown in Figure 7.12 and the resultant phase diagram in Figure 7.13.

A useful experimental parameter that we can obtain directly from Equation (7.23) is the osmotic pressure by obtaining the chemical potential for the solvent:

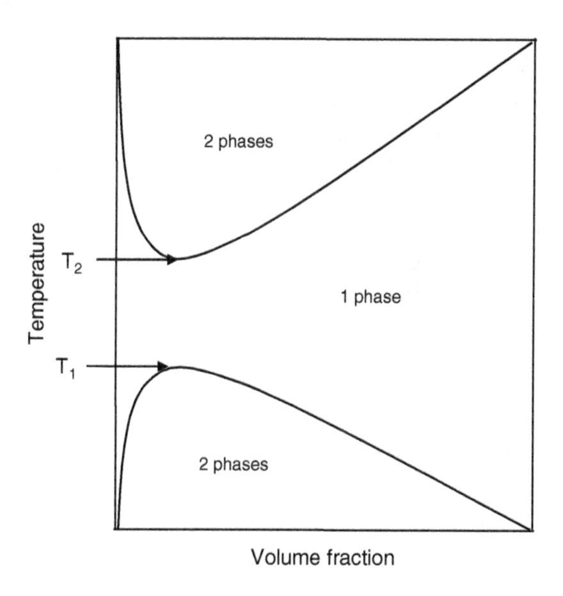

Figure 7.13 *A polymer solution phase diagram. The UCST and LCST are shown as T_1 and T_2*

$$\frac{\Pi}{c} = RT\left(\frac{1}{M} + (0.5-\chi)\frac{c}{v_0\rho^2}\right) \tag{7.24}$$

This gives another route to determine molecular weight and the χ parameter. v_0 is the molar volume of solvent and ρ is the polymer density. The equation is also the starting point for developing the theory of light scattering from polymer solutions.

Equation (7.24) also reveals another interesting fact that there is a critical value for χ. When $\chi = 0:5$, Equation (7.24) reverts to the ideal Van't Hoft equation. Under these conditions the solution is ideal and the temperature this occurs at is called the θ temperature. The balance of the osmotic pressure, which in a poor solvent compresses the chain, with the chain excluded volume, which leads to expansion, accounts for the ideal random coil behaviour of polymer chains that can be observed.

The behaviour of a polymer melt, which also shows ideal behaviour, can also be rationalised in that the osmotic pressure of the polymer in the solutions can also be calculated and shown to be ideal when extrapolated to a pure melt.

7.8.2 Polymer Melts

Mixtures of two polymers can also be described by the Flory–Huggins theory (Equation 7.22), but in this case both r_1 and r_2 can be large numbers meaning that the favourable entropy of mixing contribution to the free energy becomes very small. Mixing then is dominated by χ_{AB}, the interaction parameter between the two monomers A and B. This explains why most polymers do not mix except at relatively high temperatures.

7.8.3 Copolymers

Copolymers having more than one monomer type have more than one χ parameter. Random copolymers often have intermediate properties between the two monomers, if they are truly random, and it is possible to use a weighted χ parameter to describe some of their solution behaviour. Block copolymers such as non-ionic surfactants (Chapter 4) often aggregate as micelles in solution if one block is in a favourable solvent environment ($\chi_A < 0:5$) and one in an unfavourable solvent ($\chi_B > 0:5$). The extent of aggregation will also depend on the χ_{AB} between the blocks and the block ratio, as well as on temperature and concentration. Many block copolymer systems show very complex phase behaviour. The Flory–Huggins theory can be applied in principle to these systems to calculate the critical micelle concentrations and other solution properties.

References

(1) Flory, P. J. (1953) Principles of Polymer Chemistry. Cornell University Press, Ithaca.
(2) Flory, P. J. (1989) Statistical Mechanics of Chain Molecules. Hanser, Munich.
(3) de Gennes, P.-G. (1979) Scaling Concepts in Polymer Physics. Cornell University Press, Ithaca.
(4) Sun, S. F. (1994) Physical Chemistry of Macromolecules. John Wiley & Sons, Ltd, New York.
(5) Doi, M. (1996) Introduction to Polymer Physics. Clarendon Press, Oxford.
(6) Grosberg, A. Y., Khokhlov, A. R. (1997) Giant Molecules. Academic, San Diego.
(7) Rubinstein, M., Colby, R. (2003) Polymer Physics. Oxford University Press, Oxford.
(8) Stevens, M. P. (1999) Polymer Chemistry. Oxford University Press, Oxford.

8

Polymers at Interfaces

Terence Cosgrove

School of Chemistry, University of Bristol, UK

8.1 Introduction

In many situations we need to control the stability of dispersions, for example in pharmaceutical preparations, paints and inks. In other situations we may need to flocculate them, for example in bacterial harvesting or the drying of a paint film. Adsorbed polymers can play a key role in circumventing and controlling these situations. In order to discover how this can be achieved it is necessary to understand the basic structure of an adsorbed polymer and the importance of the solution and surface chemistry.

In this chapter we shall introduce some of the basic concepts of polymer adsorption from a theoretical point of view and then make comparisons of the background theory with experiment. The emphasis in this chapter is on discerning parameters of the adsorbed layer which are useful in constructing formulations with desired properties.

Figure 8.1 shows an atomistic simulation of an adsorbed polymer interacting with a nanoparticle of silica in water. In this case the polymer both adheres to the surface and has a long tail which protrudes into solution. This is the generally accepted picture of an adsorbed polymer with some segments attached to the surface as trains, which are joined by segments in loops. The ends of the chains which often protrude from the surface are tails. It is the balance of these three populations that gives the adsorbed layer its unique properties. Two books which contain detailed accounts of polymer adsorption are those by Fleer *et al.* (1) and Jones and Richards (2). Several review articles have also been published (3, 4).

Colloid Science: Principles, methods and applications, Second Edition Edited by Terence Cosgrove

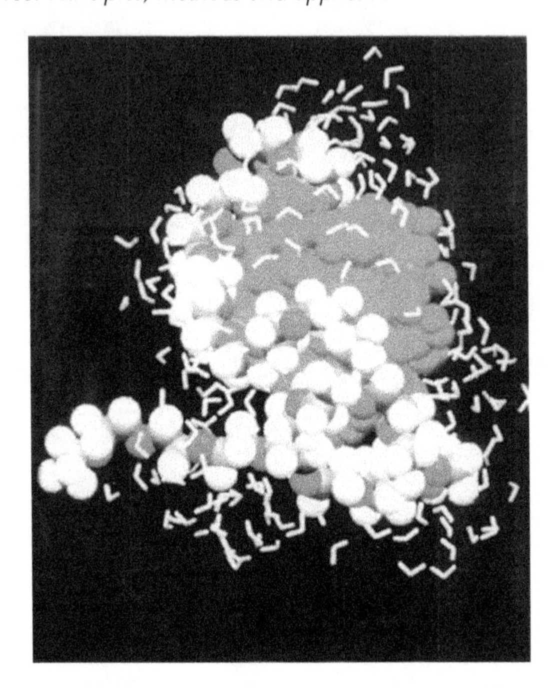

Figure 8.1 An atomistic simulation of a chain of poly(ethylene oxide) adsorbed onto a silica nanoparticle in water

8.1.1 Steric Stability

Except for some special cases (e.g. microemulsion(s)) colloidal dispersions are not thermodynamically stable but by virtue of energy barriers much greater than k_BT they are kinetically stable.

Figure 8.2 shows the interparticle potential between a pair of colloidal particles (details can be found in Chapter 3). This system is kinetically stable by virtue of a surface charge which gives rise to a repulsion (positive) which exceeds the inherent attractive potential (negative) which is due to van der Waals forces. The thermal energy must overcome this repulsion for the two particles to flocculate. However, this balance is very strongly affected by the presence of salt. By increasing the salt concentration the electrostatic repulsion is easily reduced so that the total potential becomes attractive and the sample will flocculate, as shown in Figure 8.3. In this situation and in the case of non-polar solvents, where electrostatic stabilisation is difficult, another mechanism is required to stabilise the dispersion and this is often provided by an adsorbed polymer layer. The effectiveness of this layer depends very much on the conformation of the chains involved and this is the main focus of the discussion in this chapter. The subject of steric stability *per se* is treated in more detail in Chapter 9.

8.1.2 The Size and Shape of Polymers in Solution

In ideal conditions the size of a macromolecule depends on $n^{0.5}$ where n is the number of monomers (Chapter 7). More exactly, we can define the radius of gyration, R_G, of the

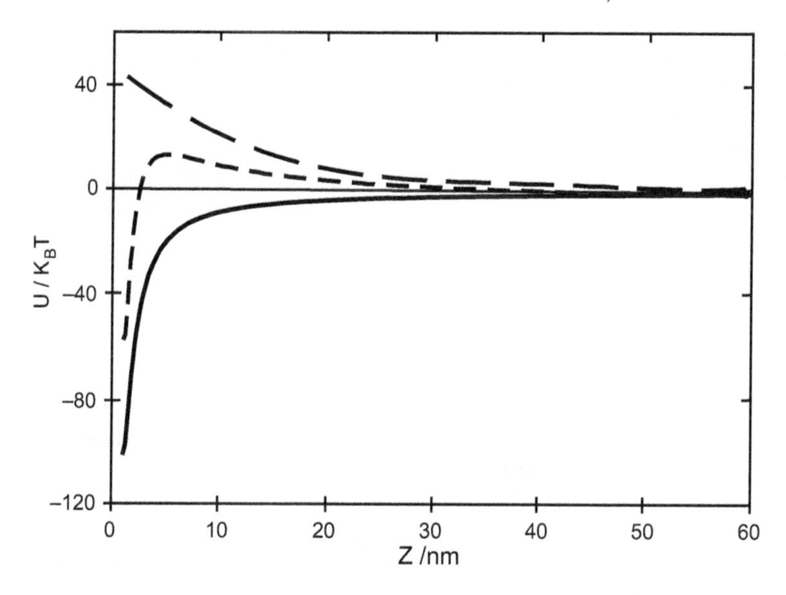

Figure 8.2 *The inter-particle potential between two AgBr particles with radius 100 nm in a 0.001 M NaCl solution long dash repulsion, solid line attraction, short dash net potential*

polymer as $R_G = C_\infty \ell n^{0.5}$ where C_∞ is the characteristic ratio and ℓ is the monomer length. This prediction is realised both in a polymer solution and in a polymer melt. In the former case the osmotic pressure of the solvent overcomes the excluded volume of the polymer chain and ideal behaviour is found. This special condition is known as a θ solvent and is

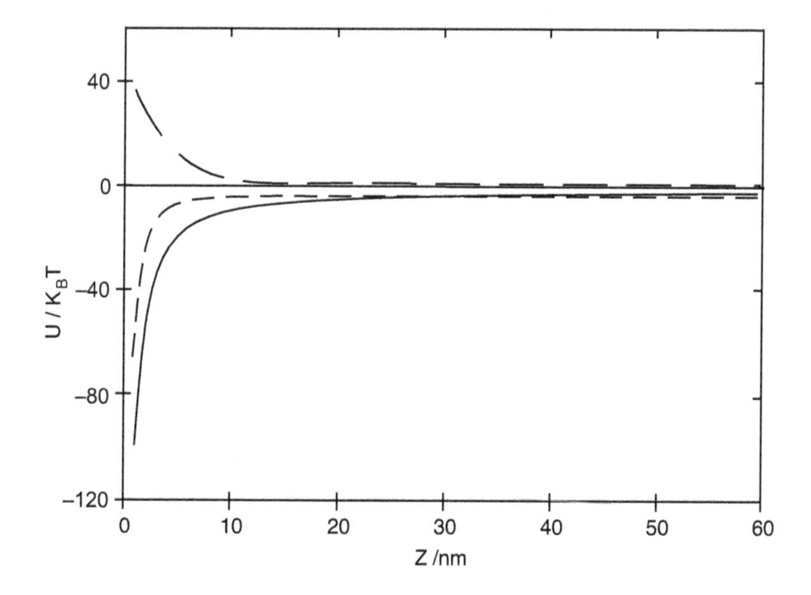

Figure 8.3 *The inter-particle potential between two AgBr particles with radius 100 nm in a 0.01 M NaCl solution long dash repulsion, solid line attraction, short dash net potential*

characterised by the Flory–Huggins parameter χ equal to 0.5. The more usual case is when the osmotic pressure does not overcome the excluded volume and the polymer chain expands (a good solvent). In this case χ is less than 0.5 and we find that the polymer coil expands leading to $R_G \sim n^{0.6}$ where R_G is the radius of gyration. The exponent in this relationship is directly related to χ (further details can be found in Chapter 7). The shape of an uncharged homopolymer in an ideal solution is approximately spherical, but for block copolymer(s) and polyelectrolytes a whole range of shapes exist and the shape of a molecule strongly influences R_G. On adsorption this picture breaks down and this is the focus of the discussion in this chapter.

8.1.3 Adsorption of Small Molecules

It is useful to start by considering the adsorption of small molecules at surfaces. These can be described in many cases by either the Langmuir or the BET isotherm(s) (5); the former being a limiting case of the latter. Equation (8.1) gives the form of the Langmuir equation where θ is the fractional surface coverage, c is the equilibrium concentration and b is a constant. The model assumes that solute molecules only interact with the surface and, hence, only a monolayer can be formed. Figure 8.4 illustrates the form of the equation. The BET model includes solute–solute interactions and, hence, a multi-layer structure can be formed, as is evident in Figure 8.4, for a particular set of interaction energies. For polymer adsorption there are examples in the literature of using the Langmuir equation to describe experimental isotherms, but, although functionally the equation often works, interpretation of the

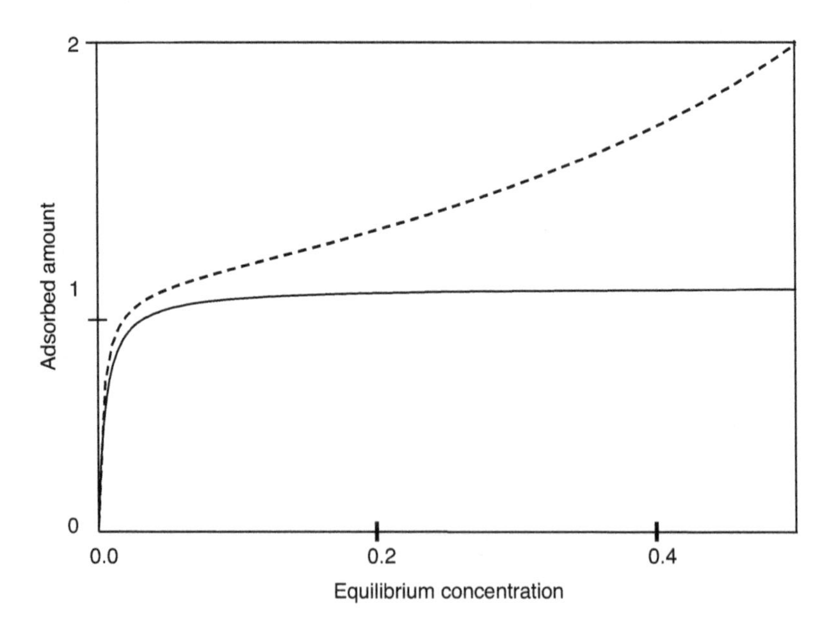

Figure 8.4 A comparison of the Langmuir (solid Line) and BET (dashed line) isotherms

thermodynamic variables such as the adsorption energy must be treated with caution. The reason for this is the effect of polydispersity (see Section 8.4.2):

$$\theta = \frac{bc}{1+bc} \quad (8.1)$$

8.2 Adsorption of Polymers

8.2.1 Configurational Entropy

Unlike small molecules, there is often a large configurational entropy penalty to pay when a polymer adsorbs from solution on to a surface. We can use a simple thermodynamic argument to estimate the change in entropy when a polymer adsorbs. For simplicity we assume that each polymer segment has 3 possible spatial orientations (Ω). That means that for n segments a polymer has 3^n possible conformations. Similarly, if the coil adsorbs completely flat then in 2 dimensions there are 2^n conformations.

We can estimate the entropy change from the third law of thermodynamics as

$$S = k_B \ln(\Omega) \quad (8.2)$$

Hence

$$S = k_B \ln(2^n/3^n) \quad (8.3)$$

For this entropy penalty to be overcome we need a critical enthalpy of at least $\sim 0.4 k_B T$. This simple analysis ignores the effect of liberating the solvent from the surface which is clearly entropically advantageous.

8.2.2 The Flory Surface Parameter χ_s

In a similar way to the definition of the Flory–Huggins solution parameter, χ, we can define a Flory surface parameter χ_s which is defined in terms of an initial state comprising a polymer melt and adsorbed solvent and a final state of bulk solvent and adsorbed polymer as shown in Figure 8.5. The reason for this particular form of this definition is that it makes χ and χ_s effectively independent, i.e. we separate the surface interactions from the solution interactions. In reality this would be very difficult to achieve.

The definition of χ_s is given by

$$\chi_s = -\left[(u_{2s} - u_{1s}) - \frac{1}{2}(u_{11} - u_{22})\right] k_B T \quad (8.4)$$

where u are the pairwise contact energies and are normally attractive (<0). S refers to the surface, 1 a solvent molecule and 2 a polymer segment. For adsorption to take place χ_s must be greater than the critical value, χ_{sc} whose value is given approximately by Equation (8.3).

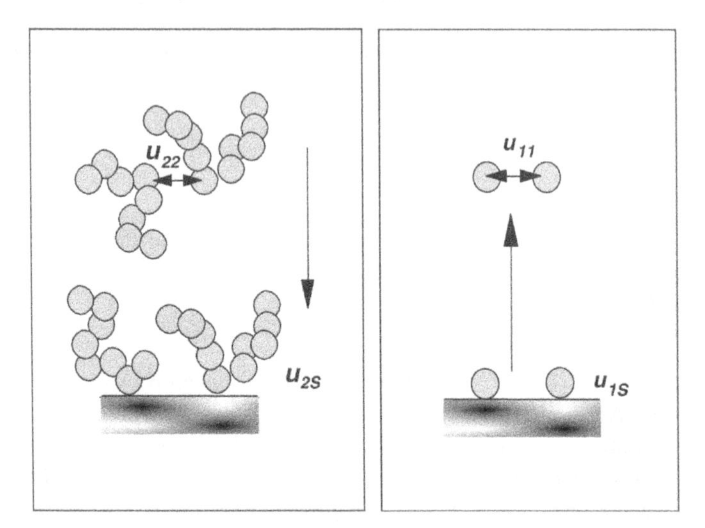

Figure 8.5 The exchange process of adsorbing a polymer segment and displacing an adsorbed solvent molecule: (1) refers to the solvent, (2) to a polymer segment and 's' refers to adsorbed

8.3 Models and Simulations for Terminally Attached Chains

Because many aspects of the adsorption of polymers depend on chain conformations, simulation methods, which explore the different ways in which a polymer can interact with an interface, are particularly revealing. The simplest example is to constrain the polymer chain to a lattice rather than free space and to attach one end of the chain irreversibly to the surface. These two constraints reduce the number of chain conformations possible by a very large amount and prevent the chain from desorbing into the solution.

The simplest of these models involves counting the number of chain conformations exactly and this is the first approach we shall use. For longer chains the approximate methods of Monte Carlo and molecular dynamics are useful both for single and multiple chains. Finally, we shall describe a full thermodynamic model of polymer adsorption where the constraint of terminal attachment is overcome and a full equilibrium of the adsorbed population of chains with those in solution can be achieved.

8.3.1 Atomistic Modelling

Ideally, to predict the interaction of a polymer with a substrate we would like to build a theory which retains the detailed atomic structures of the solvent, polymer and interface. For simple visualisations this can be done but to make a detailed study at this level for practical systems with an ensemble of long chain lengths is not currently feasible. Nevertheless, the approach is useful and in Figure 8.6 we show a simulation of polystyrene on graphite. One interesting aspect of this visualisation is that the phenyl rings do not lie flat on the surface because of steric constraints. In this case the definition of the bound sound surface layer has to be generalised to segments within a certain distance of the surface. In these situations it is the shape of the molecule, i.e. the chemistry, which is the determining factor.

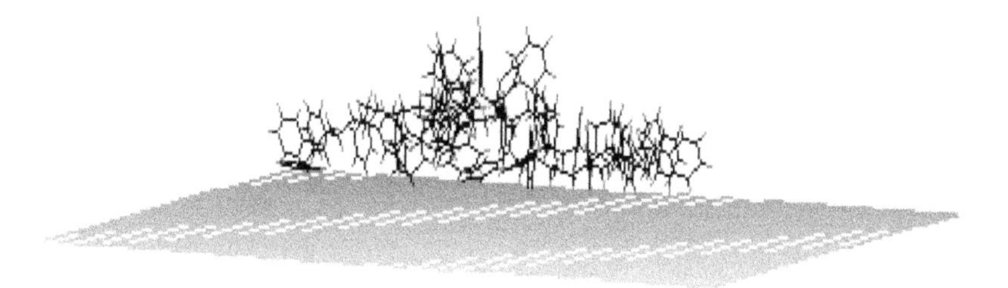

Figure 8.6 *An atomistic level simulation of a single polystyrene chain on a graphite surface. Segments in trains (touching the surface), loops and tails can be seen*

In a more realistic model we need to average over all possible chain confirmations and the result would be a volume fraction profile normal to the surface which describes the number of segments in layers starting at the interface and finishing in the bulk solution.

Given an explicit form for the volume profile we can readily calculate other important parameters using the following equations:

$$\Gamma = \rho_2 \int_0^{\text{span}} \phi_{\text{ads}}(z)dz \tag{8.5}$$

$$p = \rho_2 \int_0^{\ell} \phi_{\text{ads}}(z)dz/\Gamma \tag{8.6}$$

$$\delta_{\text{RMS}}^2 = \rho_2 \int_0^{\text{span}} \phi_{\text{ads}}(z)z^2dz/\Gamma \tag{8.7}$$

The adsorbed amount Γ is just the integral under the profile multiplied by ρ_2 the polymer density, giving units of mass/unit area. The bound fraction p corresponds to fraction of segments in the first adsorbed layer and is typically a monomer length, ℓ in width. The span in this context is the point at which the profile is zero (the maximum value is the chain length). In order to make a comparison of the experimental variables as above with a lattice model, a procedure for scaling to real space is required. There are several strategies for this, for example each lattice site could be occupied by one monomer or one statistical segment.

8.3.2 Exact Enumeration: Terminally Attached Chains

In its simplest form this approach uses a lattice with a single chain terminally attached to the interface. The basic idea is just to count every possible conformation: exact enumeration (EE). For example we shall use a cubic lattice as shown in Figure 8.7. We start at the coordinates (0, 0, 0) and reject all conformations which penetrate the surface and violate the excluded volume criterion (i.e. occupy the same lattice point). $C(n,m)$ is defined as the number of walks of length n bonds of which m are in the layer next to the interface. So, for example, $C(1,1) = 4$ and $C(2,1) = 4$. This soon becomes very difficult, and to reach more than $n = 20$ takes a very appreciable amount of computer time; even this modest chain length can have conformations of the order of 10^9. Once we have enumerated the array of numbers $C(n,m)$ we can quite easily calculate the bound fraction, p, of the walk. This is the fraction of the total number of segments that is in the layer next to the surface, i.e. as trains. For example,

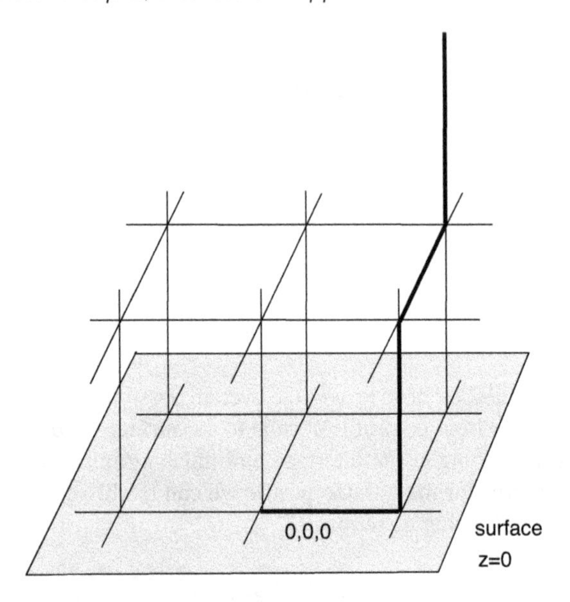

Figure 8.7 *A typical self-avoiding walk on a cubic lattice which is one of the conformations of the set C(4,1)*

for the walk shown in Figure 8.7 $p(n) = 1/4$. The results for all the conformations are combined using a Boltzmann weighting term which is just $\exp(m\chi_s/k_B T)$, and Equation (8.8) shows the statistical sum needed to find the average value of p, $\langle p \rangle$:

$$\langle p \rangle = \frac{\sum_{m=1}^{n} C(n,m) m \exp(m\chi_s/k_B T)}{n \sum_{m=1}^{n} C(n,m) \exp(m\chi_s/k_B T)} \tag{8.8}$$

It is useful to explore how $\langle p \rangle$ depends on the net adsorption energy χ_s but the calculation so far is for a finite value of N and the chain is anchored irreversibly at the surface. This means that the limiting value of $\langle p \rangle$ for $\chi_s = 0$ is $1/n$. The model can be developed further by using an extrapolation method to find the value of $\langle p \rangle$ in the limit of large n and these data are shown in Figure 8.8. At values of χ_s less than the critical value of χ_{sc} the bound fraction is zero which would correspond to no adsorption. However, below the critical value there is adsorption. In the limit as χ_s becomes much greater than $k_B T$ for an isolated chain all the segments tend to lie in the first lattice layer as trains.

This simple approach can be easily extended to work out the average number of segments in each of the lattice planes parallel to the surface. This requires some extra counting as we need to know the number of walks $C(n,m,s,z)$ which have m surface contacts and s segments in layer z. The data in Figure 8.9 show how the shape of the relative volume fraction profile, $\phi(z)$, varies as we change χ_s. At the surface we see the same picture as above. Below the critical adsorption energy, the number of segments in the first layer decreases. In this case we are dealing with a finite chain length ($n = 15$) and as these conformations all have one segment irreversibly attached at the surface, p is greater than 0. This regime has become known as a mushroom since $\phi(z)$ has a maximum. In contrast, when χ_s is 1.8, the chain collapses on the surface and this is know as a pancake.

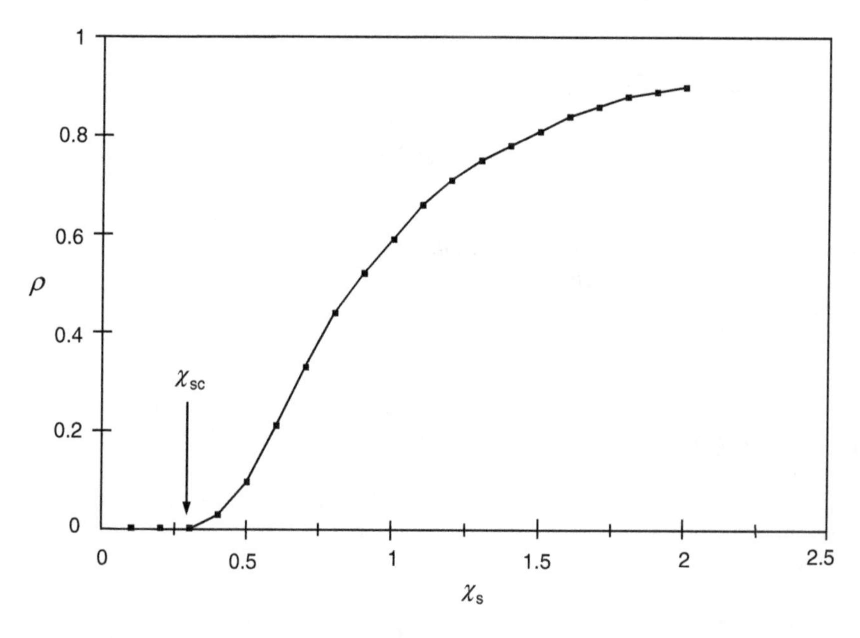

Figure 8.8 *The variation in the bound fraction ⟨p⟩ as a function of the adsorption energy χ_s*

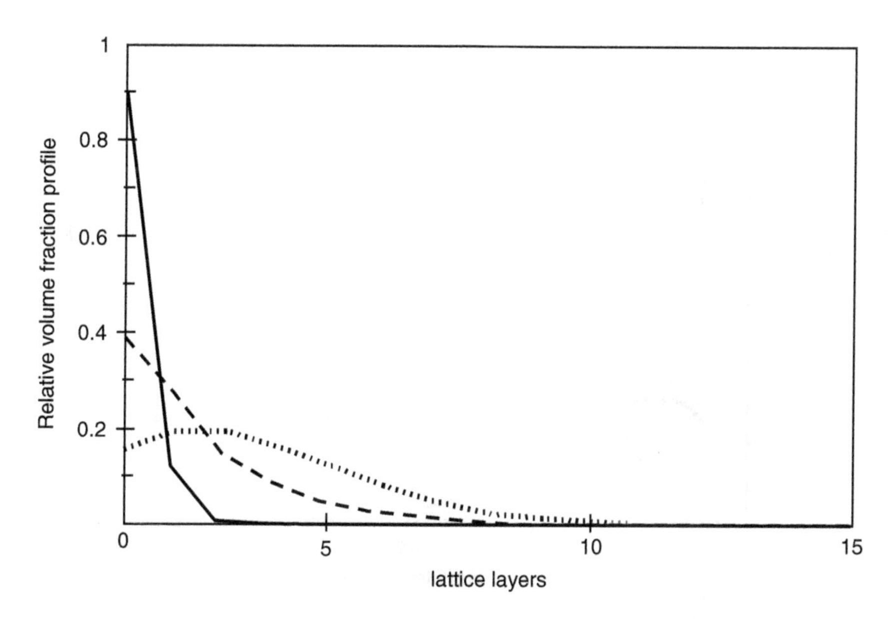

Figure 8.9 *The proportion of segments in layers normal to the surface for a terminally attached chain of 15 segments on a cubic lattice as a function of the adsorption energy χ_s. Three values are shown 0.0 (······), 0.6 (- - - - - -) and 1.8 (———)*

8.3.3 Approximate Methods: Terminally Attached Chains

The approach above is very limited as it is limited to a single short chain. Another approach is to use Monte Carlo (MC) or molecular dynamics (MD) methods (6). In these methods not all the chain conformations are generated, but a subset. By using appropriate selection criteria, this subset can be representative of the whole. The methods are not restricted to lattices and can deal with multiple chains of lengths substantially greater than with exact enumeration. A simple procedure for multiple chains is to use a periodic boundary such that any chain crossing the boundary is re-entered in the opposite side of the cell. This, however, does impose lateral coherence in the chain structures, but is still a useful procedure.

Figure 8.10 shows a chain length of 50 and a surface coverage, θ, of 0.15 which is defined as the number of adsorbed segments per surface lattice site. This approach also reproduces the profiles found above, the pancake and the mushroom, but now the volume fraction scale is absolute.

In the molecular dynamics approach, Newton's laws of motion are used to generate new conformations from old ones by evaluating the interaction forces in the system. Typically, simulations can be performed on the nanosecond timescale.

8.3.4 Scaling Models for Terminally Attached Chains (Brushes)

Another approach to discovering the structure of a terminally attached chain is through the scaling approach of de Gennes (7). In this model the chain is decomposed into g blobs, as in Chapter 7 and as shown in Figure 8.11.

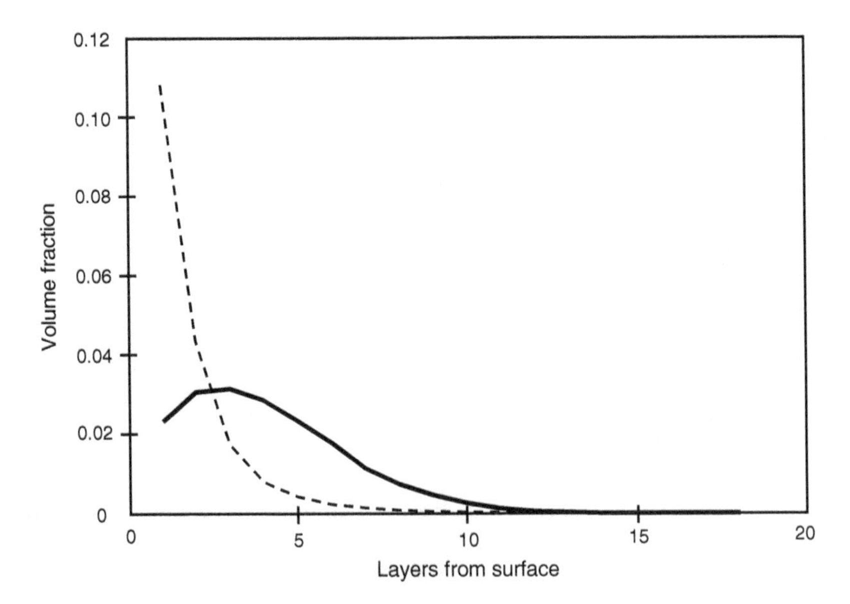

Figure 8.10 An MC simulation of a terminally attached chain with 50 segments and a surface coverage of 0.15, for two values of the adsorption energy above and below the critical value

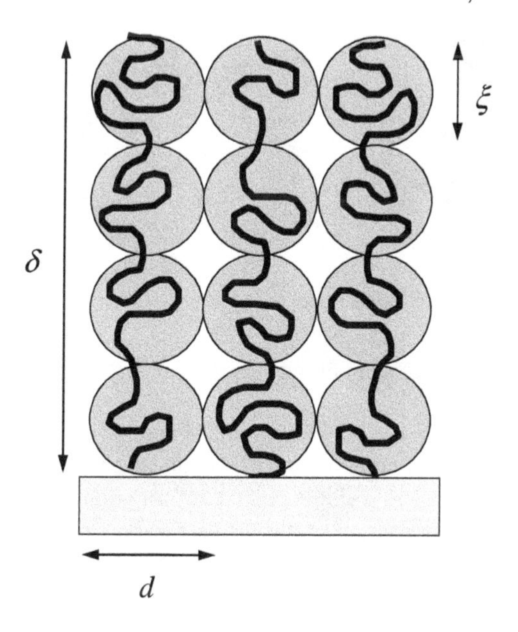

Figure 8.11 The blob representation of a series of terminally attached chains

If each blob contains a self-avoiding walk of g monomers then the blob size ξ is given by

$$\xi = g^{3/5}\ell \tag{8.9}$$

where ℓ is the length of a monomer. For n monomers the brush length is then

$$\delta = \left(\frac{n}{g}\right)\xi \tag{8.10}$$

The second, and central, assumption is that the blob size is directly related to the grafted amount σ so that $\sigma = 1/\xi^2$. Combining this result with Equations (8.9) and (8.10) we arrive at

$$\delta = n\sigma^{1/3} \tag{8.11}$$

This is a very surprising result as it predicts that the brush length is linear in chain length. This is clearly true for a rod normal to the surface but suggests that chains closely grafted together on a surface are very strongly stretched. A more sophisticated approach which confirms this result predicts that the brush volume fraction profile is parabolic (8). A further discussion can be found in Chapter 16.

8.3.5 Physically Adsorbed Chains: Scheutjens and Fleer Theory

The most successful and useful theory for polymer adsorption is that developed by Scheutjens and Fleer (SF) (1). This uses the basic ideas of the Flory–Huggins theory for polymer solutions and applies this to each layer of lattice built on a solid substrate.

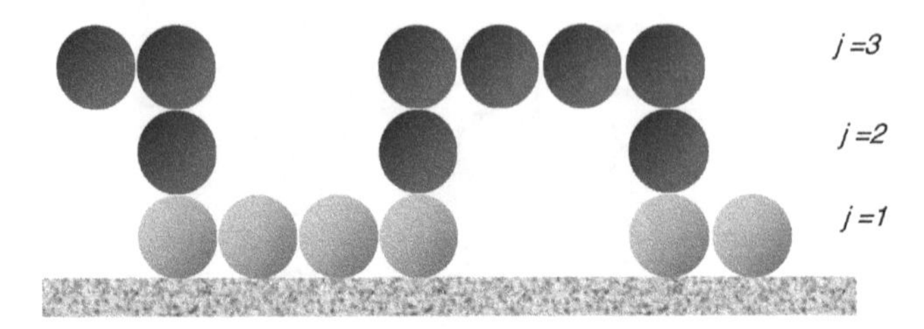

Figure 8.12 *A chain of 15 segments, six of which are at the interface*

In the SF model, layers, each containing L sites parallel to the surface, are numbered $j = 1, 2, 3$.

The goal of the model is to calculate and minimise the free energy of the system. This can be done in two stages: first calculating the energy and then the entropy.

In the example in Figure 8.12 a single chain conformation is shown which spans three layers. To find the energy U for this example we just need to find the total number of nearest neighbours for each segment. At the surface, for a cubic lattice we have 6 surface–polymer contacts and 19 polymer–solvent contacts. For three layers, $j = 1, 2, 3$ we can express the energy U as

$$U = 6\chi_s + [19\chi + 12\chi + 25\chi]/z \tag{8.12}$$

where z is the lattice coordination number.

This is a useful exercise but a more general approach is required and this can be done using the mean-field approximation; instead of using the actual number of contacts we use the probability that there is a polymer–solvent or polymer–surface contact in a given layer. This is then just the volume fraction of adsorbed (a) polymer in layer j; $\phi^a(j)$. For example, the average number of contacts in the first layer between the polymer and the surface is just $L\phi^a(1)$. More generally, Equation (8.12) can be written as

$$\frac{\Delta U}{k_B TL} = \phi^a(1)\chi_s + \sum_{j=1}^{M} \phi^a(j)\langle\lambda\rangle\chi \tag{8.13}$$

The total number of layers is M and λ is a parameter which corrects for the different number of contacts in the plane and between planes. For example, in a cubic lattice there are four nearest neighbours in the plane and one above and one below the plane and these must not be over-counted.

The calculation of the entropy is more involved and a formalism is needed that uses the mean field approach to make up explicit chains. The idea behind this is to use the concept of a free segment weighting factor G.

$G(z)$ is defined as the *weighting* in the ensemble that a monomer in layer z has compared to a segment in the bulk of the solution. For non-interacting segments if $j < 1$ then $G(z) = 0$ as no monomers can penetrate the surface. Similarly if $j > 1$ then $G(z) = 1$ as beyond layer 1 all segments are in the bulk solution (i.e. not interacting with the short range forces at the surface). If $z = 1$, however, then $G(z) = \exp(\chi_s)$ and a segment has a higher probability of being in the surface than in the bulk if $\chi_s > 0$.

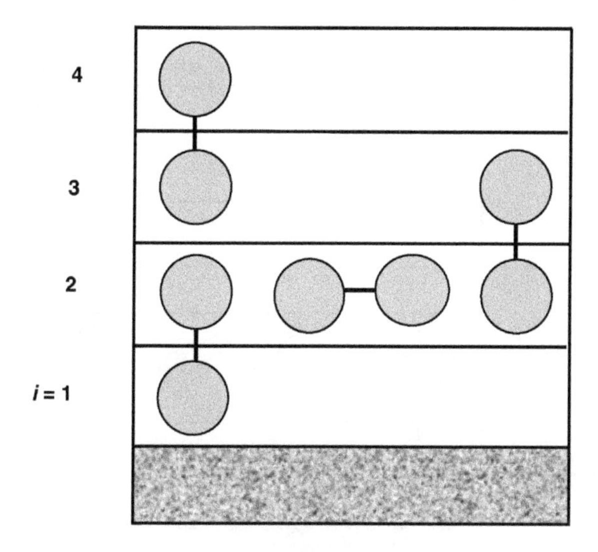

Figure 8.13 *Four dimers on a cubic lattice in different orientations*

The next step is to combine these weighting factors to form chains. Starting with the shortest chains possible, which are dimers, we can work out the weighting factors as follows. In Figure 8.13 there are four dimers in different conformations. The dimers are labelled by the lattice layer in which they exist. For example dimer (3, 4) has segment (1) in layer 3 and segment (2) in layer 4. The weighting factor for (3, 4) is given by

$$G(3,4) = \frac{G(3)G(4)}{6} = \frac{1}{6} \tag{8.14}$$

The combined weighting factor is arrived at as follows. Each segment on the lattice, not next to the surface (i.e. in layer 1) has a weighting factor of unity. The number of nearest neighbours on the lattice is six and there is only one possible way to construct the dimer (3, 4) starting at segment (3) and so the probability is 1/6.

Similarly, dimer (1, 2) has a weighting factor

$$G(1,2) = \frac{G(1)G(2)}{6} = \frac{1}{6}\exp(\chi_s) \tag{8.15}$$

Note segment (1) has an extra contribution as it is in layer (1) next to the surface.

The next step is to find an expression for the end-segment weighting factors, which is the weighting for a chain whose end segment is in layer j. This is effectively the number of chains which start in layers $(j-1)$, (j) and $(j+1)$ and end in layer j:

$$\text{i.e. } G(j;s) = G(j)\left[\frac{1}{6}G(j-1) + \frac{4}{6}G(j) + \frac{1}{6}G(j+1)\right] \tag{8.16}$$

where s is the end segment in layer j.

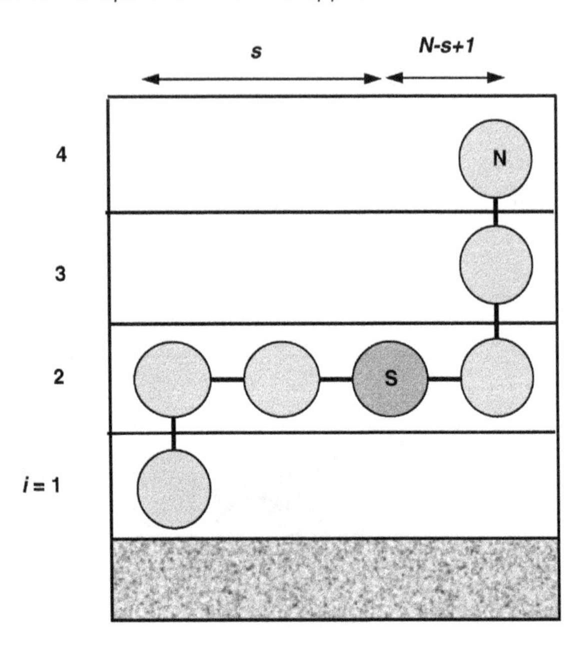

Figure 8.14 The connectivity rule for end segments forming a chain of N monomers

We can use this construction to work out the overall weighting for any chain segment in layer j by combining two chains whose end segments are in the same lattice layer. So the contribution to the volume fraction in layer j from any segment s for a chain with n segments is

$$\phi(j,s) = \frac{C}{G(j)} G(j;s)G(j,n-s+1) \tag{8.17}$$

This is shown schematically in Figure 8.14.

Hence, the total contribution to the volume fraction in layer j can be found by summing over all segments in the chain:

$$\phi(j) = \sum_{s=1}^{n} \phi(j;s) \tag{8.18}$$

This formalism can be extended to calculate explicit volume fraction profiles in equilibrium with a polymer solution. This makes it possible to estimate all the parameters needed to characterise the adsorbed layer—the adsorbed amount, the layer thickness and the bound fraction—and to make comparisons with experiment.

The volume fraction profiles in Figure 8.15 were calculated for a chain of 50 segments with an ideal solvent (the Flory parameter, $\chi = 0.5$). Two cases are evident:

- when the surface Flory parameter, χ_s, is greater than the critical value χ_{sc} we have adsorption and the profile falls monotonically to the bulk solution concentration (1000 ppm).

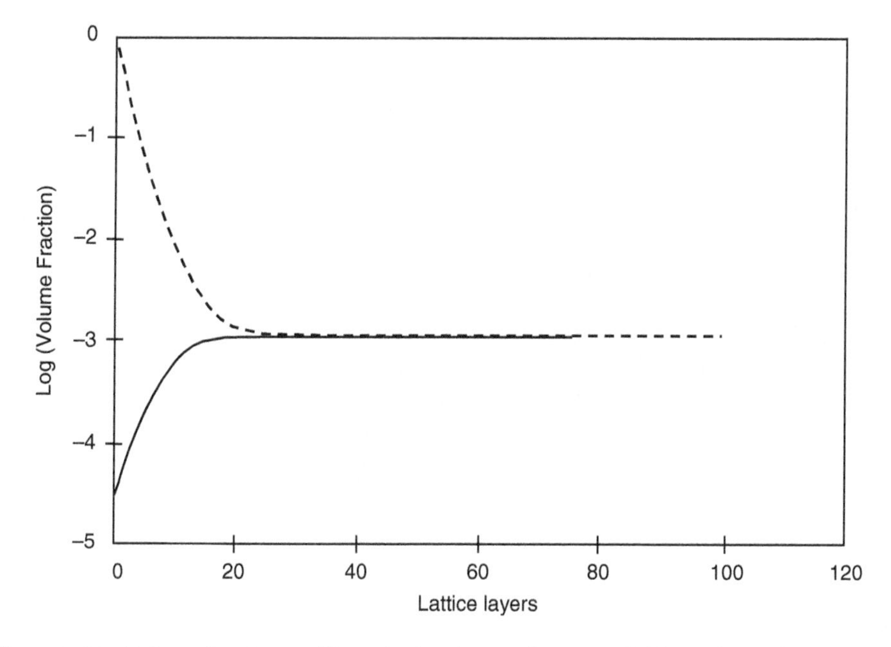

Figure 8.15 Volume fraction profiles calculated using the SF model for a chain of 50 segments with two different values of χ_s: $\chi_s < \chi_{sc}$ and $\chi_s > \chi_{sc}$. A more detailed account of this work can be found in the book by Fleer et al. (1)

- when $\chi_s < \chi_{sc}$ we have depletion and the concentration of segments at the surface is less than the bulk concentration.

Unlike the volume fraction profiles for terminally attached chains (Figure 8.10) whole chains can be completely desorbed. A more detailed account of the SF theory can be found in reference (1).

8.3.6 Scaling Theory for Physical Adsorption

The scaling approach discussed in Section 8.3.4 can also be used to model a physically adsorbed chain. The basic approach is to treat the adsorbed layer as 'self-similar' to a polymer solution. Three regimes were envisaged, as shown in Figure 8.16.

1. The proximal region which is effectively the train layer, with a width of the order of the monomer length ℓ, i.e. $\phi \sim$ constant for $z \leq \ell$.
2. The central region which is similar to a semi-dilute polymer solution. The local volume fraction of a polymer in a semi-dilute solution is given by

$$\phi \sim n/R^3 \tag{8.19}$$

and in a good solvent $R \sim n^{3/5}$ (Chapter 7) then

$$\phi \sim R^{5/3}/R^3 \sim R^{-4/3} \tag{8.20}$$

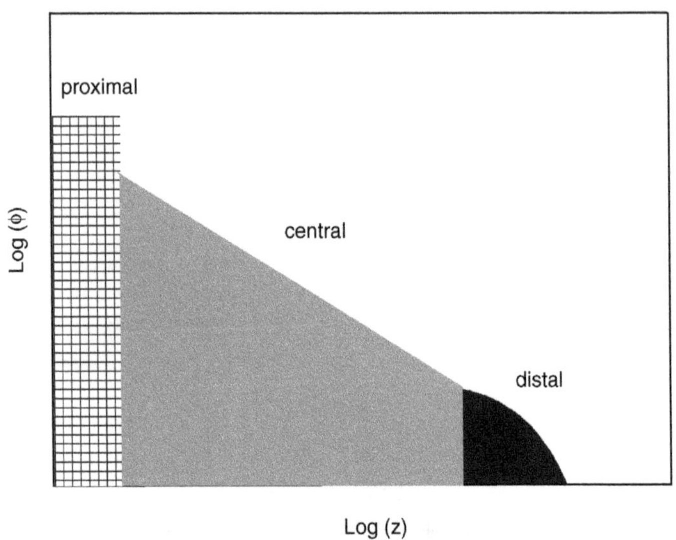

Figure 8.16 The scaling model of a physically adsorbed polymer layer showing the three regions: proximal (dark grey), central (light grey) and distal (black)

The volume fraction at a distance z from the wall can be thought of as the space in which the chains exist and hence:

$$\phi \sim z^{-4/3} \quad \text{for} \quad \ell < z < d \tag{8.21}$$

3. The distal region which is the periphery of the layer is approximated as an exponential decay:

$$\phi \sim e^{-z}(0.1) \quad \text{for} \quad d < z < \text{span.} \tag{8.22}$$

8.4 Experimental Aspects

8.4.1 Volume Fraction Profiles

There are several approaches to obtaining experimental volume fraction profiles but for particles the most successful has been to use small angle neutron scattering (SANS) and for macroscopic surfaces, neutron reflection. One approach is to fit the theoretical profile shapes directly to the scattering data and Figure 8.17 shows an example of the profiles found after fitting such data (9). The system is poly(ethylene oxide) (PEO) 110K molecular weight adsorbed on polystyrene (PS) latex of radius ~600 Å.

The two different profile shapes are very similar up to ~80 Å. Beyond this the tail region is significant and this is calculated explicitly by the SF approach but only approximately by scaling. The sensitivity of the method is ~0.001 volume fraction and so for longer chains it underestimates the chain span.

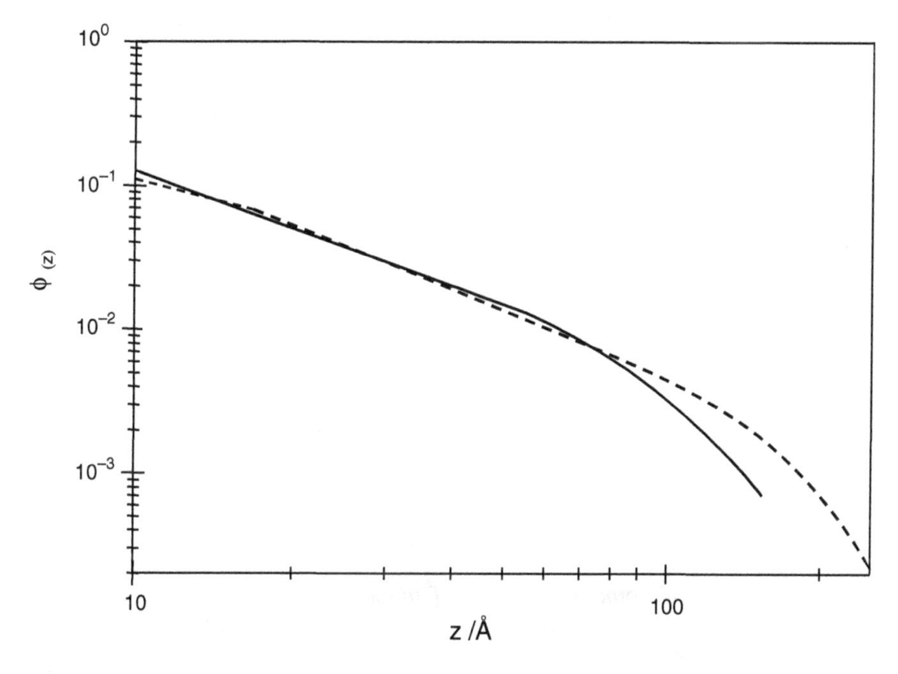

Figure 8.17 *Experimental volume fraction profiles for PEO 110K adsorbed on polystyrene latex in water, obtained by fitting SANS data to SF (- - - - - -) and scaling profiles (———)*

8.4.2 Adsorption Isotherms

The adsorption isotherm can be calculated from the SF model but not directly from scaling theory. A typical example is given in Figure 8.18.

The shorter chain has a lower affinity isotherm, similar to the Langmuir case. The longer chain has a much higher affinity which is typical for monodisperse polymers and shows that below saturation virtually all the added polymer is removed from solution. The way in which the adsorbed amount (Γ) at fixed concentration varies with molecular weight depends on solvent quality. In good solvents (see Chapter 7) Γ increases with increasing molar mass M at low M range, but for very high molar masses the plateau value becomes independent of chain length. In theta-solvents (see Chapter 7) Γ seems to increase without bounds.

From the above it is clear that adsorption is molecular weight dependent and for a polydisperse polymer there are further complications. Entropically, it is more favourable for the longer chains to adsorb though dynamically the short ones may reach the interface first. This competition means that at equilibrium shorter chains that were adsorbed initially can be displaced by longer ones and that dilution may not desorb these chains. The effects can be treated with the SF model and Figure 8.19 illustrates the important parameters. The ratio of the volume of the solution to the available surface area is the most important parameter and in dispersions with very large particles or single flat surfaces these effects can be very pronounced.

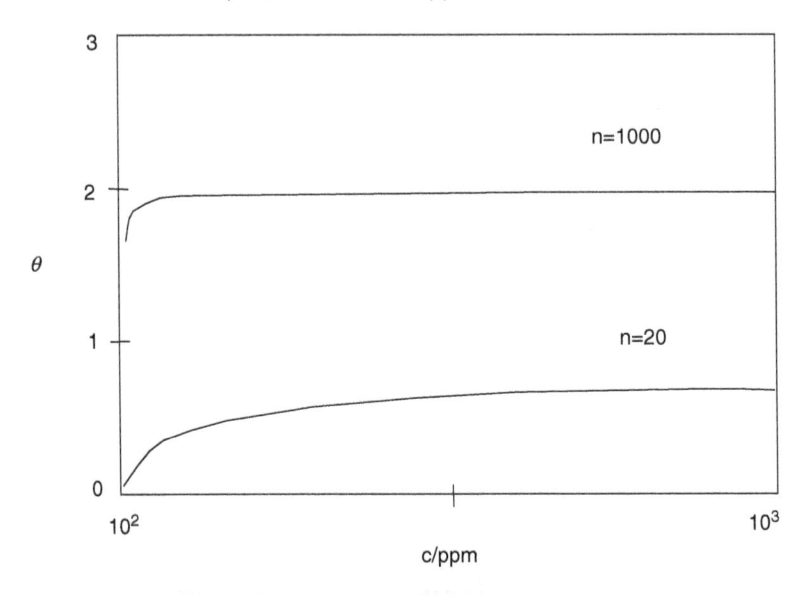

Figure 8.18 *Theoretical isotherms using the SF model for two chain lengths (n = 100 lower curve and n = 1000 upper curve) with values of χ_s of 1.0 and χ of 0.5. θ is the number of adsorbed segments per surface site*

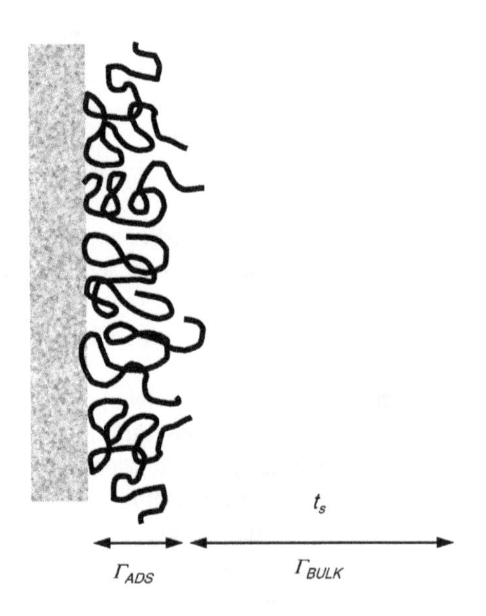

Figure 8.19 *The relation between the surface area and the volume of the solution. t_s is the thickness of the polymer solution*

The total amount of polymer in the system can be written as

$$\Gamma_{\text{total}} = \Gamma_{\text{ads}} + \Gamma_{\text{bulk}} \quad \text{and} \quad \Gamma_{\text{bulk}} = c_p t_s [\text{mg m}^{-2}] \tag{8.23}$$

where $c_p t_s$ is the volume of solution per unit area of the surface:

$$c_p t_s = V_{\text{solution}} / A_{\text{interface}} \tag{8.24}$$

This is shown schematically in Figure 8.19.

The effects of polydispersity can be seen clearly by taking just a bimodal polymer solution. The adsorption isotherms for such a situation are shown in Figure 8.20, and where the adsorbed amount is plotted against $c_p t_s$ the polymer concentration times the thickness of the solution, i.e. the total volume of polymer solution per unit area of surface. As before at low polymer solution concentrations virtually all the polymer is adsorbed, i.e. both chain lengths.

This is region I in Figure 8.21. However, as the solution concentration is increased and we approach saturation (region II) the longer chains are preferentially adsorbed and the small ones that have adsorbed are displaced into the solution. In region III, the surface is saturated. A very interesting and not entirely obvious effect of this adsorption scenario is what happens if we now dilute the polymer solution. This is shown as region IV where the longer chains are left at the surface and complete desorption is now very difficult as the equivalent polymer solution concentrations necessary to achieve this adsorbed state are very low. In some senses this can be seen as an irreversible adsorption. More details can be found in Fleer *et al.* (1).

Another aspect of polydispersity is that the adsorption isotherms become rounded and this means that any attempt to get adsorption energies from the initial slope (by a Langmuir analysis) will be fruitless.

A typical high affinity experimental adsorption isotherm is shown in Figure 8.22. The data shown are for a narrow molecular weight PEO 51K molecular weight adsorbed on PS latex. The isotherm breaks away from the vertical axis at about 0.4 mg m^{-2} which corresponds

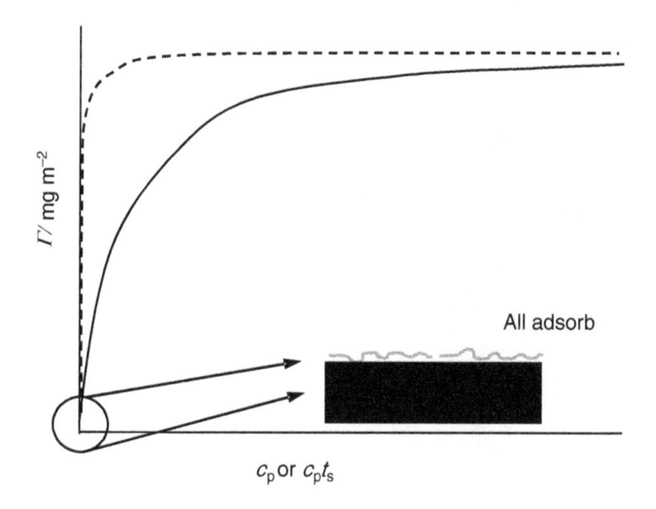

Figure 8.20 *Schematic adsorption isotherms for two different molecular weight polymers*

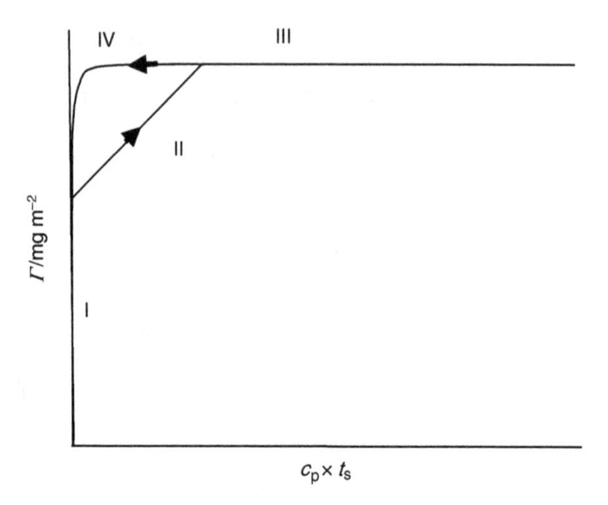

Figure 8.21 Adsorption and desorption from a bimodal mixture of polymer molecular weights

to the onset of interpolymer interactions in the layer. Values of order 1 mg m^{-2} are very typical for the adsorbed amount for uncharged homopolymers.

8.4.3 The Bound Fraction

The bound fraction describes how much of the polymer is anchored at the interface. Single chains lie flat on the surface (when $\chi_s > \chi_{sc}$), but for multiple chains this is not so. To optimise the free energy, surface sites are filled to gain energy but the chains retain a more three-dimensional structure at the surface to retain entropy. The experimental data in

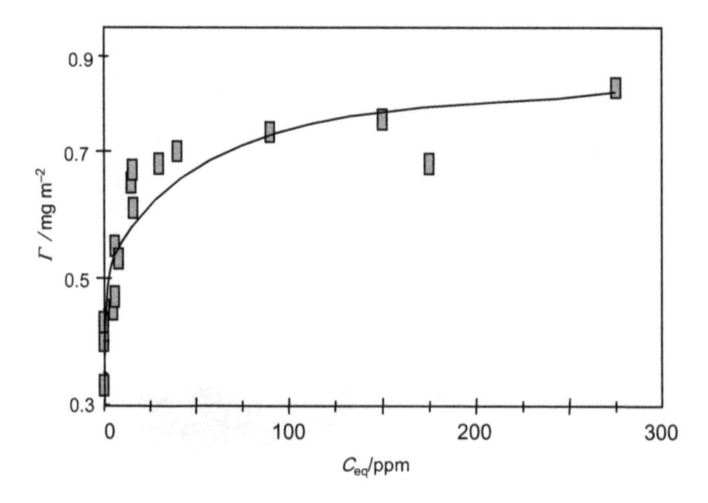

Figure 8.22 A typical high affinity adsorption isotherm for PEO 51 K g mol^{-1} adsorbed on to polystyrene latex from water

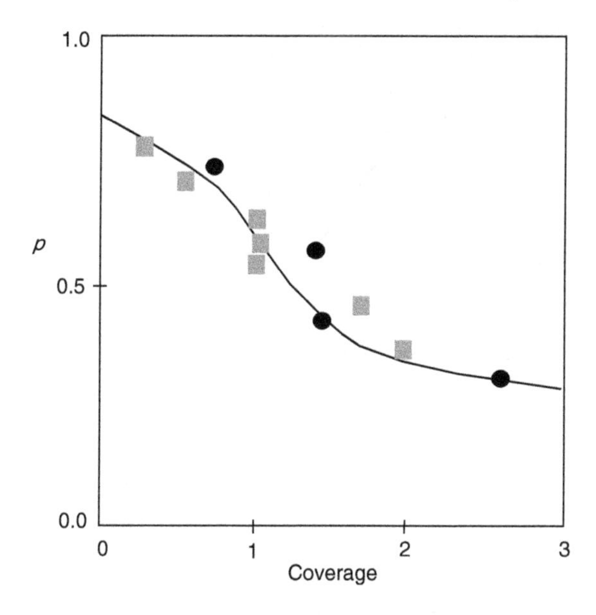

Figure 8.23 *Values of the bound fraction, p, for PVP adsorbed on silica from water. Either squares (NMR) or circles (ESR) are shown on the figure itself*

Figure 8.23 are for poly(vinyl pyrrolidone) (PVP) adsorbed on silica from water and the results have been obtained using two very similar experimental methods, nuclear magnetic resonance (NMR) and electron spin resonance (1). These methods work on the basis that segments at the interface (trains) are more restricted in mobility than those in loops and tails. Several methods can be used to estimate the bound fraction and these include FTIR, microcalorimetry, solvent NMR relaxation and SANS. For comparison, Figure 8.24 illustrates a theoretical calculation of the bound fraction using the SF model. Two different chain lengths have been used. Both theory and experiment show that at low coverages the chains lie relatively flat at the surface (a completely flat conformation would have a bound fraction $p = 1$). With increasing coverage, although the volume fraction at the surface layer will remain fairly constant to incorporate more chains, the adsorbed layer will swell, leading to the formation of loops and tails. For a flat layer the bound fraction is independent of chain length, and p values for different molecular weights will only diverge when there are sufficient number of segments in loops and tails. The SF theory shows this latter point very nicely, and both theory and experiment confirm this picture of the adsorbed layer.

8.4.4 The Layer Thickness

The layer thickness is a very important parameter for an adsorbed layer as it helps us to design steric stabilisers. Figure 8.25 shows the results of an SF calculation of the contribution of loops and tails to the hydrodynamic layer thickness, δ_H. The contribution of trains can be ignored, but the surprising factor is that tails dominate δ_H and become increasingly important as the chain length increases. The tail region is the first point of contact when two particles come together and when the repulsive/attractive steric force is felt.

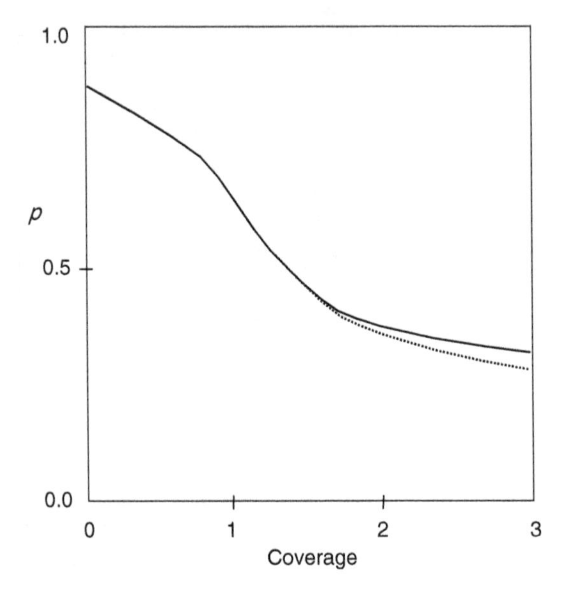

Figure 8.24 *SF calculations of the bound fraction of an adsorbed polymer as a function of the adsorbed amount for two different chain lengths (100) and (1000)*

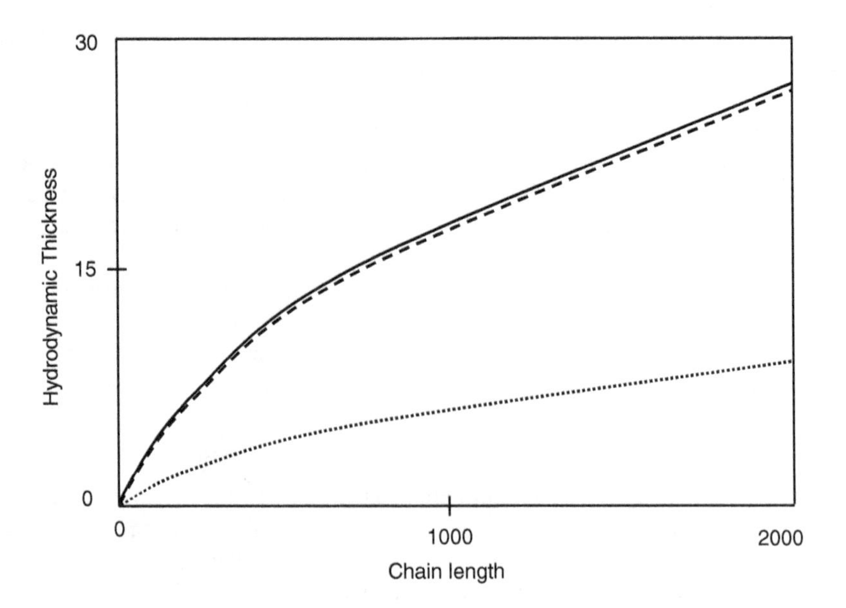

Figure 8.25 *A theoretical prediction of the contribution of loops (······), tails (- - - - - -) and the full profile (———) to the hydrodynamic thickness of an adsorbed layer as a function of chain length*

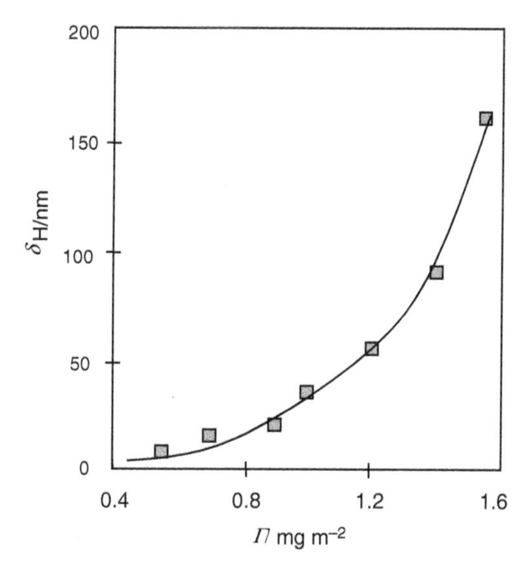

Figure 8.26 *Hydrodynamic layer thickness for PEO adsorbed on polystyrene latex as a function of the adsorbed amount*

Experimentally, dynamic light scattering (PCS) and viscosity can be used to measure δ_H and Figure 8.26 shows a set of data for PEO adsorbed on polystyrene latex as a function of adsorbed amount. The data collapse on a single curve and illustrate the increasing importance of tails.

This behaviour can be understood in that the maximum layer thickness is determined by a combination of the adsorbed amount and molecular weight. For a given molecular weight if there is a maximum in adsorbed amount (the plateau region) that will be reflected in the maximum attainable layer thickness.

The explicit molecular weight dependence of the layer thickness can be measured by taking a series of samples from the plateau region of the isotherms. Figure 8.27 shows a set of data for the same series of polymers.

The data clearly follow a scaling prediction of the form

$$\delta_H \sim n^a. \tag{8.25}$$

The exponent, a, obtained from these data was ~0.8 but other workers have reported lower values of around 0.6. For comparison, in the same figure data for the RMS (root mean square) layer thickness are also shown. These data are easily calculated from the volume fraction profile as obtained by SANS (1) or by ellipsometry. The data also show a scaling law but this time the exponent is ~0.4. The hydrodynamic thickness is determined by the tail region as seen above, whereas the RMS thickness is dominated by the central region of the profile, which is mainly loops. Figure 8.28 shows a theoretical calculation of these two thickness parameters based on the SF theory and a percolation model for the hydrodynamics (1). The agreement between the SF predictions and experiment is good and shows that the hydrodynamic layer thickness increases more strongly with n than does the RMS.

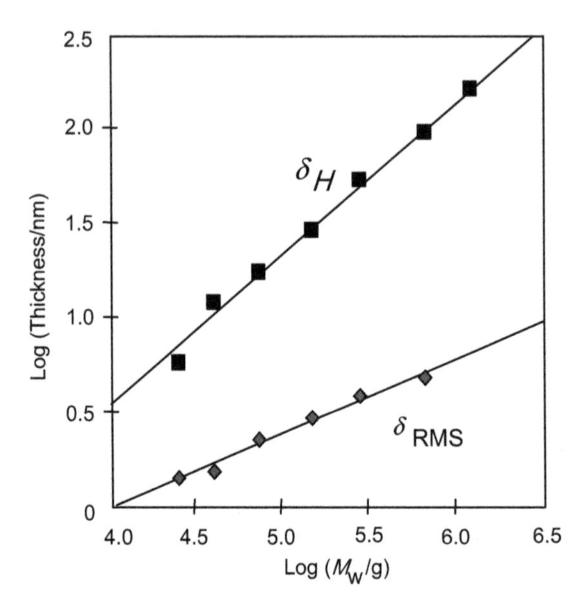

Figure 8.27 *The variation in the hydrodynamic thickness of layers of PEO adsorbed on polystyrene as a function of PEO molecular weight measured by photon correlation spectroscopy*

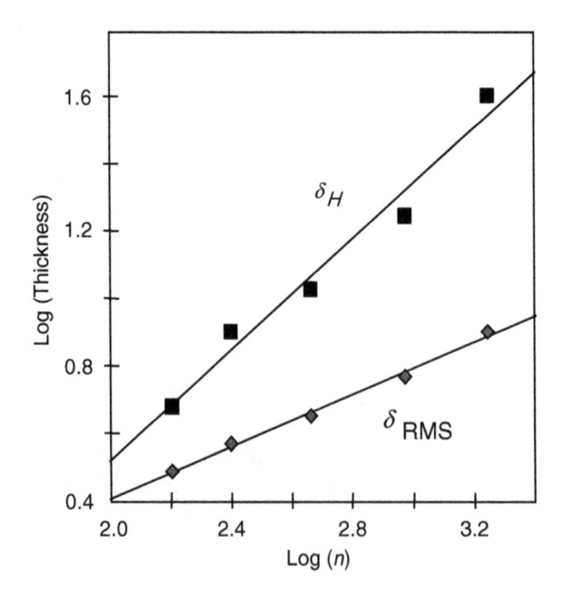

Figure 8.28 *The variation in the hydrodynamic thickness as a function of chain length calculated using the SF model and a hydrodynamic model*

8.5 Copolymers

An adsorbed homopolymer is often a compromise as a stabiliser as it needs to be strongly adsorbed and strongly solvated. These two conditions are not compatible as strong solvation implies a small χ value which reduces the adsorption preference. To optimise both these effects, the system of choice is a copolymer in which one segment type can be strongly adsorbing and poorly solvated and the other the reverse. For a block copolymer there is a strong partitioning at the interface and the adsorbing block can even form a polymer melt at the surface. Figure 8.29 illustrates this partitioning behaviour. The block copolymer forms a train region which is populated mainly by the more strongly adsorbing and less solvated block and a brush region composed of the block with the opposite properties. On the right-hand side of this figure we also illustrate the adsorption of a random copolymer. The exact surface structure that results depends on the relative ratios of the blocks and their distribution in the chain, but typically we might expect that the copolymer has an intermediate behaviour between two homopolymers made up of the respective blocks. This is shown very clearly in the SF calculation in Figure 8.30 where one constituent homopolymer does not adsorb and the other does. Both the adsorbed amount and the layer thickness behave in a very similar manner.

For a block copolymer the behaviour is quite different, as is seen in Figure 8.31 which is plotted with the same y-axis scale as Figure 8.30. In this case there is a balance between the energy gained from adsorbing the anchor block (A) and the entropy associated with stretching the buoy block (B). This leads to a maximum in the adsorbed amount and the layer thickness as a function of the fraction of adsorbing segments (ν_A) when these two effects balance. So in choosing a block copolymer there is an optimum block ratio at approximately $\nu_A \sim 0.2$ to get the maximum steric effect. For a random copolymer the adsorption varies from depletion when $\nu_A = 0$ to that of homopolymer A when $\nu_A = 1$.

The experimental data in Figure 8.32 are for a series of ABA block copolymers of PEO and poly(propylene oxide) (PPO) on PS latex which show clearly the trend of increasing

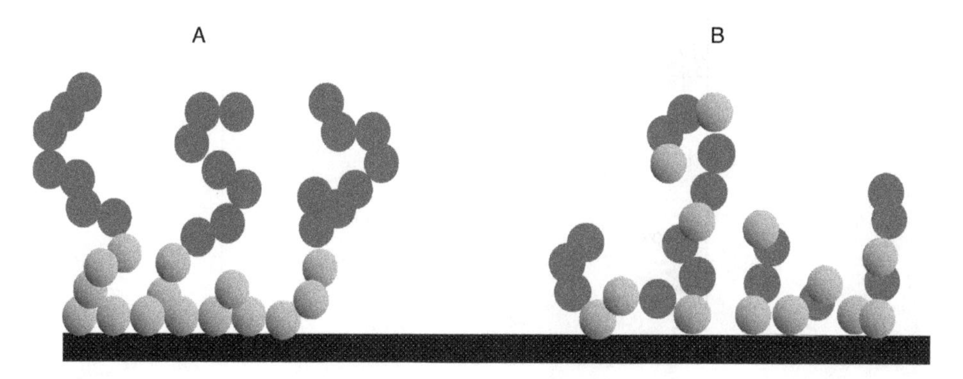

Figure 8.29 A schematic representation of the adsorption of a block copolymer (A) and a random copolymer (B). The adsorbing segments are in grey and the non-adsorbing ones in black

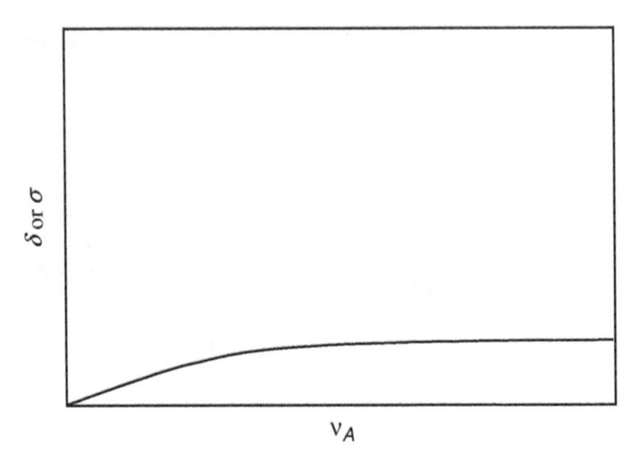

Figure 8.30 Theoretical thickness or adsorbed amount for an adsorbed random copolymer based on the Scheutjens–Fleer theory. This figure has the same y-axis scale as Figure 8.31

adsorbed amount up to a certain critical fraction of the adsorbing monomer as predicted by theory. The general results are quite typical of block copolymer systems.

8.5.1 Liquid/Liquid Interfaces

The SF formalism can be used to investigate block copolymers adsorbed at penetrable interfaces, for example an emulsion where the hydrophobic block penetrates into the oil phase and the hydrophilic one into water (1). Figure 8.33 shows how this partition can take place. The experimental system shown in Figure 8.34 is for a pluronic PEO–PPO–PEO block copolymer with 96 monomers in each of the two PEO chains and 69 PPO units. The data were obtained using neutron reflection. In this instance the PPO is not very soluble in the oil and the penetration is not as large as would be found, for example, with an alkyl chain.

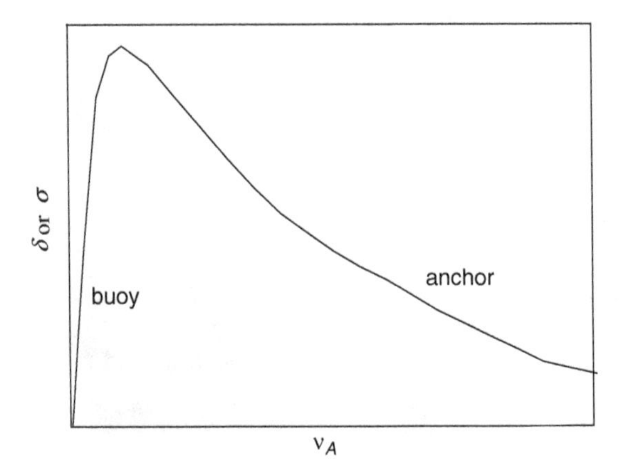

Figure 8.31 Theoretical thickness or adsorbed amount for an adsorbed block copolymer based on the Scheutjens–Fleer theory

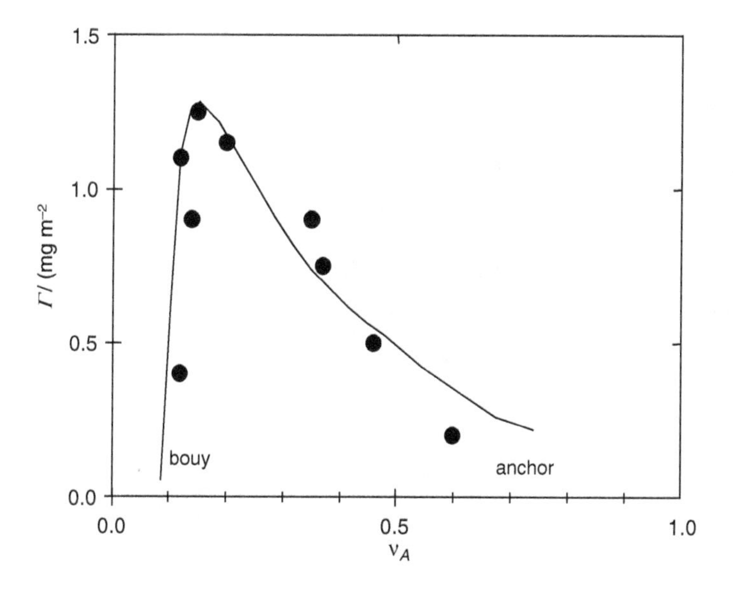

Figure 8.32 The adsorbed amount for a series of pluronic (ABA) block copolymers adsorbed on polystyrene latex

8.6 Polymer Brushes

The terminally attached chain discussed at the beginning of this section can also be investigated using the basic SF method and for comparison Figure 8.35 shows examples similar to the ones in Section 8.3.

The brush system has been studied extensively and in Figure 8.35 we show an experimental example of polystyrene (PS) grafted onto silica and dispersed in dimethylformamide (DMF) which is a good solvent and from which PS does not physically adsorb. Under

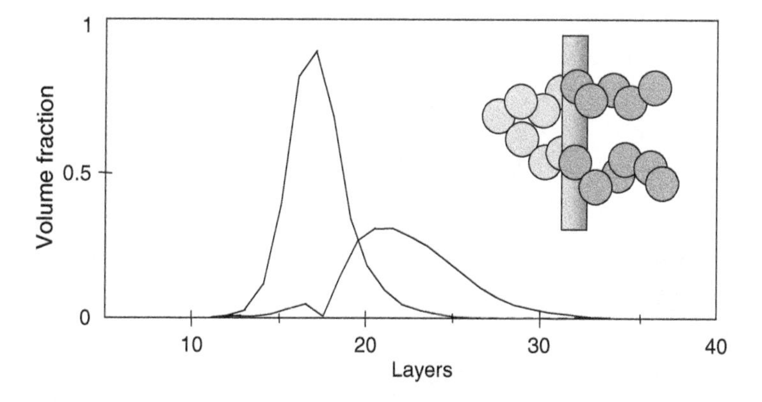

Figure 8.33 SF calculation for the adsorption of an ABA block copolymer at the liquid/liquid interface. The inset shows schematically how the segments could arrange

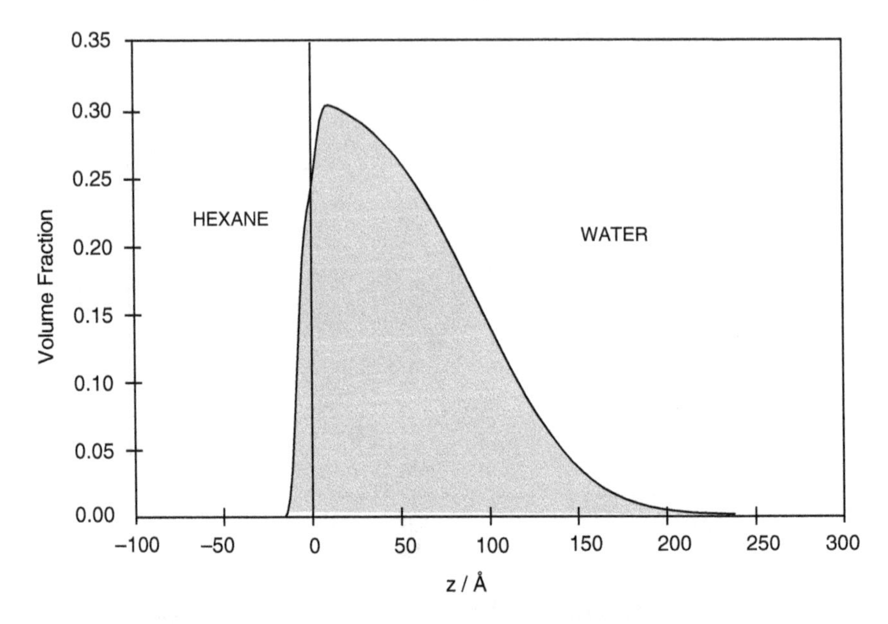

Figure 8.34 *Neutron reflectivity derived volume fraction profiles for an ABA block copolymer of PEO and PPO adsorbed at the hexane/water interface*

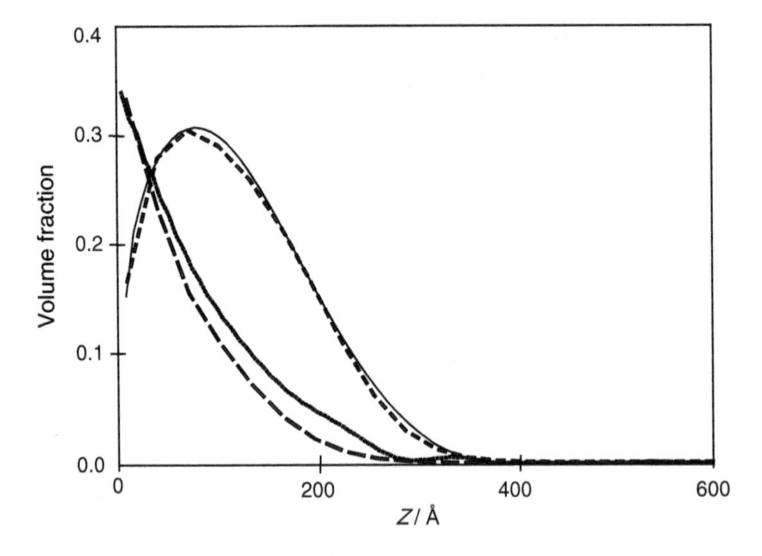

Figure 8.35 *A comparison of the volume fraction profiles obtained from small-angle neutron scattering and SF calculations for polystyrene chemically grafted to silica. Short dashes and solid lines (theory) correspond to a pure toluene solvent ($\chi_s = 1:0$) and the dotted and long dashes (theory) correspond to the effect of pre-adding dimethylformamide ($\chi_s = 0.0$)*

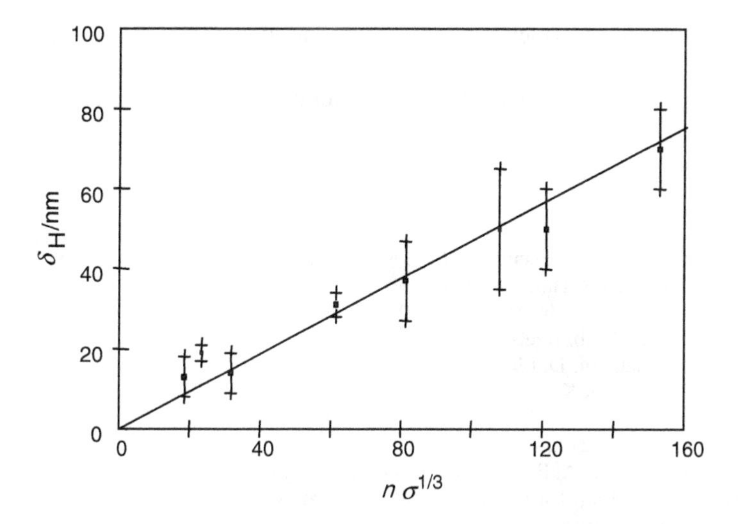

Figure 8.36 *The hydrodynamic thickness of terminally attached brushes of polystyrene on silica as a function of chain length and grafting density*

these conditions we would expect the mushroom picture. In toluene where physical adsorption does take place, we expect a pancake. The comparison of the SANS data with the SF (solid line and long dashes) prediction is startling (1).

The same brush system as above has been studied with PCS and SANS to compare with the scaling prediction for a brush (Equation 8.11). From Figure 8.36 it can be seen that for a series of samples of different molecular weights and grafting densities we get a universal plot. The height of the brush surprisingly scales with the number of segments ($\sim n$) which is in marked contrast to a polymer in a good solvent solution where the coil size $\sim n^{0.6}$.

8.7 Conclusions

In this chapter we have learnt that the structure of a polymer at an interface is very different from that in solution. Two very different scenarios can occur: adsorption and depletion. In the adsorbed state the polymer chain no longer has spherical symmetry and there is an entropic penalty, but this is offset by a gain in energy through adsorption and the gain in entropy on displacing the solvent from the interface. The ends of the adsorbed chain stretch away from the surface by a considerable distance compared to the solution radius of gyration and these farthermost tail segments are responsible for the particle hydrodynamics and the onset of interactions between particles. Polydispersity can lead to rounded isotherms and effective irreversible adsorption. Random copolymers can show behaviour intermediate between those of the two respective homopolymers whereas block copolymers can segregate at surfaces giving an anchor and a buoy layer. There is an optimum value of block size to give the maximum thickness for a given overall chain length. For terminally attached chains, mushroom and pancake layer shapes can be found, depending on the attraction of the polymer to the surface, and for this system the thickness is linear in chain length.

For depletion interactions and its effects on colloid stability see Chapter 9.

References

(1) Fleer, G., Stuart, M. C., Scheutjens, J. H. M. H., Cosgrove, T., Vincent, B. (1993) Polymers at Interfaces. Chapman & Hall, London.
(2) Jones, R. A. L., Richards, R. W. (1999) Polymers at Surfaces and Interfaces. Cambridge University Press, Cambridge.
(3) Netz, R. R., Andelman, D. (2003) Phys. Rep., 380: 1–95.
(4) Granick, S., Kumar, S. K., Amis, E. J., Antonietti, M., Balazs, A. C., Chakraborty, A. K., Grest, G. S., Hawker, C., Janmey, P., Kramer, E. J., Nuzzo, R. H., Russell, T. P., Safinya, C. R. (2003) J. Polym. Sci. B, 41: 2755–2793.
(5) Adamson, A. W., Gast, A. P. (1997) Physical Chemistry of Surfaces. Academic Press, New York.
(6) Binder, K. (1995) Monte Carlo and Molecular Dynamics Simulations in Polymer Science. Oxford University Press, Oxford.
(7) de Gennes, P.-G. (1979) Scaling Concepts in Polymer Physics. Cornell University Press, Ithaca.
(8) Milner, S. T. (1991) Science, 251: 905–914.
(9) Marshall, J. C., Cosgrove, T., Leermakers, F., Obey, T. M., and Dreiss, C. A. (2004) Langmuir, 20: 4480–4488.

9

Effect of Polymers on Colloid Stability

Jeroen van Duijneveldt

School of Chemistry, University of Bristol, UK

9.1 Introduction

Many colloidal suspensions in practice also contain polymers. These may be adsorbed to the particles, chemically attached to their surfaces, or they may be free in solution. Whichever form they take, their presence has a major effect on the stability of colloidal suspensions. This chapter explores some of those possibilities. The effect of polymers on colloid stability is an area of active research interest and a few examples of recent results are given below.

9.1.1 Colloid Stability

What is meant by colloid stability? As discussed in Chapter 1, in practice there are several levels at which the stability of suspensions can be considered. First of all, we focus on the interaction between a pair of particles in suspension. Generally van der Waals forces (see Chapter 3) operate between any two colloidal particles in suspension and these can lead to strong attractions at close contact of the particles. If no stabilisation mechanism is provided, the particles will aggregate rapidly (a non-equilibrium process).

Once a stabilisation mechanism has been provided to avoid irreversible aggregation, the suspension as a whole may still have a tendency to phase separate, into a dilute and a dense phase for instance, as a result of weak particle attractions. A phase equilibrium may be obtained, or long-lived non-equilibrium (such as gel) states may form.

Yet another meaning of the term colloid stability relates to the tendency of a suspension to undergo sedimentation. If particles are well stabilised in terms of their pair interaction, the stabilised suspension may still undergo sedimentation so the final suspension may not be

Colloid Science: Principles, methods and applications, Second Edition Edited by Terence Cosgrove

considered stable. This chapter shows how polymers can be used to affect colloidal stability at all these levels.

9.1.2 Limitations of Charge Stabilisation

The charge stabilisation of colloidal particles is discussed in detail in Chapters 2 and 3. However, there are a number of limitations in relying on charge stabilisation (alone). Effective stabilisation requires the use of a polar solvent, for instance. Charge-stabilised particles are very sensitive to addition of salt, in particular high valency counter-ions. In fact, charge-stabilised suspensions are essentially unstable (there is only a kinetic barrier to aggregation), which can create difficulties at high concentrations or when the suspension is sheared. Finally, if a suspension is likely to undergo freeze–thaw cycles, they may not recover from the freezing cycle.

9.1.3 Effect of Polymers on Interactions

The outline of this chapter is as follows. After an introduction of particle interactions, three key mechanisms are discussed by which polymers can affect interactions between colloidal particles, and hence their overall stability:

- steric stabilisation, where a polymer is adsorbed or chemically attached to the surface of the particles and the presence of the polymer gives rise to particle repulsions
- depletion interactions, where non-adsorbing polymers in solution induce attractions
- bridging interactions, where adsorbing polymers induce particle attractions by adsorbing to two particles at the same time.

For more detail about colloids and in particular the role of polymers in controlling colloid stability, the following textbooks are recommended.

- The main text on polymeric stabilisation still is the monograph by Napper (1).
- Some more recent developments are summarised in the texts by Fleer and co-workers and by Jones and Richards (2, 3).
- The description of steric interactions below is based on the approach presented by Russel *et al.* (4).
- Surface forces are discussed in detail in the book by Israelachvili (5).

9.2 Particle Interaction Potential

First we consider the total interaction potential between two particles. This may consist of a number of different contributions. For instance, van der Waals attractions will in general have to be taken into account. In particular in polar solvents (but not exclusively), charge interactions need to be considered. Finally there are interactions due to the presence of polymer chains. In the general case all of these may be of importance simultaneously. However in this chapter we will focus on the polymer-mediated interactions alone. By studying suspensions close to refractive index matching, van der Waals attractions are minimised (see Chapter 3). This is most easily realised in non-aqueous systems.

Due to their high molecular weight, polymer chains have a large number of degrees of freedom and structural correlations between polymer segments complicate their theoretical description. This chapter only includes a few simplified approaches to gain a qualitative understanding, whilst providing references to more detailed descriptions. A further point to note is that the timescales involved in the structural relaxation of polymers in solution can be significant, which means that colloidal particles do not necessarily experience the equilibrium interaction potential described by theory. Examples of this are given below.

9.2.1 Measuring Surface Forces

In developing a fundamental understanding of the role of polymers on interaction between surfaces (and hence particles) a key role was played by the surface force apparatus (SFA), developed by Israelachvili and co-workers (5). In the SFA two crossed cylindrical mica surfaces are used, with mica being chosen because this material can be cleaved to give a molecularly smooth surface. The surfaces are immersed in a solution and brought very closely together. The surfaces are mounted on a stiff cantilever and an interferometric technique is used to determine the distance D between the two surfaces, and the deflection of a laser beam is used to detect the bending of the cantilever, from which the surface force can be obtained.

This technique gives detailed information on the interaction as a function of distance between two surfaces, or nanometre length scales. It measures $F(D)/R$, where R is the mean radius of curvature of the mica cylinders. The Derjaguin approximation then relates this force to the interaction energy between flat surfaces per unit area, $E(D)$, as

$$F(D)/R = 2\pi E(D) \tag{9.1}$$

This and other methods for measuring surface forces are discussed in detail in Chapter 16.

9.3 Steric Stabilisation

An important mechanism for providing colloid stability is that of steric stabilisation. It involves covering the colloidal particles with a dense polymer layer. In a good solvent for the stabilising polymer, this gives rise to steep repulsions between the particles.

9.3.1 Theory

Perhaps the most common method to coat particles with a polymer layer is to allow the polymer to adsorb from solution. A detailed description of the resulting adsorbed layer is given in Chapter 8. Polymers at a surface may either adsorb or deplete, depending on the value of the surface χ parameter χ_s. If it is larger than a critical value, $\chi_s > \chi_{sc}$, then adsorption will result, if it is less the polymer will be depleted from the surface instead. Figure 9.1 summarises the adsorption process.

The left-hand panel shows an adsorption isotherm, giving the adsorbed amount per unit of surface, Γ, as a function of the equilibrium concentration of polymer in solution, c_{eq}. As the equilibrium concentration is increased, in this case Γ quickly reaches a plateau. In practice this plateau is typically a few milligrams per square metre of surface of particles.

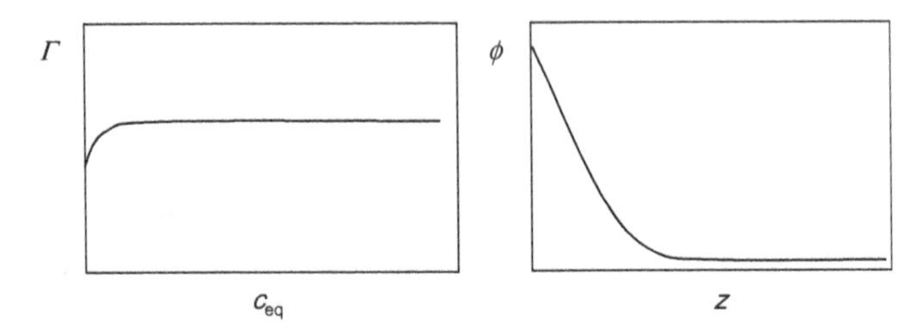

Figure 9.1 *Summary of polymer adsorption behaviour. The left-hand panel shows an adsorption isotherm, and the right-hand panel the polymer volume fraction profile*

In the right-hand panel the polymer volume fraction is sketched as a function of distance from the surface, z, summarising the structure of the polymer layer. Predicting such properties analytically is not easy but there are numerical methods available which we can use to calculate such properties, for instance the Scheutjens–Fleer method detailed in Chapter 8.

Figure 9.2 illustrates the principle of steric interactions. The particles are covered with a dense layer of polymer, either through adsorption or through chemical grafting. As a first approximation each polymer layer is represented as having a constant polymer volume fraction φ throughout the layer.

When two particles approach, the polymer layers will touch as soon as the particle surface distance D becomes less than twice the layer thickness. Whilst maintaining the polymer concentration in each layer as a step function, two possible scenarios now arise: (a) interpenetration, in which the two polymer layers gradually intermingle as the surfaces

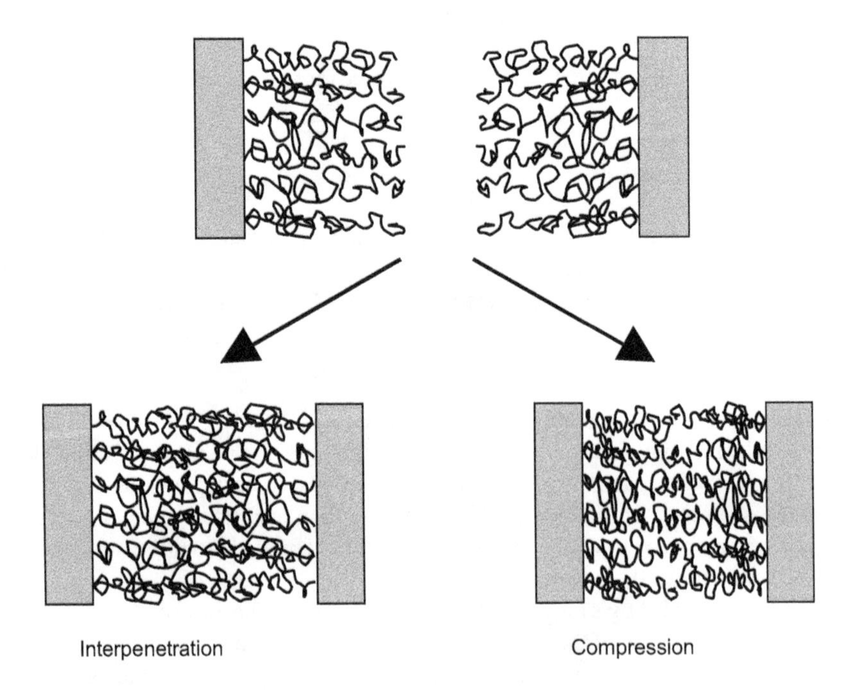

Figure 9.2 *Principle of steric interactions*

approach, locally doubling the polymer segment concentration, or (b) compression, in which the concentration of polymer segments between the surfaces gradually increases from its initial value but is at the same value across the gap between the surfaces.

In a good solvent (Chapter 7) for the polymer the local increase in polymer concentration carries a free energy penalty and this gives rise to a repulsive steric interaction between the particles. This repulsion will be felt as soon as the surface separation D becomes less than twice the unperturbed layer thickness L_0, $D < 2L_0$, and it will increase steeply as soon as $D < L_0$, when compression of the polymer layers is inevitable. This steep repulsion provides the steric stabilisation of suspensions. Whereas van der Waals attractions still operate between the colloidal particles, effective stability may ensue provided the attractions are weak (compared to k_BT) at separations $D \approx L_0$, where the steric repulsion sharply increases (see also Chapter 3).

A simple analytical expression for the steric interaction energy V_{ster} between two polymer-coated particles of radius a was derived by Fischer (6):

$$V_{ster}/k_BT = 4\pi a \Gamma^2 N_A \frac{\bar{v}_2^2}{\bar{V}_1}\left(\frac{1}{2}-\chi\right)\left(1-\frac{D}{2L_0}\right)^2 \quad (9.2)$$

The derivation assumes a constant segment density in the polymer layers and assumes linear superposition of these densities and is therefore only valid for weak overlap between the layers, in the regime $L_0 < D < 2L_0$. The partial specific volume of the polymer chains is written $\bar{v}_2$ and the molar volume of the solvent molecules as $\bar{V}_1$. In this expression the interactions scale as the square of the surface coverage Γ, thus highlighting the importance of a high surface coverage for effective steric stabilisation. For repulsive interactions, $V_{ster} > 0$, good solvent conditions are required, i.e. the Flory–Huggins parameter $\chi < 0.5$. The role of solvent quality will be explored further in Section 9.6. A more detailed theory was proposed by Hesselink, Vrij and Overbeek (7). An extensive overview of theories of steric stabilisation is given in the text by Napper (1).

For interaction between polymer brushes, the Alexander–de Gennes scaling theory yields for the repulsive pressure (8):

$$P(D) \approx \frac{k_BT}{s^3}\left[(2L_0/D)^{9/4}-(D/2L_0)^{3/4}\right] \quad (D < 2L_0) \quad (9.3)$$

In this expression s is the mean distance between grafted chains. As $\Gamma = 1/s^2$ this implies that $P \propto \Gamma^{3/2}$, a slightly weaker dependence than the Fischer theory suggested. Equation (9.3) compares very well with data on aqueous surfactant bilayers (9). This result is discussed in detail in Chapter 16.

In the case of adsorbing polymers the interaction will also depend on the equilibrium conditions chosen, i.e. whether the adsorbed amount is able to adjust itself during the approach of the surfaces (full equilibrium) or not. The text by Fleer *et al.* (2) gives a much more detailed account of the different possibilities and of theoretical approaches, such as the Scheutjens–Fleer method, to tackle them.

An example of steric interactions between two surfaces due to adsorbed polymer is shown in Figure 9.3. A surface force apparatus was used to study the force–distance curve between mica sheets to which poly(ethylene oxide) (PEO) was adsorbed with $M_w = 160\,000\ \text{g mol}^{-1}$,

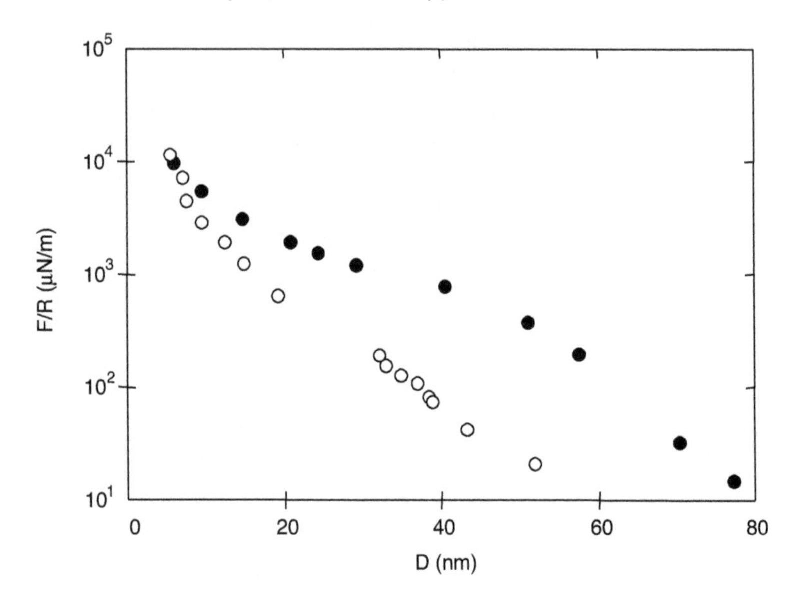

Figure 9.3 *Force–distance curve for PEO ($M_w = 160\,000\,g\,mol^{-1}$) in 0.1 M KNO_3. Full circles: initial compression. Open circles: compression after rapid decompression. Reprinted with permission from Ref. (10). Copyright (1984) American Chemical Society*

from an aqueous solution containing 10 mg L^{-1} PEO and 0.1 M KNO_3. For this polymer $R_g = 13$ nm and adsorption was essentially irreversible (after Klein and Luckham, 10).

Initial compression follows the upper curve (full circles). Repulsions become visible at a surface separation of $6R_g$. Rapid decompression followed by compression follows the lower curve (open circles). The force is now less than before, which is ascribed to the slow adjustment of the adsorbed polymer concentration profile between the plates (on a typical timescale of minutes to hours).

9.3.2 Steric Stabiliser Design

In order to achieve effective steric stabilisation, a number of conditions need to be met. For adsorbing polymers we require:

- high surface coverage, Γ
- strong adsorption, $\chi_s > \chi_{sc}$
- good solvent for stabilising chain, $\chi < 0.5$
- low free polymer concentration, $c_{eq} \approx 0$

Strong repulsions are only obtained at high surface coverage, and in fact at low surface coverage bridging interactions might otherwise result (see below). The adsorbing polymer needs to be in the plateau region of the adsorption isotherm (Figure 9.1). Strong adsorption is required to ensure high surface coverage, and the good solvent condition is needed for effective repulsions. Finally, a low concentration of free polymer is desirable to avoid the depletion attractions discussed below. Although it is sometimes possible to achieve successful steric stabilisation using homopolymers, a more promising route in general is to employ copolymers.

In practice copolymers are used, built up from two different monomers, which we denote here as A, which adsorbs strongly to the particles, and B, which does not adsorb and which dissolves well in the solvent. Different architectures can be considered:

- random copolymers A/B
- BAB block copolymers
- A-$(B)_n$ graft copolymers
- chemically grafted polymer B
- surfactant with tail group B.

Although an A/B random copolymer is straightforward to synthesise, this architecture is not very effective at creating a thick, dense stabilising layer. It is more effective to use a BAB type block copolymer where the A portion would adsorb and the two B portions would stick out into the solution and provide steric stabilisation (Chapter 8.5). Further along the same line, an A-$(B)_n$ type graft copolymer can be used, where the A backbone adsorbs and the B portions would stick out into solution again. Finally, B type polymers can be end-grafted onto the particle surfaces. The advantage is that the polymer is then bound covalently and hence will not desorb. Removal of the unreacted polymer, if required, is time consuming, however. Finally, surfactants may be used in the same way, where the head groups of the surfactants would adsorb to the particles, and the tail groups of the surfactants would provide the steric stabilisation layer.

Model suspensions in which the suspended particles have near-hard sphere interactions can be obtained by relying on steric interactions, using a polymer layer which is thin compared to the particle core. Two well-studied systems are (a) poly(methylmethacrylate) (PMMA) stabilised with a graft copolymer of poly(hydroxystearic acid) (PHS) (11) and (b) silica stabilised with a grafted layer of stearyl alcohol (12). In both cases non-aqueous solvents are used, with near-refractive index matching of the particles with the solvent minimising Van der Waals attractions.

9.3.3 Marginal Solvents

As mentioned above, for effective steric stabilisation a good solvent for the stabilising polymer chains is required. In other words, steric stabilisation is expected to break down around the Θ temperature (where $\chi = 0.5$) (1). An example which has practical importance is the stability of PEO-stabilised particles in water, where aggregation may be observed on temperature increase (depending on salt concentration) as aqueous solutions cease to be a good solvent for PEO at elevated temperatures.

To describe the interactions between polymer layers the approach of Russel *et al.* (4) is followed here. It starts from a model due to de Gennes and Alexander, representing each polymer layer as having a constant polymer concentration, i.e. a step function of distance (more realistic density profiles are presented in Chapter 8). In this approach the Helmholtz free energy of a polymer solution of a polymer of N statistical segments is written:

$$\frac{A}{Mk_{\mathrm{B}}T} = \ln\frac{\delta n}{N} - 1 + \frac{1}{2}Nv\delta n + \frac{1}{6}Nw\delta n^2 \tag{9.4}$$

where δn is the polymer segment density, $M = \delta n V/N$ the number of polymer chains per unit of volume, $w^{1/2}$ is the physical volume per polymer segment and the so-called excluded volume per segment $v = w^{1/2}(1-2\chi)$ takes into account the solvent quality. Starting from this an approximate expression is derived for the free energy per chain in a grafted polymer layer of thickness L_0,

$$\frac{A_{\text{chain}}}{k_{\text{B}}T} = \frac{3}{2}\left(\frac{L_0^2}{Nl^2} + \frac{Nl^2}{L_0^2} - 2\right) + \frac{1}{2}Nv\delta n + \frac{1}{6}Nw\delta n^2 \qquad (9.5)$$

The thickness of a polymer layer can be determined from this equation under the constraint that the total amount of polymer in the layer is fixed

$$\delta n L_0 = N\Gamma N_{\text{A}}/M_{\text{w}}$$

In the approach by Russel *et al.* linear superposition of two polymer layers is again assumed which allows the steric interaction energy per chain to be evaluated as a function of the surface separation D. To illustrate the role of solvent quality on steric interactions this theory is used here to obtain a few specific results. Figure 9.4 shows predicted interaction curves (expressed as interaction energy per polymer chain, A_{chain}).

The calculations were carried out to model the behaviour of a suspension of silica spheres (diameter 88 nm), to which polystyrene (PS) of $M_{\text{w}} = 26\,600\ \text{g mol}^{-1}$ was end-grafted to a moderately high coverage, $\Gamma = 2.3\ \text{mg m}^{-2}$. The particles were suspended in cyclohexane, a marginal solvent for PS (13). In this solvent the R_{g} of PS of this size is 4.4 nm.

Figure 9.4 shows interaction curves for different values of the χ parameter. In a very good solvent ($\chi = 0$) steeply repulsive interactions are predicted which already set in at a surface

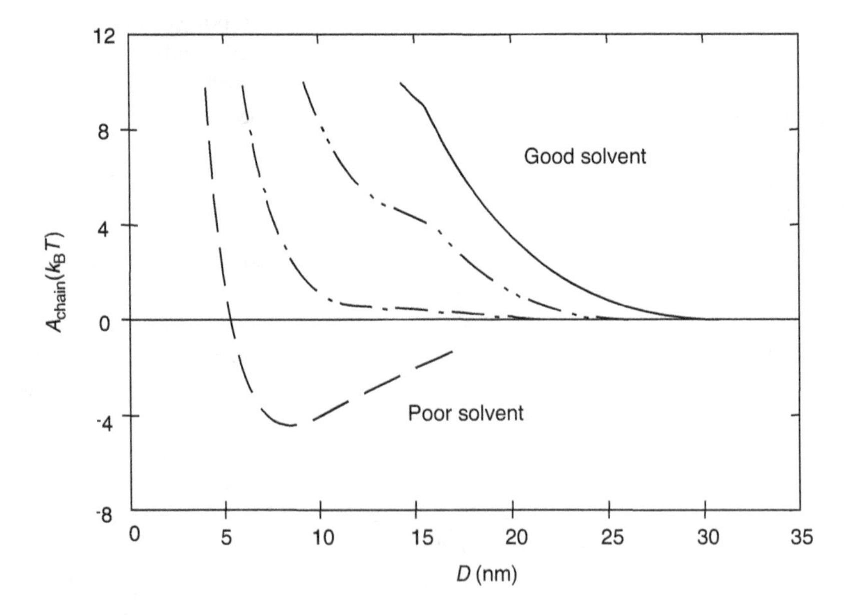

Figure 9.4 *Effect of solvent quality on steric interactions obtained from mean field theory. Polystyrene stabilised silica, see text for details. Solvent χ parameter (top to bottom) 0.0, 0.31, 0.57, 0.79*

separation of 30 nm. On approaching Θ solvent conditions the stabilising layers shrink somewhat and as a result repulsions only set in at smaller separation D, but still rise steeply at short separations. Finally, in this case for $\chi = 0.6$, at intermediate distances an attractive interaction is obtained. The fact that purely repulsive interactions are still predicted under slightly worse than Θ conditions is due to the high physical packing fraction of polymer segments in the stabiliser layers (represented by w in the theory) compensating for the negative excluded volume (represented by v).

Experimentally the particles discussed here were found to aggregate at temperatures a little below the Θ temperature for PS in cyclohexane of 34 °C (13). At low particle concentrations fractal, open aggregates were formed, whereas space-filling gels were obtained at particle volume fractions around 0.1.

Theoretical curves such as those in Figure 9.4 suggest that one would have a rather delicate control of particle interactions, through the temperature. Once they have aggregated, it should be possible to redisperse particles by reheating, and thus improving solvent quality again. Indeed with short stabilising chains (stearyl alcohol grafted silica for instance) this tends to be the case. However, particles treated with longer stabiliser chains (such as the PS treated particles discussed here) are often hard to redisperse after they have been aggregated for some time. This is possibly a result of the polymer chains having formed bridges across to the next particle (see Section 9.5), or a result of physical entanglement of the polymer chains. In Figure 9.3 time-dependent effects in steric interactions were also encountered for adsorbed polymers.

9.4 Depletion Interactions

Addition of free (non-adsorbing) polymer in solution induces so-called depletion interactions between colloidal particles. To gain an understanding of this we will discuss the Asakura–Oosawa (AO) model of depletion interactions. The particles are considered as hard spheres of diameter d and the polymers are represented by little spheres of diameter $2L_0$. Within this description the polymer coils do not interact, and hence the osmotic pressure of the polymer solution, Π, can be calculated from their number concentration n_{pol} using the Van't Hoff law,

$$\Pi \approx n_{\mathrm{pol}} k_{\mathrm{B}} T \tag{9.6}$$

The polymer coils do have hard sphere interactions with the colloidal particles, however, and hence they are excluded from a depletion layer with thickness L_0 around each particle, shown as dotted circles in Figure 9.5.

When two particles approach to a surface separation less than $2L_0$, the depletion layers overlap and the available free volume for the polymer is increased. Due to the polymer osmotic pressure this results in an effective attraction between the particles ('attraction through repulsion'). The size of the polymer molecules therefore sets the range of these attractions. Their strength can be controlled by varying the polymer concentration. Following this argument the depletion potential is given by

$$V_{\mathrm{dep}} = -\Pi V_{\mathrm{ov}}, \quad (d < r < d + 2L_0) \tag{9.7}$$

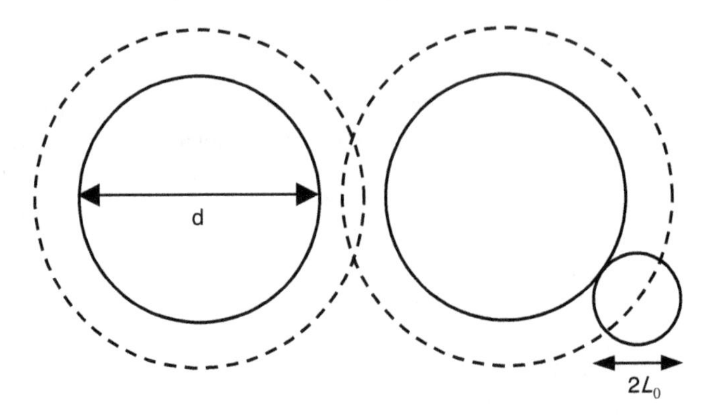

Figure 9.5 *Asakura–Oosawa model for depletion interactions*

where r is the particle centre–centre distance and the overlap volume of the depletion layers, V_{ov}, is given by

$$V_{ov} = \left(1 - \frac{3r}{2d(1+\xi)} + \frac{1}{2}\left[\frac{r}{d(1+\xi)}\right]^3\right)\frac{\pi}{6}d^3(1+\xi)^3 \tag{9.8}$$

with $\xi = L_0/a$ the polymer/colloid size ratio.

Examples of this potential are shown in Figure 9.6 as $V^*_{dep} = V_{dep}/\Pi v_0$ for a few values of ξ. The interaction potential is normalised by dividing by the osmotic pressure and by the

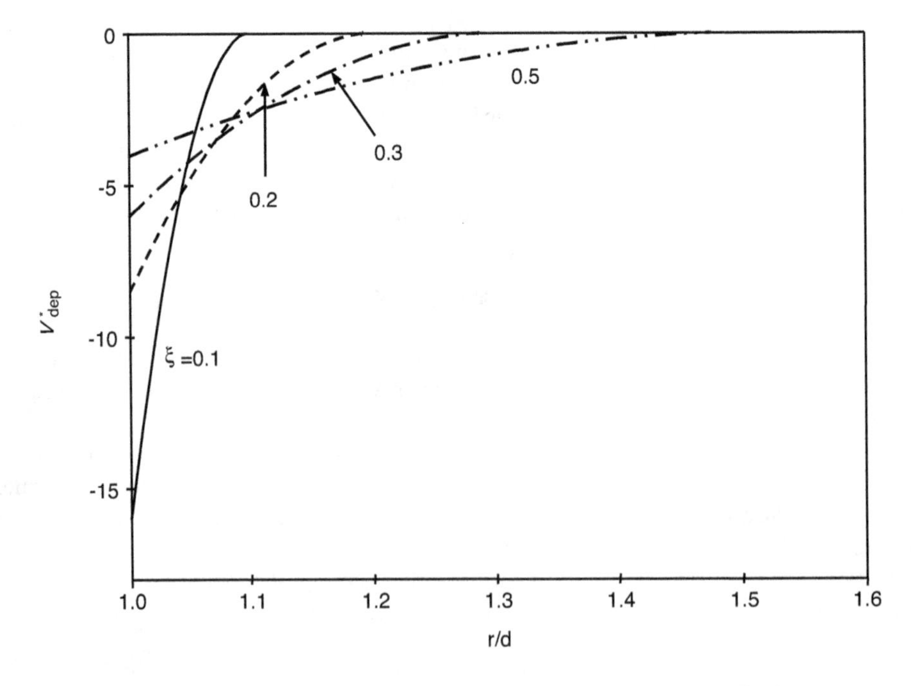

Figure 9.6 *Depletion potential (Equation 9.8) for different polymer/colloid size ratios ξ*

volume swept out by a polymer coil, $v_0 = 4\pi L_0^3/3$, and the particle separation is normalised by the particle diameter, *d*. An almost triangular attraction function is obtained, which becomes narrower as the polymer size is reduced. Typically for the size of the polymer coil (thickness of the depletion layer) the radius of gyration of the polymer is taken ($L_0 = R_g$) (4).

Although the AO model is rather approximate it has led to useful predictions, some of which are discussed below. A more detailed description of depletion interactions is given in Fleer *et al.* (2). This is an area of active research, see for instance, references (14, 15) and references therein.

The depletion mechanism allows for the switching on of particle attractions in a very controlled manner and the range and strength of the attractions can be varied independently. As a result phase transitions can take place where different phases are formed, which differ in the concentration and ordering of the colloidal particles. A few theoretical predictions are shown here, obtained by combining the AO model with scaled particle theory for the polymer-free volume (16). In Figure 9.7 calculated phase diagrams are plotted for size ratios $\xi = 0.1$ and $\xi = 0.4$. The polymer concentration is expressed as the volume fraction of coils, $\phi_{\text{coil}} = v_0 n_{\text{pol}}$. Monodisperse hard spheres form colloidal crystals at particle volume fractions $\phi_{\text{p}} > 0.5$ so the pure colloidal suspension already has a fluid to solid transition. This was verified experimentally using sterically stabilised PMMA colloids (17). It is assumed here that the particles have a narrow size distribution; if the relative polydispersity exceeds about 10% crystals are not obtained.

On addition of a small polymer (compared to the colloid) (short-range attractions, $\xi = 0.1$) a wide immiscibility gap opens up, with a dilute fluid coexisting with a dense solid. Using somewhat larger polymers (long-range attractions, $\xi = 0.4$) a qualitatively different behaviour emerges with a gas–liquid transition at intermediate polymer and colloid concentrations, and an area where all three phases (gas, liquid and solid) coexist. Note that the requirement of equal chemical potentials implies that not only the colloid but also the polymer concentration is different in the three phases.

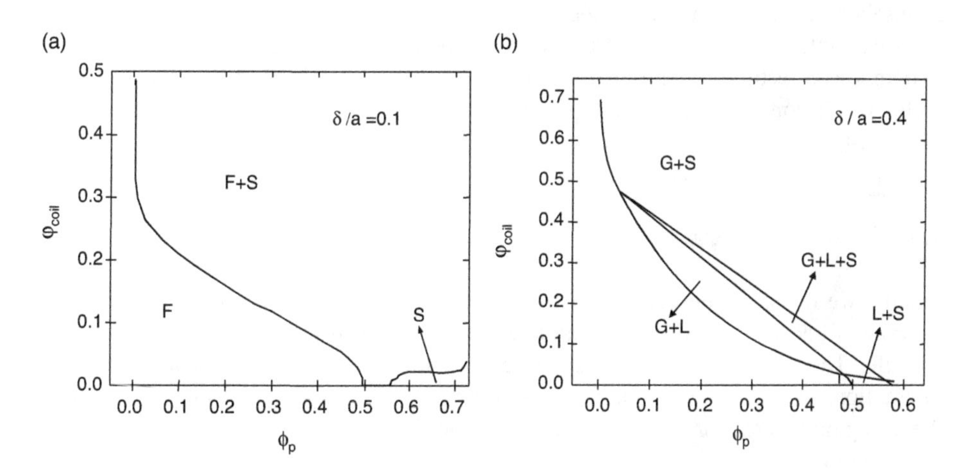

Figure 9.7 *Theoretical phase diagrams for colloid–polymer mixtures based on the AO model. Fluid (F), solid (S), gas (G) and liquid (L) phases are predicted to form. Size ratio (a) $\xi = 0.1$ and (b) $\xi = 0.4$. Reprinted with permission from Ref. (16)*

Experimental work confirms these predictions (18). However, experimentally at high attraction strength (polymer concentration) non-equilibrium states are often found, such as aggregates (at low particle concentration) or gels (at higher particle concentration). These non-equilibrium states can be long lived and may actually be the state a product is formulated in (for instance a weak gel). At moderate polymer concentration the resulting gels may be weak enough to undergo syneresis, with a sudden collapse observed in some cases (19).

An important point to note about depletion interactions is that low polymer concentrations, often only a few mg ml^{-1}, can induce phase separation. When sterically stabilised particles are used with a thick stabilising polymer layer, the depletion effects are significantly reduced as the free polymer is now able to penetrate the stabilising layer. Also, the depletion effect may be lost on increasing polymer concentration further, resulting in restabilisation (2).

The AO model discussed here is suitable to describe what is referred to as the 'colloid limit' with $\xi < 1$. However, one can also consider the opposite limit, $\xi > 1$ where small particles are added to solutions of large polymers. This is relevant for instance in protein crystallisation experiments where it is common practice to add polymer in solution to aid the crystallisation. Hence this is referred to as the 'protein limit'. Phase separation in dilute (gas) and dense (liquid) phases can still occur. However, a different theoretical description is required as the polymer concentrations involved are typically above the overlap concentration, $\phi_{coil} > 1$ (20–23).

This chapter focuses on the role of polymers in controlling interactions between colloidal particles. However, as an aside, it is worth noting that depletion-type interactions occur more generally. In suspensions containing more than one suspended species it can be useful to think of one (typically the smallest) species as inducing a depletion attraction between the larger particles. The depletant can for instance consist of small particles, or perhaps of surfactant micelles. However, a theoretical description of such systems is more difficult than the idealised polymer case discussed here, because interactions between the small objects themselves can usually not be neglected.

An example is that of hard sphere mixtures, which for size ratio larger than 5 are predicted to demix as a result of such depletion interactions. For such a phase separation to occur high overall concentrations are typically needed, in the region of 50% by volume of solids (24).

9.5 Bridging Interactions

Finally, we consider the case of bridging interactions (see Figure 9.8). At low surface coverage Γ, adsorbing polymers may attach themselves to the surface of more than one particle. The effect of this on the particle interactions is referred to as the bridging interaction. Whereas steric interactions are repulsive under good solvent conditions, the bridging interaction can lead to attractions in this case when homopolymers are adsorbed (4, 5).

Polymeric flocculants can be designed to maximise bridging interactions, in cases where aggregation is desired. ABA tri-block copolymers where the A block adsorbs are types of molecules which lend themselves for this type of behaviour. Often polyelectrolytes are used,

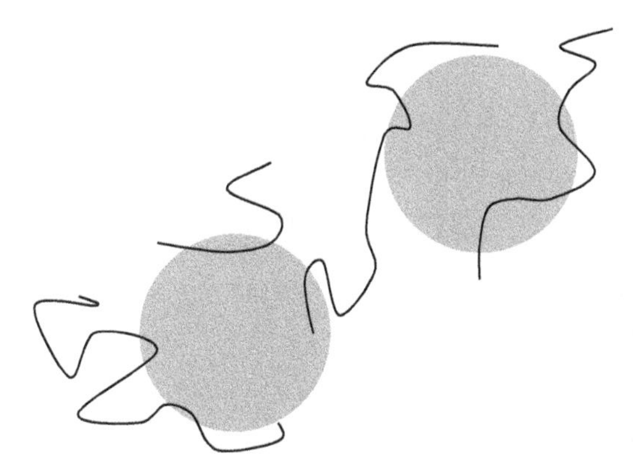

Figure 9.8 *Principle of bridging interactions*

where charges on the polymers are used to enhance adsorption to the target particles. Counter intuitively, it is possible to achieve bridging flocculation also with polymers with the same charge sign as the particles; the adsorption is then mediated by counter-ions such as Ca^{2+}.

Polymeric flocculants are used in a variety of applications, for instance waste water treatment, clarification of drinks such as beer and wine, and also in processing of minerals. Following aggregation, suspended fine particles will settle down more easily or the resulting flocs may be filtered off.

So far the effects of steric interactions, depletion and bridging have been discussed separately here. In practice more than one mechanism may have to be taken into account. The same adsorbing polymer may give rise to both steric and bridging interactions, or a weakly adsorbing polymer may cause both bridging and depletion attractions for instance.

As an illustration of the complex scenarios possible see Figure 9.9. A stability map is shown for aqueous polystyrene (PS) latex particles, carrying terminally grafted poly(ethylene oxide) (PEO) chains, to which poly(acrylic acid) (PAA) is added ($M_w = 14\,000\,\mathrm{g\,mol^{-1}}$), as a function of pH (25). W_2 is the weight fraction of added PAA. At low pH the PAA coacervates with the PEO and at low PAA concentration this results in bridging flocculation. At higher pH the PAA does not adsorb onto the PEO coated particles, but instead depletion flocculation may occur at sufficiently high concentrations of added PAA. An intermediate range of pH values is observed where neither effect is sufficient to cause flocculation.

9.6 Conclusion

Many products and processes involve colloidal suspensions and often polymers are also involved in solution or attached to the particles. The presence of these polymers has a major impact on the aggregation state and therefore on the flow properties (rheology, see Chapter 12) of these suspensions. Whereas the influence of polymers on the behaviour of solid, spherical particles is discussed in this chapter, the types of interaction are general and would also apply to the behaviour of emulsion droplets or of non-spherical (rod-like or plate-like) particles.

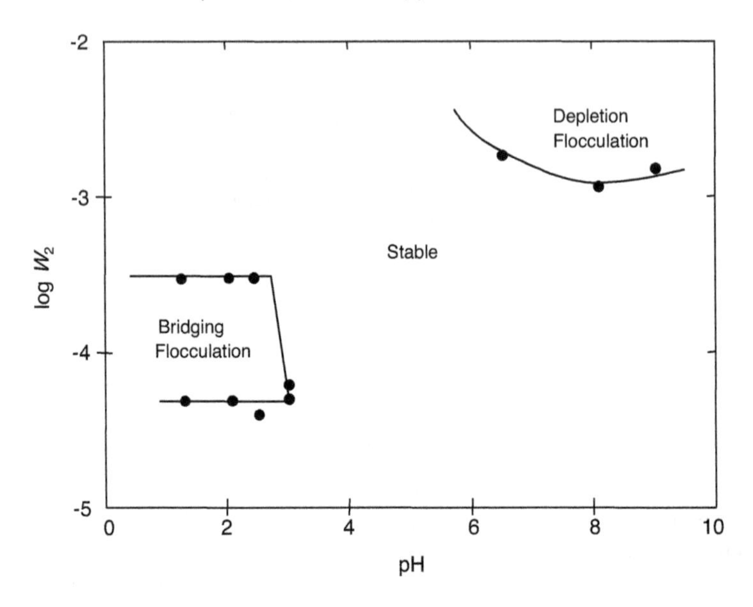

Figure 9.9 *Stability map of PS latex particles stabilised with PEO chains on addition of PAA as a function of pH. W_2 is the weight fraction of PAA. Reprinted with permission from Ref. (25)*

Depending on the application, there is a range of possibilities for what the practical aim might be. When a stable suspension of finely suspended small particles is required, as in inks, a good steric stabilisation is needed. However, larger particles would typically have a tendency to settle down even if well-stabilised in terms of their particle interaction potential. Equivalently oil-in-water emulsion droplets would tend to cream up. A weak flocculation induced by depletion interactions might help prevent this settling or creaming by building a weak gel network. Finally there are applications such as water clarification where a rapid and strong aggregation of the particles is needed, and polymeric flocculants could be used.

References

(1) Napper, D. H. (1983) Polymeric Stabilization of Colloidal Dispersions. Academic Press, New York.
(2) Fleer, G. J. *et al.* (1993) Polymers at Interfaces. Chapman & Hall, London.
(3) Jones, R. A. L., Richards, R. W. (1999) Polymers at Surfaces and Interfaces. Cambridge University Press, Cambridge.
(4) Russel, W. B., Saville, D. A., Schowalter, W. R. (1989) Colloidal Dispersions. Cambridge University Press, Cambridge.
(5) Israelachvili, J. (1992) Intermolecular and Surface Forces. 2nd edn. Academic Press, London.
(6) Fischer, E. W. (1958) Elektronenmikroskopische Untersuchungen zur Stabilität von Suspensionen in makromolekularen Lösungen. Kolloid Z., 160: 120–141.
(7) Hesselink, F. T., Vrij, A., Overbeek, J. T. (1971) Theory of stabilization of dispersions by adsrobed macromolecules. 2: Interaction between two flat particles. J. Phys. Chem., 75(14): 2094–2103.
(8) de Gennes, P. G. (1987) Polymers at an interface – a simplified view. Adv. Colloid Interface Sci., 27(3-4): 189–209.

(9) Israelachvili, J. N., Wennerström, H. (1992) Entropic forces between amphiphilic surfaces in liquids. J. Phys. Chem., 96(2): 520–531.
(10) Klein, J., Luckham, P. F. (1984) Forces between two adsorbed poly(ethylene oxide) layers in a good aqueous solvent in the range 0–150 nm. Macromolecules, 17: 1041–1048.
(11) Antl, L., *et al.* (1986) The preparation of poly(methyl methacrylate) lattices in nonaqueous media. Colloids Surfaces, 17(1): 67–78.
(12) van Helden, A. K., Jansen, J. W., Vrij, A. (1981) Preparation and characterization of spherical monodisperse silica dispersions in non-aqueous solvents. J. Colloid Interface Sci., 81(2): 354–368.
(13) Weeks, J. R., van Duijneveldt, J. S., Vincent, B. (2000) Formation and collapse of gels of sterically stabilized colloidal particles. J. Phy. -Condensed Matter, 12(46): 9599–9606.
(14) Tuinier, R., Rieger, J., de Kruif, C. G. (2003) Depletion-induced phase separation in colloid--polymer mixtures. Adv. Colloid Interface Sci., 103(1): 1–31.
(15) Fleer, G. J., Tuinier, R. (2008) Analytical phase diagrams for colloids and non-adsorbing polymer. Adv. Colloid Interface Sci., 143(1-2): 1–47.
(16) Lekkerkerker, H. N. W., *et al.* (1992) Phase-behavior of colloid plus polymer mixtures. Europhys. Lett., 20(6): 559–564.
(17) Pusey, P. N., van Megen, W. (1986) Phase-behavior of concentrated suspensions of nearly hard colloidal spheres. Nature, 320(6060): 340–342.
(18) Ilett, S. M., *et al.* (1995) Phase-behavior of a model colloid–polymer mixture. Phys. Rev. E, 51 (2): 1344–1352.
(19) Starrs, L., *et al.* (2002) Collapse of transient gels in colloid–polymer mixtures. J. Phys. -Condensed Matter, 14(10): 2485–2505.
(20) Bolhuis, P. G., Meijer, E. J., Louis, A. A. (2003) Colloid–polymer mixtures in the protein limit. Phys. Rev. Lett., 90(6): art. no. 068304.
(21) Sear, R. P. (2001) Phase separation in mixtures of colloids and long ideal polymer coils. Phys. Rev. Lett., 86(20): 4696–4699.
(22) Mutch, K. J., van Duijneveldt, J. S., Eastoe, J. (2007) Colloid–polymer mixtures in the protein limit. Soft Matter, 3(2): 155–167.
(23) Mutch, K. J., *et al.* (2009) Testing the scaling behavior of microemulsion–polymer mixtures. Langmuir, 25(7): 3944–3952.
(24) Dijkstra, M., van Roij, R., Evans, R. (1999) Phase diagram of highly asymmetric binary hard-sphere mixtures. Phys. Rev. E, 59(5): 5744–5771.
(25) Cawdery, N., Milling, A., Vincent, B. *et al.* (1994) Instabilities in dispersions of hairy particles on adding solvent-miscible polymers. Colloids Surfaces (A): Physicochem. Eng. Aspects, 86: 239–249.

10

Wetting of Surfaces

Paul Reynolds

Bristol Colloid Centre, University of Bristol, UK

10.1 Introduction

The sciences of surfaces and colloids are inextricably linked since the chemistry and physics governing the surface properties of both large and small areas are identical in that they describe the interactions between molecules and collections of molecules. As a result the interaction between a material placed on a surface and an extensive solid surface can be adequately described using our knowledge of both surfaces and colloids. In this chapter we outline the established theories and explanations of wetting and use these to provide a more practical appreciation of aspects of these phenomena. The basic principle is simple in that when a liquid is placed on a solid surface it will spread to some extent. Here both liquid and solid can be described as bulk. However, in the region of interaction where the bulk phases meet, local interactions are of primary importance. Consider the adsorption of a macromolecule or a surfactant or indeed a grease or oil stain on a surface and the effect that this has on a liquid placed on the surface. Clearly this surface modification made with only a molecular, or macromolecular, dimension has a dramatic influence on the macroscopic observation of the liquid drop behaviour on that surface. Wetting, the coverage of a surface with another material, is an important industrial and academically interesting and often challenging area.

The basic premise developed in this chapter is that a liquid in contact with a solid in the presence of a given vapour exhibits behaviour which depends on all three components. The rules that govern this behaviour may not be well developed and are empirical in many instances. However, there is a scientific rationale behind these observations. It is the

Colloid Science: Principles, methods and applications, Second Edition Edited by Terence Cosgrove

intention here to illustrate how these rules are developed and expose the scientific background to the interactions.

10.2 Surfaces and Definitions

We would normally associate wetting with the coverage of a solid surface with a liquid material. In doing so we would assume that the liquid is mobile on the solid surface. The liquid surface is a dynamic surface whereas the solid surface is not. Interestingly, perhaps what can be thought of as a fundamental property of a solid and a liquid, the surface energy and the surface tension, respectively, have identical units, i.e. they are dimensionally the same. An interesting discussion of the units and dimensions is given in Adamson and Gast (1). It is important to note that vapour has not been omitted from the definitions of surface tension and surface energy and is an important component of the overall interaction. Moreover, given the necessity to include vapour it becomes clear that this will change depending upon the temperature, pressure and partial pressure of the vapour phase.

There are four surfaces that we can consider in the absence of a vacuum:

- liquid/vapour – surface tension
- liquid/liquid – interfacial tension
- liquid/solid – surface energy or interfacial tension
- solid/vapour – surface energy.

The units of both surface tension and surface energy are Newton per metre ($N\,m^{-1}$) and Joule per square metre ($J\,m^{-2}$). Although dimensionally the same and equivalent in value, they are defined differently – as stated below.

Each of the surfaces identified above contributes to the overall picture of the wetting of a surface, which in the case of a liquid on a solid leads to a well defined shape of droplet for any given materials and conditions. In describing this behaviour we shall determine which properties of the materials are important in controlling wetting.

10.3 Surface Tension

Consider first the liquid/vapour surface. It is a very dynamic surface where molecules from the liquid phase are leaving the surface and molecules from the gaseous phase are entering the liquid phase continuously. The interface is a difficult region to model and requires the development of models based upon an imaginary surface known as the Gibbs dividing surface. This gives the basis for developing equations for surface excesses (see Section 4.3.2). Unfortunately many elementary texts prefer to show a liquid schematically as a set of spheres with interactions radiating in all directions and being equivalent to one another. At the surface of the liquid we lose the equivalence in the direction of the surface, simply because there are very few molecules above the surface. As a result it is 'shown' that the force, or tension, arising in a surface is due to the 'missing forces' of interaction of a molecule in the surface of the liquid phase in the direction of the vapour phase. The picture is schematically satisfying but lacking in scientific rigor, yet surface tension can be shown to depend only on the interactive forces between molecules.

However, we can make measurements in the surface which relate to surface properties. These lead to definitions of the surface tension. It is possible to write two definitions of surface tension and both will take some mental agility to fully appreciate what is meant. Thus common definitions are: *surface tension is the force acting at right angles to a line of unit length in the surface of a liquid, or surface tension is the force per unit length, which when multiplied by the distance moved to create a new area is equal and opposite to the work done in extending the area of surface by unit amount.*

The latter expression shows that the surface tension can be interpreted as the energy per unit surface area and that this has a tendency to reduce its area to a lower free energy arrangement. A full discussion of surface tension is given in Chapter 4.

10.4 Surface Energy

We can think about a surface as being created by the work required in bringing a molecule from the bulk to the surface. Thus, *the work that is required to increase that surface area by unity is the surface free energy*. Of course there is a tendency of that surface to contract and this is where the picture of molecules in the surface at least illustrates the ideas here because we can visualise that state of tension in the surface and we would define that as the surface tension.

10.5 Contact Angles

When a liquid is placed on a solid surface it can spread to form a continuous film or form discrete droplets. In the case of the latter a range of different behaviours can be observed. It has been stated previously that for a given solid and liquid and a defined set of conditions, temperature and pressure for instance, a liquid drop will form a well-defined shape on the solid surface. Figure 10.1 shows this schematically but additionally shows that the shape subtended by the liquid at the three-phase line of contact, where the solid (s), liquid (l) and vapour (v) meet, has a defining angle, called the contact angle (θ).

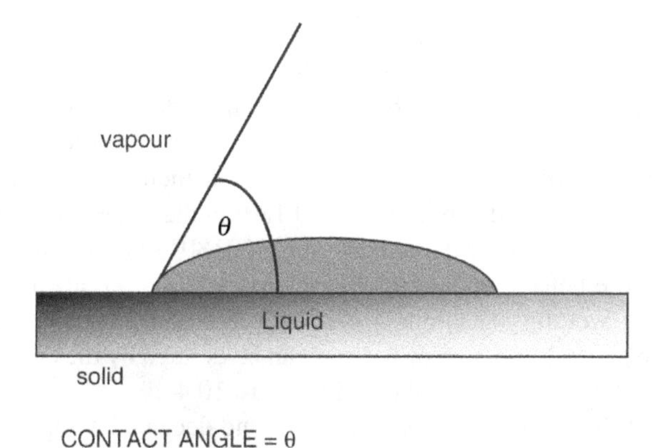

Figure 10.1 *Contact angle at the three-phase line of contact of solid, liquid and vapour*

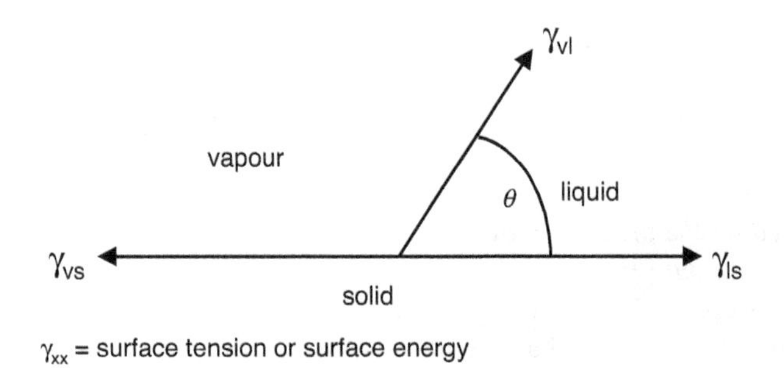

Figure 10.2 *Equilibrium balance of forces for the tensions of liquid drop on a solid surface*

Interpretation of this into more fundamental forces reveals the nature of the contact angle. This is done by considering the tensions in each of the surfaces, as shown in Figure 10.2.

Taking the tensions at the three-phase point of contact and putting them into a balance of forces resolved in the plane of the solid surface, it is seen that there are three tensions (γ): one acting in the direction of the vapour/solid (γ_{vs}) and this is opposed by the two tensions, the surface tension in the liquid/solid surface (γ_{ls}) and a component of the surface tension of the vapour/liquid (γ_{lv}). When resolved in the plane of the solid surface the function becomes the cosine of the contact angle. The resolution of these tensions gives Young's equation (2) (Equation 10.1)

$$\gamma_{ls} + \gamma_{lv} \cos\theta = \gamma_{vs} \tag{10.1}$$

It must be remembered, however, that the contact angle whilst giving useful and fundamental information about the solid/liquid interaction is defined in the presence of the vapour phase. This is often forgotten and a change in vapour (partial) pressure and composition can have very profound effects on the subsequent contact angle.

10.6 Wetting

Armed with the previous definitions, wetting *per se* can now be discussed. For some liquids a zero contact angle is obtained and might also be called perfect wetting and hence spontaneous spreading. Another possibility is partial wetting, where a contact angle is subtended somewhere between 0° and 90°. The 90° may well be thought of as an arbitrary distinction between wetting and non-wetting but nonetheless we find it an important distinction. An angle subtended between 90° and 180° in the liquid would be a non-wetting condition and finally of course if the contact angle is 180° then we have a perfectly non-wetted surface. These behaviours are shown schematically in Figure 10.3.

When discussing wetting behaviour it is most often convenient to speak in terms of contact angle. The general principle of wetting can be defined by the characteristic surface properties of the solid and liquid as shown in Figure 10.4.

Consider three different liquids, mercury, water and decane, having surface tensions of 484, 72 and 24 $\mathrm{mN\,m^{-1}}$ respectively, that are placed on a planar surface of a material. In the first instance we choose a high energy surface, magnesium oxide with a surface energy of

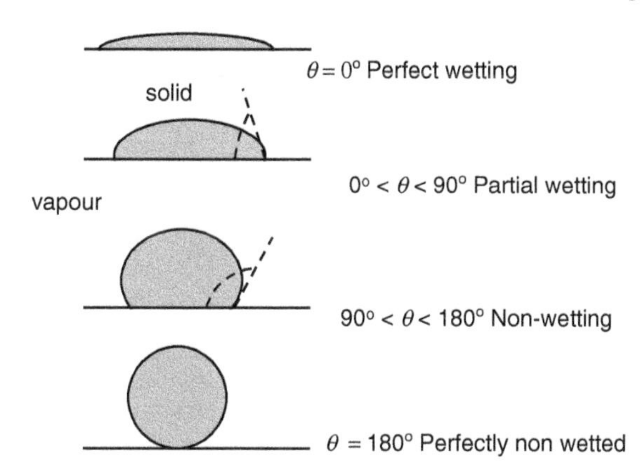

Figure 10.3 *Wetting as described by different values of the contact angle*

1200 mN m^{-1}. We would expect a finite contact angle and therefore wetting. However, the three liquids would show a different interaction with a solid with a reduced surface energy. For example if we use silica with a surface energy of 307 mN m^{-1}, water and decane would be seen to wet the surface whereas mercury would not. Thus, mercury with a higher surface tension than the silica surface energy does not wet the surface whereas water and decane do. Reducing the energetics of the solid surface further, using polyethylene with a surface energy of 31 mN m^{-1}, shows that only decane wets the surface; mercury and water do not. Again we will see that mercury and water have higher surface tensions than the surface energy of polyethylene. Finally, a low-energy solid surface, polytetrafluoroethylene (PTFE), shows non-wetting of all three liquids, and all three liquids have higher surface tensions than the surface energy of PTFE alone.

It is tempting using the above observations to conclude that a useful rule is: *liquids with low surface tensions wet solids with high surface energies.*

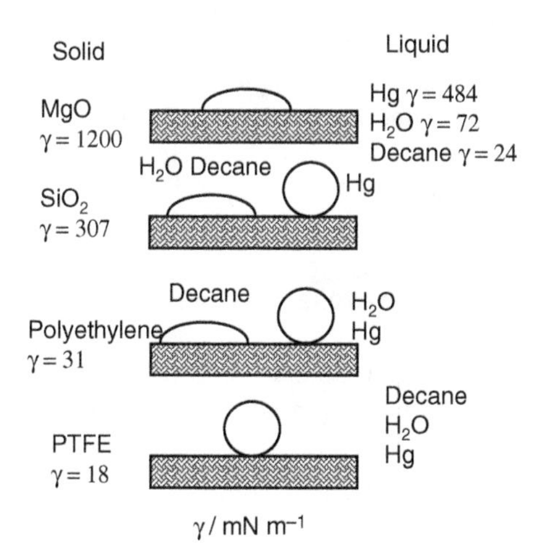

Figure 10.4 *Differential wetting of three liquids on solid surfaces of different surface energy*

Table 10.1 *Liquid surface tension and contact angles on solid surfaces*

Liquid	γ (mN m^{-1})	Solid	θ (degree)
Hg	484	PTFE	150
Water	73	PTFE	112
		Paraffin wax	110
		Polyethylene	103
		Human skin	75–90
		Gold	66
		Glass	0
Methyl iodide	67	PTFE	85
		Paraffin wax	61
		Polyethylene	46
Benzene	28	PTFE	46
		Graphite	0
n-decane	23	PTFE	40
n-octane	2.6	PTFE	30
Tetradecane/water	50.2	PTFE	170

Evidently, the magnitude of the contact angle follows some rationale governed by the magnitude of surface tensions and energies. A more complete description of the governing scientific principles is beyond the scope of this chapter; however, it would entail a consideration of Hamaker constants which describe the interactions between molecules in adjacent materials (3) (see Chapters 3 and 16). This idea is adopted later to describe the observed behaviour from an empirical standpoint.

There are other generalisations which start to appear when studying comparative contact angles measured for liquids in contact with materials. For example, Table 10.1 identifies some materials in contact with a variety of liquids.

From Table 10.1 it can be seen for instance that water has a range of contact angles when in contact with a low-energy solid surface like PTFE going through to higher energy solid surfaces like glass. There is a clear trend in these values.

There are other trends apparent in these data and one can be seen for PTFE; high contact angles (150°) are found for mercury, through to *n*-octane which has a contact angle of 30°. The difference in the surface tensions of the liquids ranges from 484 mN m^{-1} for mercury to 2.6 mN m^{-1} for *n*-octane. The above observations can be used to conclude another tentative rule: *high surface tension liquids tend to have large contact angles, while polar solids tend to have smaller contact angles.*

10.7 Liquid Spreading and Spreading Coefficients

The idea that there is a rational organisation to our observations can be taken a little further with the specific case of perfect wetting (zero contact angle).

It is apparent from Young's equation (Equation 10.1) that when the contact angle, θ, is zero it reduces to

$$\gamma_{sv} - \gamma_{ls} - \gamma_{lv} = 0 \quad (10.2)$$

For real wetting we can evaluate Equation (10.2) as the spontaneous spreading coefficient, *S*. If *S* is positive then spontaneous wetting occurs whereas a negative value of *S* would result in a finite contact angle.

The coefficient *S* has some interesting properties when examined practically. One example is the wetting properties of the homologous series of paraffins in contact with a given material. As the carbon number is changed *S* can become progressively less negative until spreading occurs. For example hexane spreads on water but decane does not. Perfect wetting occurs at the point at which the surface energy and surface tension are equivalent and this observation can be used for surface characterisation (4).

10.8 Cohesion and Adhesion

Surface energies can also be used to define cohesion and adhesion from the condition of spontaneous spreading. Adhesion and cohesion simply define the interactions which lead to the cohesive nature of materials and the adhesive nature of one material when in contact with another material. There is no implication in these definitions of any chemical bonding. Adhesion between a solid and liquid is defined as the work required to separate the solid from the liquid. Thus, the work of adhesion involves the creation of new surfaces of vapour/liquid and solid/vapour, and the destruction of the old surface of liquid/solid. The Dupré equation (Equation 10.3) defines this and Figure 10.5 shows a schematic of the process:

$$W_{\mathrm{ls}} = \gamma_{\mathrm{sv}} + \gamma_{\mathrm{vl}} - \gamma_{\mathrm{ls}} = W_{\mathrm{a}} \tag{10.3}$$

where W_{a} is the work of adhesion and W_{ls} is the work required to separate the solid from the liquid surface. It is worth noting that all the quantities are defined per unit surface area.

In the same way cohesion can be seen to be the creation of two liquid/vapour interfaces and the destruction of a liquid/liquid interface. The cohesive nature of a liquid (or solid) involves the separation of the liquid from itself. Clearly from this construction the work of cohesion, W_{c}, can be seen to be

$$W_{\mathrm{c}} = 2\gamma_{\mathrm{lv}} \tag{10.4}$$

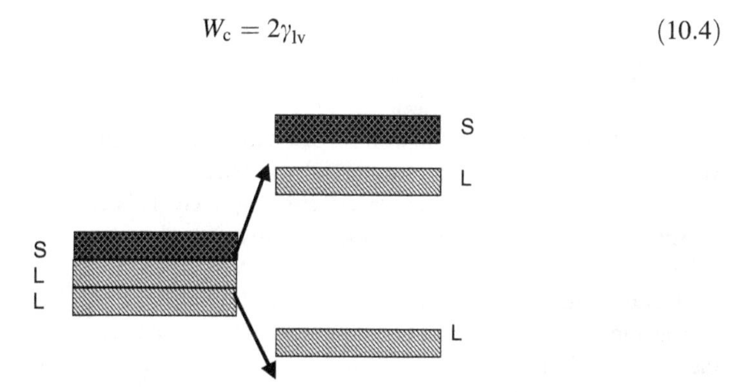

Figure 10.5 *Solid–liquid adhesion and cohesion. Adhesion is the separation of the solid from the liquid. Cohesion is the separation of the liquid from itself*

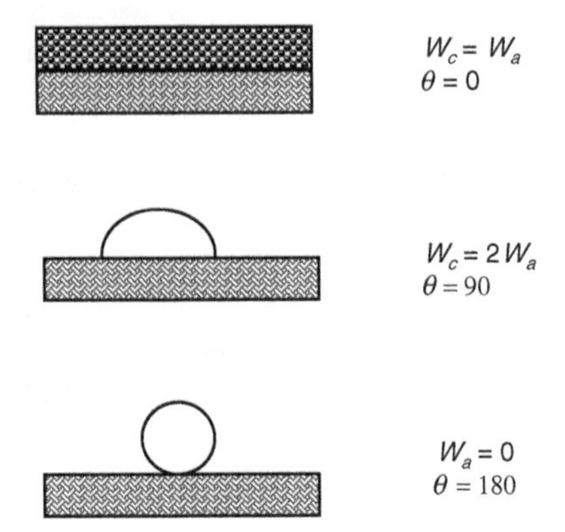

Figure 10.6 *Simple relationships between the work of cohesion and the work of adhesion for a solid and a liquid*

Adhesion and cohesion have another quite simple relationship which can be developed by taking into account the contact angle, shown as

$$\cos\theta = -1 + 2\left(\frac{W_a}{W_c}\right) \tag{10.5}$$

The result of this is that the contact angle is seen to be governed by the competition between cohesion of a liquid to itself and adhesion of a liquid to a solid. Schematically Figure 10.6 shows some simple results. At a 0° contact angle the work of cohesion equals the work of adhesion; at a 90° contact angle, the condition just between wetting and non-wetting, the work of cohesion equals twice the work of adhesion. For perfect non-wetting the work of adhesion is zero.

Cohesive failures and adhesive failures in materials are more easily appreciated with this understanding.

10.9 Two Liquids on a Surface

So far we have considered the wetting of a surface by a single liquid in equilibrium with its vapour. There is, however, a commercially important case where two immiscible liquids sit on a solid surface and differentially wet the surface. This case is shown schematically in Figure 10.7 where the solid S is differentially wetted by liquid A and liquid B.

This is a commercially important case since very often there are two immiscible liquids in contact on a surface and it is not difficult to find examples.

The appropriate tensions and the directions in which they are acting have been drawn in the diagram, from which it is clear that they can be resolved in the same way as done previously to recover Young's equation. However, this time the equation (Equation 10.6) applies to the two liquids in contact with each other and in contact with the solid surface:

$$\gamma_{BS} = \gamma_{AS} + \gamma_{AB}\cos\theta_A \tag{10.6}$$

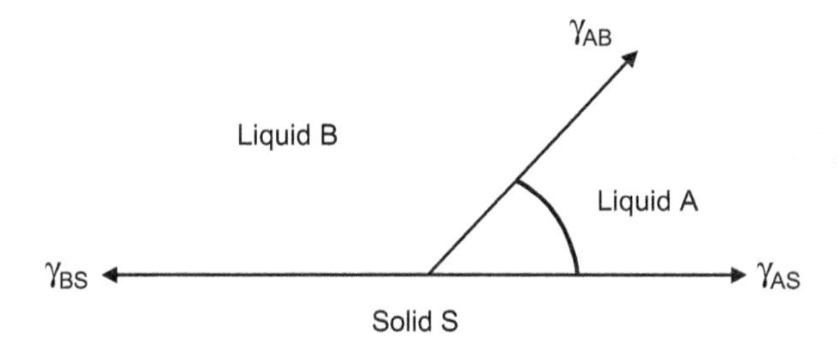

Figure 10.7 *Two immiscible liquids on a solid surface*

From the diagram it can be seen that liquid A preferentially wets the solid, and that is because the contact angle is smaller for liquid A than it is for liquid B, i.e. $\gamma_{BS} > \gamma_{AS}$.

This leads us to another rule of wetting: *for two immiscible liquids on a solid surface, in general the liquid with the smaller solid/liquid surface tension or surface energy wets the solid preferentially.*

Table 10.2 shows various results of studying two immiscible liquids wetting the surface of a solid. The solids exemplified here are alumina, PTFE, mercury (note that mercury is also taken as a liquid in the same table) and glass. It can be seen how they are differentially wetted. In Table 10.2 the angle θ_{SA} is the angle subtended by liquid A on the solid surface. Thus in the first case of water and benzene on alumina, liquid A is water having a contact angle of 22° when benzene is present as the second liquid phase. Hence here water is better wetting, or the one that is preferentially wetting the solid (alumina). Another example is that of a PTFE surface which can be seen to be completely dewetted by liquid A, water, and thus it will not wet PTFE in the presence of decane. With water and benzene on the surface of mercury, water has a contact angle of 100° and so is dewetted preferentially. These two results may not be surprising in that they show water to be the dewetted phase but it is the magnitude that illustrates that the phenomena are not entirely predicted by our knowledge of a single liquid on a solid surface. The value of 100° for water on mercury, in the presence of benzene, is not far off the 90° distinction between wetting and non-wetting. Complete dewetting, 180°, for water on PTFE in the presence of decane is very rarely observed for a single liquid. Perhaps the most surprising result is that mercury and gallium as liquids in contact with glass show mercury has a zero contact angle and therefore spreads ideally.

Consider, for instance, detergency where initially oil or grease adheres to a fabric surface, or perhaps a ceramic surface. It is clear that we need to change the energetics, surface energies, to be able to remove the oil from the surface, and it is generally a water solution that will be used, in conjunction with a surfactant. This case will be considered later.

There are other important commercial cases which can be identified and some occur in the manufacturing industry sector. The preparation of emulsions using small solid particles to

Table 10.2 *Differential wetting of a surface by two immiscible liquids*

Solid	Liquid A	Liquid B	θ_{SA}/(degree)
Al_2O_3	Water	Benzene	22
PTFE	Water	Decane	180
Hg	Water	Benzene	100
Glass	Hg	Gallium	0

stabilise the emulsion necessarily has differential wetting of the solid surface by the two immiscible liquids. These are called Pickering emulsions. Essentially packing of the small particles around the liquid droplet determines the type of emulsion, water in oil (W/O) or oil in water (O/W). The particles therefore stabilise against aggregation and coalescence. The rule is: *the liquid phase which better wets the particle is going to be the external phase.*

This we can compare this with Bancroft's rule (1), which states that the continuous phase should be the phase in which the surfactant is the most soluble. This follows from a simple geometrical argument based on the ratio of the head group area to that of the hydrophobic chain cross-section of a simple surfactant and dictates the type (or stability) of the emulsion. Thus if particles are better wetted by water than oil they will have a greater surface area covered by water. As a result the oriented wedge of water-wetted surface fits together better in between a set of spheres surrounding a larger sphere, where the oil only wets a small area of the particles surface. The result is an oil in water emulsion. This is shown schematically in Figure 10.8. Silica is a good example of this effect whereas carbon black, which is wetted better by oil than water, results in a water in oil emulsion.

There are many other important examples that can be found in the electronics and personal care industries. In the process of soldering there is a liquid metal on a surface in contact with a liquid flux, which is there to solubilise impurities and remove them from the solid/liquid metal interface. Clearly this is complex since the temperatures are high and fluctuate and this will lead to changes in the surface energies and the surface tension of the flux. Also the properties of the flux/molten metal interface and flux/metal surface will change as the level of impurities solubilised in the flux changes. In personal care applications excess foam production can be a serious problem, for example in washing machines. Antifoamers, which are often liquid silicones containing particulates, principally silica, are commonly used (5, 6). These also work by differential surface wetting of the silica by water and silicone in order to make the thin foam films unstable.

Figure 10.8 *A Pickering emulsion is a liquid droplet stabilised by particles which are better wetted by the continuous phase than the internal phase. More efficient packing leads to a more stable emulsion arrangement*

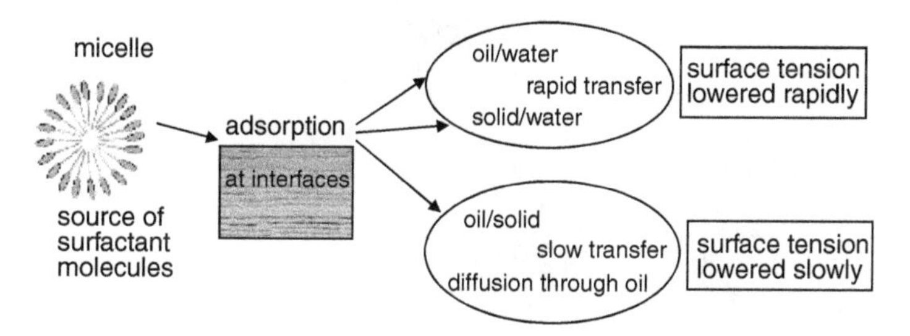

Figure 10.9 *Schematic of the detergency process*

10.10 Detergency

One clear example of wetting of a surface with two immiscible liquids is that of everyday washing. An oil stain (liquid) on a fabric (solid) surface is in contact with an immiscible phase when immersed in washing water. The detergent action at the oil/water/solid interface is influenced by adsorption of surfactants. The line of tensions may be analysed in much the same way as previously described. The great difference here is that the surface energies and the surface tensions are constantly being modified with time as surfactant from the washing water adsorbs to the oil and fabric surfaces. The dynamics of the mechanism become very important. What is critical, however, is the rate at which the fabric/oil surface is modified by surfactant adsorption. This is a slower process than adsorption at the other surfaces because of the amount of time necessary to allow surfactant to diffuse through the oil droplet. This process leads to the roll-up mechanism, leading to oil droplet detachment. This is shown schematically in Figure 10.9.

The source of surfactant molecules is the micelles in the formulated detergent. These diffuse to the appropriate surfaces and adsorb. The oil drop as it rolls up, because of the contact angle change driven by the change in the surface tension balance of forces, exposes a new surface which then adsorbs surfactant to a greater extent. This stabilises the oil from re-establishing an adsorbed state on the fabric surface and allows the drop to ultimately detach (Figure 10.10). This process is, however, aided considerably by the agitation occurring in the washing process, as well as possibly the elevated temperature, although in some parts of the world washing takes place at ambient temperatures and no hot water is used.

10.11 Spreading of a Liquid on a Liquid

Spreading can also be observed for two immiscible liquids when one is placed on the surface of the other. The lens formed from one liquid on the surface of the other can be described by what may be termed a 'generalised Young's equation' (Equation 10.7). This is necessary since all the surfaces involved are deformable, as shown in Figure 10.11. Under these conditions we need to allow for the contact angles of each to be resolved in a horizontal direction,

$$\gamma_{wv}\cos\theta_3 = \gamma_{ov}\cos\theta_1 + \gamma_{ow}\cos\theta_2 \tag{10.7}$$

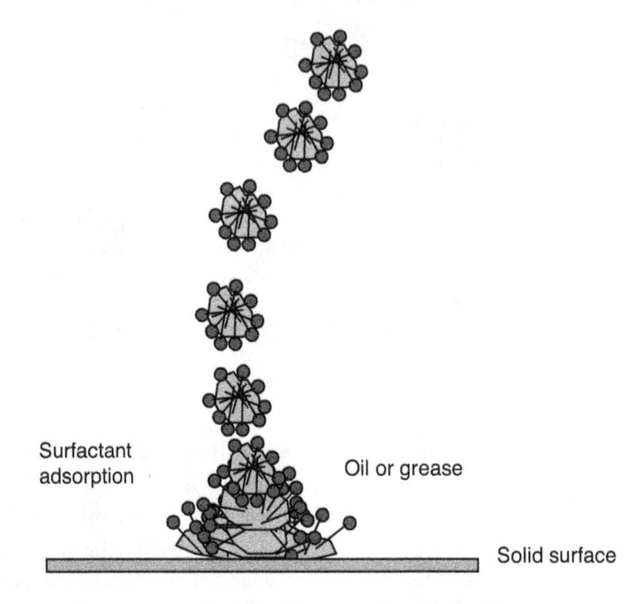

Figure 10.10 *The roll-up mechanism in detergency*

The picture is relatively complicated, since, as spreading develops with time the angles all change because of the finite size of the spreading liquid reservoir. Again, simply using the definitions and expressions used previously, the balance of forces, leading to work of adhesion (W_{AB} between liquids A and B) and work of cohesion ($= 2\gamma_A$ of the spreading liquid), can be used to evaluate the spreading coefficient of one liquid on the other:

$$S = W_{AB} - 2\gamma_A = \gamma_B - \gamma_A - \gamma_{AB} \tag{10.8}$$

Clearly for the spreading of liquid A on liquid B the work of adhesion between A and B must exceed the work of cohesion of the spreading liquid A, and the difference between the two quantities is the spreading coefficient, S, of A on B. As shown previously, if the value of S is positive, spreading results and if it is negative there is no spreading.

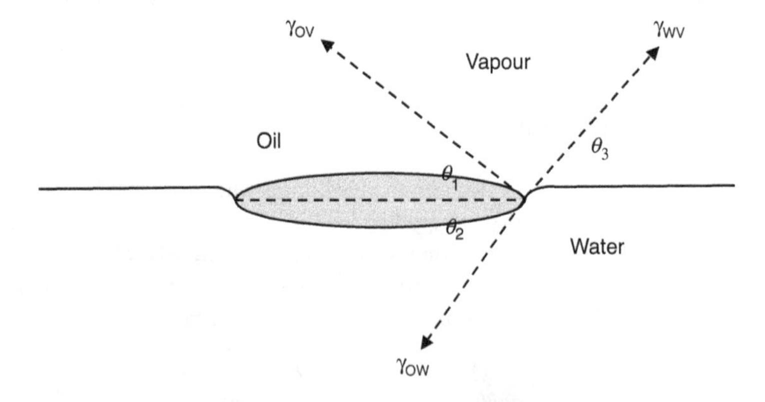

Figure 10.11 *Surface tensions for a liquid lens on an immiscible liquid*

Table 10.3 *Spreading coefficients of one liquid on another (liquid A on water, $\gamma = 72.5\,mN\,m^{-1}$)*

Liquid A	γ_A (mN m^{-1})	γ_{AB} (mN m^{-1})	S (mN m^{-1})
Octyl alcohol	27.5	8.5	36.5
Oleic acid	32.5	15.5	24.5
Bromoform	41.5	40.8	−9.8
Liquid paraffin	31.8	57.2	−16.8

Table 10.3 shows some examples of the spreading of one liquid on another. By taking liquid A as the spreading liquid (octyl alcohol, oleic acid, bromoform, liquid paraffin) on water, we can look at the interfacial tensions and the surface tensions. The surface tension of water in each case is 72.5 mN m^{-1}, and the liquid surface tensions of A vary only from 27.5 to 31.8 mN m^{-1}. However, the interfacial tensions change dramatically (from 8.5 to 57.2 mN m^{-1}). The spreading coefficients are obtained from the previous relationships, and two are positive values (spreading) and two are negative (non-spreading).

Whilst there is a systematic change in spreading coefficient, there are clearly some complicating issues. This is shown by the time-dependent behaviour of the spreading of benzene on water.

When a drop of benzene is placed on water a lens forms, which expands and spreads. After a period of time it retracts back to reform the original lens. It turns out that there is a limited solubility of benzene in water, and of water in benzene. It takes time for this equilibrium to be established after the drop of benzene is placed on the water surface. From Table 10.4 the initial values of the surface tensions of water and benzene give rise to a spreading coefficient of 8.9 mN m^{-1}, the interfacial tension being 35 mN m^{-1} and constant. However, after a period of time, water uptake in benzene reduces the surface tension of benzene very slightly, which also increases the spreading coefficient. After a longer period of time the benzene re-equilibrates in the water and the interfacial surface tension of water drops to 62.2 mN m^{-1}. The final sum for spreading coefficient is now negative (−1.6 mN m^{-1}) so it is now non-spreading. The time-dependent dissolution of benzene in water and vice versa alters the behaviour from benzene being spreading to non-spreading on water (1).

The vapour phase composition and (partial) pressures clearly have a significant influence on the spreading behaviour. The role of solubilised impurities in the case of detergency and for two immiscible liquids on a solid surface can also influence the balance of forces markedly. The results may not be as expected. Consider these processes when put into a real situation such as the soldering example; the fine balance of behaviour to create the desired adhesion/cohesion conditions becomes very difficult to predict.

Table 10.4 *Spreading of benzene on water*

	γ_{water} (mN m^{-1})	$\gamma_{benzene}$ (mN m^{-1})	S (mN m^{-1})
Initial	72.8	28.9	8.9
	72.8	28.8	9
Final $\gamma_{WB} = 35$ mN m^{-1}	62.2	28.8	-1.6

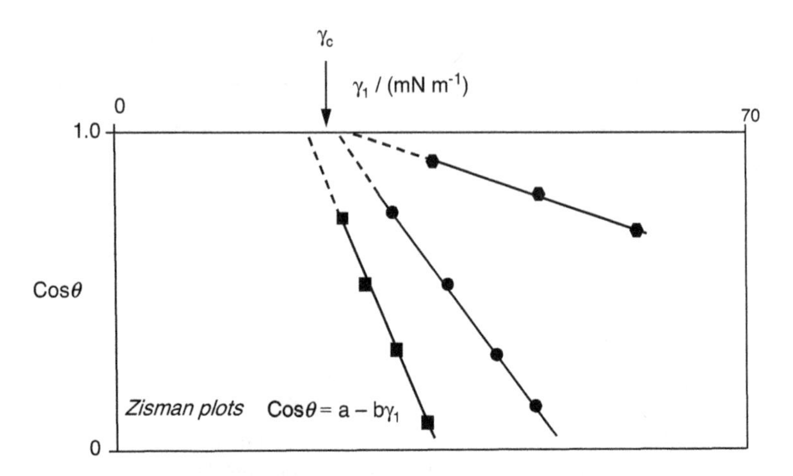

Figure 10.12 Zisman plot for homologous series of liquids on a solid surface

10.12 Characterisation of a Solid Surface

One method of characterising the solid surface is to use a series of liquids of varying surface tension, which can be obtained in a homologous series and observe those which wet and those which do not. Our general rule is that for wetting, the surface energy of the solid/vapour interface is going to be greater than the surface tension of the liquid/vapour interface. For the fluids, a ranking order can be arranged, which will give a good indication of the solid surface energy from the null spreading condition. However, we can go further than this using the homologous series of liquids, as drops, on the surface.

A plot of the cosine of the contact angle against the liquid surface tension gives us a characterisation tool for the surface. An example is shown in Figure 10.12 and is linear. This is called a Zisman plot (4). The extrapolation of the cosine of the contact angle for different homologous series tends to the same intercept. This suggests that the surface energy can be characterised by this extrapolated value, which can be termed a 'critical surface tension' (γ_C). The results, however, should be treated as semi-empirical.

10.13 Polar and Dispersive Components

Hamaker constants, like surface tensions, are related to intermolecular forces. For non-polar materials when dispersion forces dominate, the surface tension properties of liquids can be calculated directly using Hamaker constants (1, 3) using the following equation (1).

$$\gamma_i^d = \frac{A_{ii}}{24\pi (r_i^0)^2} \tag{10.9}$$

A_{ii} is the Hamaker coefficient and $r^0{}_i$ is a distance parameter used in integrating the Hamaker expression (7). Some values are given in Table 10.5.

Based only on the dispersion force, it is clear that for non-polar materials (*n*-octane, *n*-hexadecane, PTFE and polystyrene) the predicted values obtained from calculations match

Table 10.5 *Comparison of calculated (calc) and experimentally (exp) measured values of surface tension*

Material	γ^{d}_{calc} (mN m^{-1})	γ^{d}_{exp} (mN m^{-1})
n-octane	21.9	21.6
n-hexadecane	25.3	27.5
PTFE	18.5	18.3
Polystyrene	32.1	33
Water	18	72.4

almost exactly the experimentally obtained values. When the material has a polar nature, such as water, there is a huge disparity between the calculated value and the experimental values. Furthermore, we can also calculate contact angles as in Equation (10.10) (7):

$$\cos\theta = -1 + 2\phi\left(\frac{\gamma_s}{\gamma_{lv}}\right)^{1/2} - \frac{\pi_{sl}}{\gamma_{lv}} \tag{10.10}$$

ϕ is a constant which accounts for the relative molecular sizes, polarity, etc. and depends upon the molecule under consideration, but under conditions of non-polarity approximates to 1. There is an additional term in this expression, the surface pressure π_{sl}, which essentially reduces the solid surface energy through adsorption of vapour from the liquid at the solid/vapour interface. Some caution must be exercised using these equations since significant polar forces clearly affect the results.

10.14 Polar Materials

Polar materials can be treated in a similar manner by adding extra contributions to the total surface tension (8). The surface tensions can be divided into separate independent terms, the most simple being the addition of the polar component and the dispersive component, γ^{p} and γ^{d}; thus

$$\gamma = \gamma^{p} + \gamma^{d} \tag{10.11}$$

It is also possible, if necessary, to break the surface tension down into other components, hydrogen bonding for instance. We could add the hydrogen bonding term to the dispersive term, and add in a polar term to that and so on. The total surface tension is subdivided into individual contributions. For a non-polar material, of course, the surface tension is simply equal to the surface tension of the dispersion component.

An alternative approach is to use a set of semi-empirical expressions. These generally contain a term in, or derived from, the geometric mean theorem which is used to combine the Hamaker constants of different materials. Equation (10.12) is a useful expression which combines the contributions from two components A and B:

$$\gamma_{AB} = \gamma_A + \gamma_B - 2\left(\gamma^{d}_{A}\ \gamma^{d}_{B}\right)^{1/2} - 2\left(\gamma^{p}_{A}\ \gamma^{p}_{B}\right)^{1/2} \tag{10.12}$$

Table 10.6 *Surface tension components for solid surfaces*

Surface	γ(mN m^{-1})	γ^d(mN m^{-1})	γ^P(mN m^{-1})
PTFE	18–22	18–20	0–2
Polyethylene	33	33	0
PMMA	41	30	11
PET	44	33	11

Thus the surface tension components of liquids can be obtained from measurement of interfacial tension against an immiscible probe. Other semi-empirical expressions exist for relating the polar and dispersive components of a liquid and solid with the contact angle. These models are also based on methods of combining Hamaker constants such as the geometric mean theorem for instance (7). The Owens–Wendt model, Equation (10.13), is one example:

$$1+\cos\theta = 2(\gamma_s^d)^{1/2}\left[\frac{(\gamma_l^d)^{1/2}}{\gamma_{lv}}\right] + 2(\gamma_s^p)^{1/2}\left[\frac{(\gamma_l^p)^{1/2}}{\gamma_{lv}}\right] \tag{10.13}$$

This expression can be used to estimate the solid surface energy and the contributions from both the dispersive and polar components. The most simple method is to measure contact angles with two probe liquids, one polar and one non-polar, but whose individual components are known. The simultaneous equations can be solved to give estimates of the solid surface properties.

Table 10.6 shows the numerical values obtained for the dispersive and polar components from a series of materials, and we can see there is a reasonable fit. PTFE, for instance, which is a very non-polar material, is more or less completely dispersive. Polyethylene is non-polar and again the only contributing component is dispersive. For poly(methyl methacrylate) (PMMA) there is a contribution from a polar component, as is the case for polyethylene terephthalate (PET) (9).

10.15 Wettability Envelopes

It may be imagined that up until now we have a *rule* which tells us if wetting will occur when placing a liquid on a solid. This comes from knowing the surface tension of the liquid and the surface energy of the solid. Thus the rule which has been formulated states that 'a liquid, having a lower surface tension than the solid surface energy, will wet that solid'. In practice it is found that this is not always the case and it is not, therefore, an immutable rule.

A 2D map of wetting can be constructed by using the components of surface tension and a plot produced which is designed to show where wetting will occur. To illustrate this the Owens–Wendt model described previously has been used to construct a plot.

It is suggested above that there are other ways of understanding how contact angles, hence degrees of wetting, arise from an understanding of the forces existing in the materials and between the materials. Rather than considering surface tension (or surface energy) as a single component it can be seen that surface tension is the sum of individual components, dispersive and polar components for instance, and these can be summed to yield the surface tension. It is possible to take these components which have been generated by the empirical expressions and draw (2D) maps of wetting.

The experimental programme required to produce the necessary information is relatively simple and follows the rationale described in Section 10.14. The contact angles for a solution on two standard substrates, for instance glass and poly(vinyl chloride) (PVC), are determined. One of the surfaces is polar and the other is non-polar. Following this the contact angles of two standard liquids on the substrates are determined (iodomethane and water, for instance). The contact angle on the substrate of the liquid of particular interest is subsequently determined. From this information the polar and dispersive components are derived by using the Owens–Wendt equation. For the unknown sample these components are plotted against each other. An example plot is shown in Figure 10.13.

The dispersive component is plotted along the y-axis against the polar component along the x-axis. Four points labelled A, B, C and D are shown on the plot. An envelope, the 'wettability envelope', is also plotted. The envelope is created when the Owens–Wendt model is solved for the case of a contact angle of 90°; so the area bounded by the axes and the curve is less than 90° and that outside this boundary is greater than 90°. For each of the four liquids it can be seen that A and B will wet the substrate. From this it is clear that the two materials A and B, which have different overall surface tensions, can be plotted on a map, and both can be seen to be wetting. The values of polar and dispersive contributions have been calculated previously. This starts to give an understanding of wettability. It is also clear that D is dewetted, since it sits outside that envelope. In the case of liquid C it has a contact angle of 90° and so is on the border of wetting and dewetting. It is possible to make a comparison here between this type of approach and the calculation of solubility parameters.

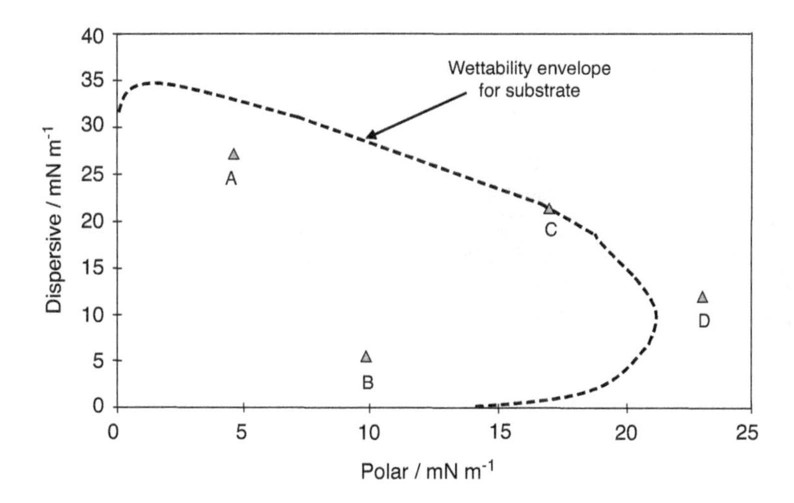

Figure 10.13 Wettability envelope for a substrate with a series of liquids

The Hildebrandt solubility parameter (3) gives a single value of solubility, whereas it is also possible to break the solubility parameter into components, as is done in the Hansen solubility parameters (3, 10), or partial solubility parameters. In the latter a three-component coordinate set of parameters can be identified, which map out the solubility in a more detailed manner and give a greater insight into the solubility of materials, polymers for instance, in different solvents. Thus the solubility of a material is put onto a 3D map which has axes of polar, dispersive and hydrogen bonding. The same principle is adopted here, but using surface tensions, to see how wettability is influenced by the components of surface tension. This gives another view and method of predicting the wetting behaviour of substrates with liquids.

10.16 Measurement Methods

Clearly it is of major importance to be able to make measurements of contact angles, interfacial tensions and surface tensions, and this itself is quite a broad subject and topic to consider. What is presented below is an identification of some of the more important techniques. It is suggested that if the reader requires a more in-depth description, further reading can easily be found on each of the techniques (1).

Capillary rise experiments can be used for both surface tension and contact angle. In each case the experimental elegance employed can potentially offer a high degree of accuracy producing contact angles with $\pm 0.1°$ accuracy. This is fine in an academic sense; however, new commercially available equipment gives rapid capabilities for industrial applications. Contact angle can be routinely measured by imaging techniques, essentially computer frame grabbing. The Du Nouy ring is a reasonable method for surface tension measurement and additionally is still the most appropriate method for interfacial tensions. The spinning drop is another interfacial tension measurement technique but is useful only for ultra-low interfacial tensions. Maximum bubble pressure techniques are able to study the development of surface tension with time. At small times a bubble blown in surfactant solution will give a value close to the pure solvent, whilst at long times the equilibrium value will be given.

Techniques requiring drop or bubble shape analysis have been advanced over recent years because of greatly improved computational methods when applied to an image of the drop or bubble profile. The method, known as 'axisymmetric drop shape analysis' (11), can be performed on the measured profiles of captive bubbles, sessile drops, hanging drops and hanging bubbles, as shown in Figure 10.14. The principle of the technique is that the pressure drop across a curved surface given by the Laplace equation is in balance with the gravitational pull on the bubble or the drop, as shown by

$$\gamma\left(\frac{1}{R_1}+\frac{1}{R_2}\right) = \Delta p_0 + \rho g z \qquad (10.14)$$

From this we can develop some idea of the drop shape from the balance between the Laplace pressure and the gravitational force, since it is a function of surface tension of the

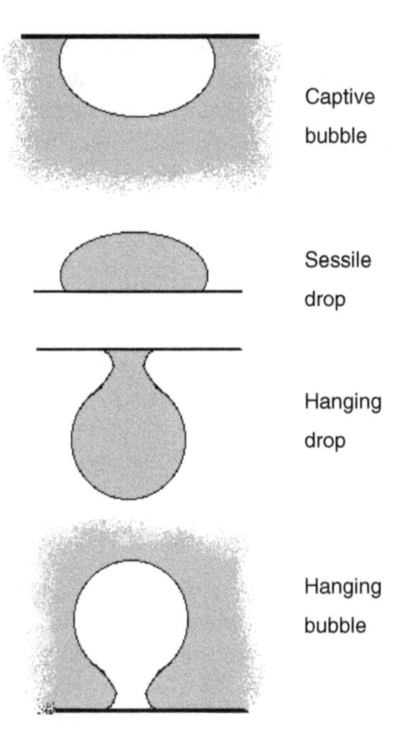

Figure 10.14 *Axisymmetric drop shape analysis measures the profiles of symmetrical drops in bubbles*

interface, γ. The appropriate expressions are written in terms of surface tension, the principal radii of the bubble or the drop, R_1 and R_2 and the pressures associated with them, Δp, the pressure drop across the interface. This has two contributions: Δp_0 the pressure difference at a reference plane and an additional pressure which arises from the gravitational component at any point on the surface. This is given by the density difference (ρ) between the bubble or drop and the continuous phase and its vertical distance z from the reference plane. The reference plane is a datum line from which values of z can be established, hence its position needs to be established and a calibration taken to ensure determining accurate values of z. The principal radii can be used to calculate the surface tensions from knowledge of the density differences, the height to z above the reference plane and the curvature at that height. The analysis involves a complex set of first-order differential equations that are solved numerically to give the results. From the measured profiles it is possible to obtain parameters such as surface tension, contact angle, the drop radius, drop volume and surface area.

The method itself is not new and goes back many years to a set of tabulated calculation results of Bashforth and Adams (12). The functions are look-up tables which make corrections to the observed values of diameter, height and radius of the drop or bubble. Using these values the surface tensions can be calculated.

There are a number of commercially available machines and because of the nature of the experiment it is rapid, simple and produces a wealth of information. Because the method observes the drops or bubbles optically the technique requires the magnification to be calibrated so that absolute dimensions are known. This can be done in a number of ways but the preferred method is to measure an object of known size using a sapphire ball. It is worth noting that the methods require an axially symmetric drop or bubble. This condition would be expected to be fulfilled for all the cases with the possible exception of the sessile drop. It is often noticed that a drop spreading on a surface does so in a series of jumps. These we can argue are due to a number of causes, surface roughness, oil or contaminant on the surface, different exposed surface energy domains, for instance. The consequence of this is that the drop may not be axially symmetric. Therefore, a degree of care must be exercised in performing these experiments.

An additional advantage of using this type of experimental equipment is that previously obtained images can be presented to the analysis system. Hence, photographs of high-temperature molten metals in contact with a surface, for instance, can be analysed to give information regarding the surface tension and contact angle.

10.17 Conclusions

The topic of wetting as outlined in this chapter is seen to be a fascinating study. Only a glimpse of the complete topic is shown here, but as more advanced reading into the subject is undertaken, it will be appreciated that there is still much to understand. Many of the expressions and arguments shown in the text can at best be described as semi-empirical. This confirms our belief that wetting is still a developing science. Perhaps it is because the subject has such immense utility to industrial processes that our full understanding has lagged behind the experimental development. Naturally, within the text here there are many missing components of the subject, and one which deserves further reading is that of the dynamics of wetting (13). This is a rich topic for investigation with tremendous implications for industrial processing.

References

(1) Adamson, A. W., Gast, A. P. (1997) Physical Chemistry of Surfaces, 6th edition. Wiley, New York.
(2) Young, T. (1805) Phil. Trans. R. Soc. (London), 95: 65.
(3) Barton, A. F. M (1991) CRC Handbook of Solubility Parameters and Other Cohesion Parameters, 2nd edition. CRC Press, Boca Raton, FL, Chapters 1 and 2.
(4) Zisman, W. A. (1964) Adv. Chem. Ser., A43: 1.
(5) Garret, P. R. (ed.) (1993) Defoaming, Theory and Industrial Applications, Surfactant Science Series vol. 45. Dekker, New York.
(6) Schulte, H. G., Hofer, R. (2003) In Karsa, D. R. (ed.), Surfactants in Polymers, Coatings, Inks and Adhesives. Applied Surfactant Series vol. 1. Blackwell, Oxford.
(7) Girifalco, L. A., Good, R. J. (1957) J. Phys. Chem., 61; Good, R. J. (1964) Adv. Chem. Ser., 43: 74.
(8) Fowkes, F. M. (1962) J. Phys. Chem., 66: 382; Fowkes, F. M. (1964) Adv. Chem. Ser., 43: 99.
(9) Van Krevelen, D. W. (1990) Properties of Polymers, Their Correlation with Chemical Structure; Their Numerical Estimation and Prediction from Additive Group Contributions, 3rd edition. Elsevier, Amsterdam.

(10) Patton, T. C. (1979) Paint Flow and Pigment Dispersion, 2nd edition. Wiley-Interscience, New York.
(11) Neuman, A. W., Spelt, J. K. (eds.) (1996) Applied Surface Themodynamics, Surfactant Science Series vol. 63. Dekker, New York.
(12) Bashforth, F., Adams, J. C. (1883) An Attempt to Test the Theories of Capillary Action. Cambridge University Press, Cambridge.
(13) Blake, T. D. (1993) In Berg, J. C. (ed.), Wettability. Dekker, New York.

11
Aerosols

Nana-Owusua A. Kwamena and Jonathan P. Reid

School of Chemistry, University of Bristol

11.1 Introduction

Aerosols play important roles in a wide range of disciplines including colloid science, pharmacy, combustion, and atmospheric chemistry and physics. An aerosol is defined as a mixed phase system that is composed of a dispersed phase, either solid or liquid particles, and a dispersion medium, the surrounding gas. Although the dispersed phase may represent a very small fraction of the aerosol volume, the terms aerosol and particle are often used interchangeably to refer to the particle phase. Particle diameters can span five orders of magnitude, extending from aerosol nuclei at the nanometre scale to cloud droplets, dust particles and sea salt spray at the 100 μm scale. This corresponds to a range in volume and mass of fifteen orders of magnitude.

Aerosols that are generated directly through suspension by mechanical action are referred to as primary aerosols; common atmospheric examples include dust and sea salt. Secondary aerosols are formed by chemical processes or the condensation of low volatility compounds onto pre-existing aerosols, commonly described as gas-to-particle conversion. Natural atmospheric aerosols are largely primary in origin with an estimated flux of 3100 teragrams (10^{12} g) per year (Tg yr^{-1}), arising from desert sands, sea spray, rock weathering, soil erosion, biomass burning and volcanoes. Chemical constituents include aluminosilicates, ores, clays, metal halides, sulfates, and elemental and organic carbon. Anthropogenic sources are estimated to contribute ~450 Tg yr^{-1} to the atmospheric aerosol burden. Emissions are both primary and secondary and are more localised geographically than

Colloid Science: Principles, methods and applications, Second Edition Edited by Terence Cosgrove

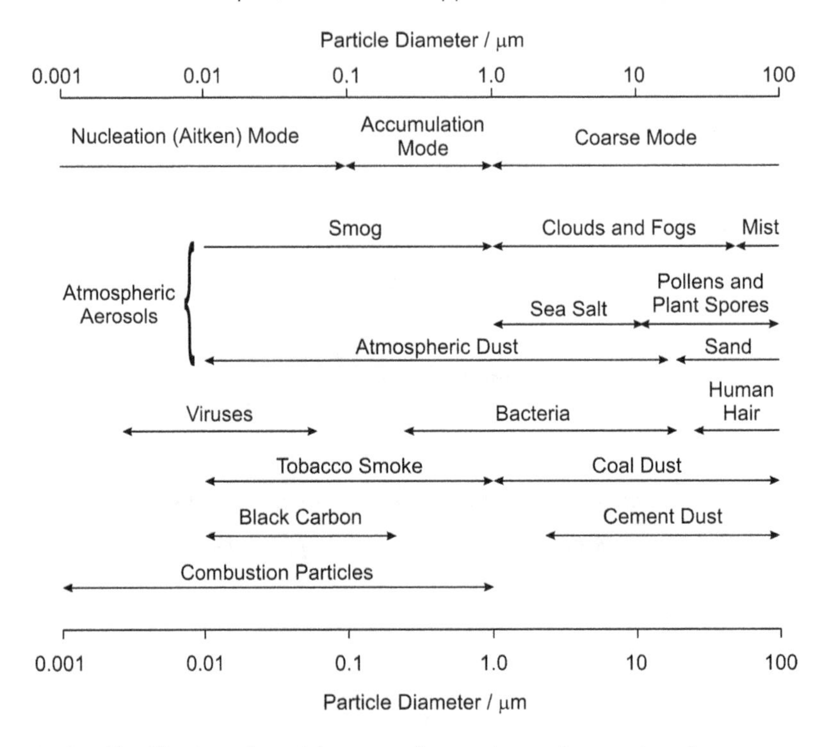

Figure 11.1 *Classification of particles according to size and examples of common aerosols*

natural sources, arising from fossil fuel combustion, heavy industry and transportation. Chemical constituents include sulfates, nitrates, organics, soot and heavy metals.

Particles are commonly classified according to their size (Figure 11.1). Particles smaller than 0.1 μm in diameter are referred to as nucleation mode particles (or Aitken nuclei) and are generated by gas-to-particle conversion. Growth of the nucleation mode by condensation and coagulation leads to accumulation mode particles, which range from 0.1 to 1 μm in size. Particles generated by incomplete combustion fall into the nucleation and accumulation modes. Organic components as well as sulfate and nitrate ions are commonly found in particles less than 1 μm in diameter. Particles larger than 1 μm are referred to as coarse mode particles and include pollens, dust and sea salt aerosol. Particle concentrations are generally expressed as the number of particles per cubic centimetre (particles cm^{-3}) or a mass per cubic centimetre (usually $\mu g\, cm^{-3}$). Typical particle concentrations in the atmosphere are 10–1000 cm^{-3} for nucleation and accumulation mode particles, falling to $<1\, cm^{-3}$ for particles in the coarse mode.

Aerosol size distributions and concentrations are commonly indicative of their source. For example, particle concentrations are usually lowest over the oceans and largest in urban areas with average concentrations of $100\, cm^{-3}$ and $>1 \times 10^5\, cm^{-3}$, respectively. Although the maximum particle concentration usually occurs in the accumulation mode, the coarse mode can dominate the aerosol volume distribution ($\mu m^3\, cm^{-3}$) and mass concentration ($\mu g\, cm^{-3}$). The larger surface-to-volume ratio of smaller particles and their high number densities lead to a surface distribution ($\mu m^2\, cm^{-3}$) that peaks in the accumulation mode.

Indeed, the accumulation mode plays a dominant role in mediating important heterogeneous atmospheric chemistry, for example in the origin of acid rain and the formation of the ozone hole. The accumulation mode also influences the radiative balance of the earth through direct scattering of solar and terrestrial light and through the impact of aerosols on cloud albedo and lifetime (1).

Epidemiological studies have highlighted the significant impact that aerosols can have on human health, with higher mortality and morbidity rates at particle concentrations lower than had previously been thought (2). Although chemical composition is important in determining toxicity, with particles acting to adsorb and concentrate potentially carcinogenic species, physical properties are also important. Toxicity can be correlated with particle size, shape, electrical charge and solubility, with fine particles ($< 1\,\mu m$ in diameter) able to penetrate deep into the respiratory tract. Particles are divided into size classifications for regulatory air quality standards. Particles less than $2.5\,\mu m$ in diameter are known as $PM_{2.5}$ and fall within the respirable range allowing them to penetrate deep into the respiratory tract, increasing the rate of progression of respiratory, pulmonary and cardiovascular diseases (2, 3).

The lifetime of an aerosol can range from seconds to days. During this time the aerosol may interact with other gas- or condensed-phase species and undergo chemical and physical transformations or ageing. These transformations can occur by condensational growth or evaporative loss, or by heterogeneous chemistry with trace gas-phase reactant species. Further, variations in relative humidity or coagulation of particles through collisions can lead to changes in aerosol size distributions and mixing states. The cross section of particles greater than $10\,\mu m$ results in coagulation contributing appreciably to the rate of growth. Conversely, the high diffusion constant of nucleation mode particles leads to rapid diffusional loss through coagulation. Solid inclusions may be incorporated within a liquid host by coagulation, leading to a transition from an externally mixed to an internally mixed particle. An external mixed aerosol population is observed when the compositions of individual particles are homogeneous, but particles in the population have radically different compositions. This is distinct from an internal mixture in which any one particle has a composition that is representative of the entire population.

Dry deposition through impaction, diffusion or gravitational settling provides the dominant loss mechanism for most particles. Small particles are buoyant and are lifted by upwelling air currents. Loss for these smaller particles occurs when air currents direct the particles close to surfaces leading to impaction. In contrast, the dominant loss process for coarse particles is gravitational settling or sedimentation. For particles of intermediate size, particles move by Brownian motion and are eventually deposited by sedimentation. Impaction occurs for particles of high mass and inertia that are not able to adapt their path with the deflected air flow around an object. A gradient in particle concentration in the atmosphere is thus established, with lower particle concentrations near the surface as a result of deposition. In the atmosphere, wet deposition through incorporation into existing rain droplets and wash out is another important loss mechanism for particles 0.1–$10\,\mu m$ in diameter.

As a result of the wide variation in size, concentration, composition, morphology and phase, a detailed characterisation of the state of an aerosol is challenging (4–6). Ideally, all of the physical and chemical properties of a particle would be measured, examining, for example, the variation in particle composition with particle size. To complicate aerosol analysis further, the highly dynamic nature of aerosol particles can lead to pronounced and rapid temporal variations in particle size and composition, arising from the large surface area interacting with the surrounding gas phase. Clearly, compromises must be made in the analysis of aerosols. In order to better understand the thermodynamic and kinetic factors that control the composition and reactivity of aerosols across a range of disciplines, methods for probing the size, concentration and composition of aerosols are imperative. Therefore, we first provide an overview of some of the different techniques that are available for generating and sampling aerosols, characterising particle size distributions and determining the chemical composition of aerosols. We then introduce the thermodynamic factors that govern the equilibrium size and composition of an aerosol, before considering the kinetics of aerosol transformation.

11.2 Generating and Sampling Aerosols

Generating an aerosol of a well-defined size distribution and composition is crucial for calibrating and testing aerosol analysis techniques and for performing detailed experiments under controlled laboratory conditions (7). Particle size distributions may be described as being either monodisperse or polydisperse. A monodisperse aerosol of known size, shape, density and/or composition is required to test sampling systems and to calibrate particle sizing instruments. A polydisperse aerosol contains particles of a wide range of particle sizes, shapes, densities and/or compositions. A general requirement of any aerosol generating method is to yield a reproducible output of particles of uniform concentration, composition and size. Recent reviews by McMurry (8) and Finlayson-Pitts and Pitts (7) provide additional information on the different methods of generating and sampling aerosols that are currently in use.

11.2.1 Generating Aerosols

The most common method for generating a polydisperse liquid aerosol is the compressed air nebuliser, as illustrated in Figure 11.2. Compressed air at a pressure of 30–250 kPa exits at high velocity from a nozzle or orifice. The low pressure resulting from the Bernoulli effect in the exit region of the nozzle draws liquid from a reservoir into the gas stream and disperses the liquid jet into droplets (4). The resulting aerosol spray is directed onto a surface where large particles are lost by impaction and small particles remain in the gas flow. This results in a polydisperse stream of droplets with mass median diameters between 1 to 5 μm and particle concentrations as high as 10^7 particles cm^{-3}. Ultrasonic nebulisers are able to produce a higher volume of a polydisperse aerosol at higher concentrations than compressed air nebulisers. Ultrasonic waves, generated by a piezoelectric crystal, are focused onto a liquid surface. Capillary waves form at the surface and then break to form a dense aerosol cloud.

To generate liquid droplets larger than 5 μm in diameter, a vibrating orifice aerosol generator (VOAG) is generally used (4). A liquid jet is unstable to a mechanical disturbance and by applying a regular mechanical vibration to a jet, formed by forcing liquid through a small orifice, droplets of monodisperse size can be generated. The droplet size is dependent on the frequency of the modulating vibration, provided by a piezoelectric crystal, and the

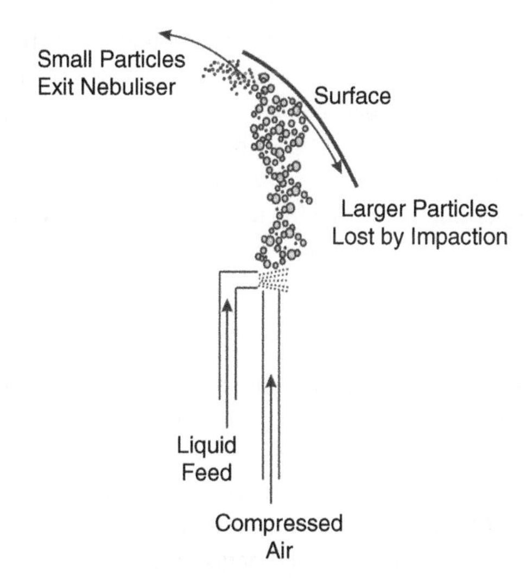

Figure 11.2 *A schematic of a compressed air nebuliser*

size of the orifice. Initial droplet diameters of 5–200 μm can be readily achieved with concentrations of 10–500 cm^{-3}. Coagulation can cause rapid increases in particle size unless the aerosol is dispersed in a flow of air.

Homogeneous nucleation can be used to generate polydisperse liquid organic aerosols < 1 μm in size. A vessel is filled with a low volatility liquid and heated. As an inert gas passes through the heated vessel the gas becomes saturated with the vapour of the liquid. Upon leaving the heated region, the gas cools and becomes supersaturated, with the vapour condensing to form an aerosol.

Solid particles can be obtained by nebulising a solution of a salt dissolved in a volatile solvent using a nebuliser or VOAG. Evaporation of the volatile solvent leads to the formation of solid particles with a final diameter that is determined by the initial volume of the liquid droplet and the mass fraction of the solid material. Nebulising a liquid suspension of solid particles can also lead to a controlled source of monodisperse solid particles. For example, monodisperse polystyrene and poly(vinyl toluene) latex spheres, 0.01–100 μm in diameter, can be commercially obtained with standard deviations in size of only a few percent. Once dispersed in a solvent and atomised, the solvent evaporates leaving monodisperse spheres.

The simplest method for generating solid particles is to disperse a dry powder by gravity feed into a high-velocity air stream. Concentrations of up to 100 g m^{-3} can be achieved, although this is dependent on the particle size range, shape and the moisture content of the powder. Particle sizes between 1 to 100 μm can be dispersed in this fashion. Dry and hydrophobic powders disperse more readily than moist and hydrophilic ones. Agglomeration in the powder to be dispersed is a major problem and this can be overcome by introducing the powder into a fluidised bed consisting of 200 μm diameter beads.

Multicomponent aerosols may be generated by atomising solutions of mixed composition. Mixed composition aerosols may also be obtained by generating an aerosol of low volatility and coating the aerosol with a second compound. An aerosol of low volatility is usually generated by means of an atomiser and is then passed through a heated vessel containing another low volatility compound. On leaving the heated region, the low volatility gas condenses onto the pre-existing aerosol. Whether the resulting aerosol is core-shell in structure or internally mixed depends on the relative difference in the vapour pressures of the components and the temperature of the heated region. Further, the thickness of a coating can be adjusted by varying the temperature of the heated vessel.

11.2.2 Sampling Aerosol

In sampling an aerosol it is important to minimise particle loss in the sampling inlet. Therefore, regardless of particle size or inertia, isokinetic sampling must be achieved. The sampling inlet must be aligned parallel to the gas flow and the gas velocity in the inlet must be identical to the velocity of the gas approaching the inlet. If this is not maintained, distortions in the size distributions passing through the inlet may result. Sampling particles through an inlet into an instrument at reduced pressure can have a significant impact on the partitioning of volatile components between the condensed and gas phases. Loss of a volatile component, such as an organic component or water, can lead to changes not only in composition, but also to particle size and phase.

It may be advantageous to select a specific size range for analysis when sampling an aerosol. Inlets can be designed to deliver a sharp cut off in the size distribution, allowing particles smaller than a certain aerodynamic diameter to pass through, as illustrated in Figure 11.3(a). Impactors classify particles according to their inertia (Figure 11.3b). Particles are accelerated through a nozzle or circular jet toward a substrate positioned at 90° to the gas flow. Particles larger than a limiting size have sufficient inertia to cross the flow streamlines and impact on the substrate, while smaller particles follow the deflected streamlines avoiding impaction on the substrate. When a number of impactors are operated in series with progressively smaller cut off sizes, the aerosol sample can be size fractionated. This is referred to as a cascade impactor (Figure 11.3b).

Impactors can allow an accurate determination of particle concentrations for particle sizes smaller than 10 μm and typical fractionations include the separation of 10, 2.5 and 1 μm sized particles. A low-pressure cascade impactor can enable the selection of particles in the nanoparticle range and particle sizes down to 50 nm can be studied routinely. A fraction of particles may bounce on encountering the substrate, particularly if the substrate is solid, and become re-entrained in the aerosol flow. This can be minimised by coating the substrate with oil or grease. Alternatively, this problem may be avoided entirely by replacing the substrate with a receiving tube, known as a virtual impactor.

A collimated flow of particles can be achieved by passing the aerosol through a series of axisymmetric contractions and enlargements, known as an aerodynamic lens (9), before a nozzle expansion. Particles smaller than a critical size can be confined very tightly to the axis of the flow, with beam waists of as little as a few millimetres.

Aerosol samples are commonly collected on fibrous or porous membrane filters. Filters are made from a wide range of materials and have collection efficiencies exceeding 99%. Particle removal occurs as a result of collision and adsorption of the particle to the fibre surface through interception, inertial impaction, diffusion or electrostatic interaction (10). Fibrous filters are composed of glass, plastic or cellulose fibres and the fibres range from <1

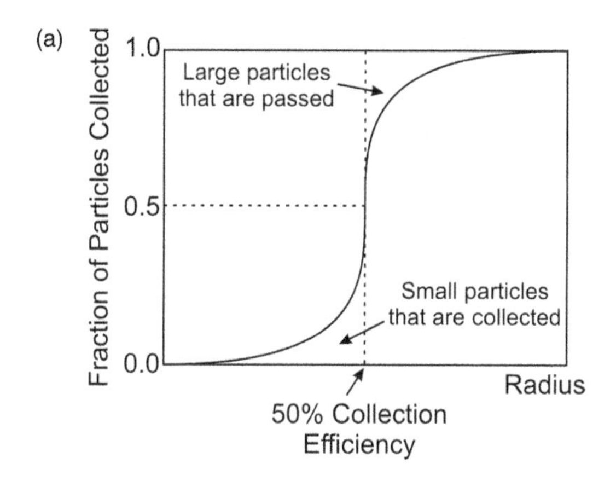

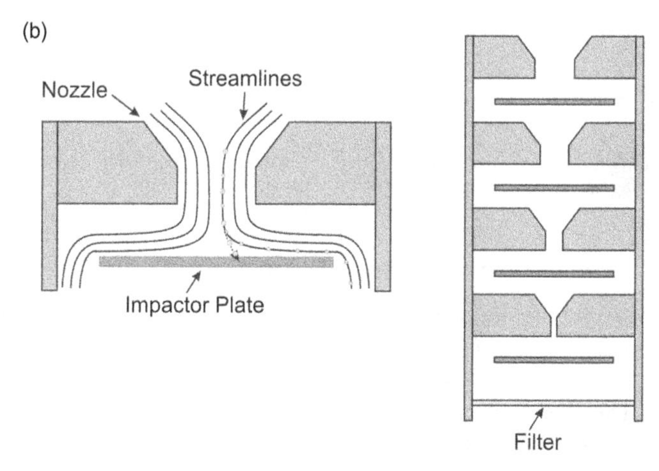

Figure 11.3 *(a) Inlets are designed to collect particles larger than a specified size with a sharp cut point; (b) An impactor and cascade impactor*

to 100 μm in diameter, forming an entangled mat. Examples of materials used for porous membrane filters include Teflon, polyvinyl chloride and sintered metals.

Diffusion denuders are used for sampling semi-volatile compounds. The sampled aerosol is passed through a conduit and gaseous components are removed by diffusion to the conduit walls and subsequent reaction with a co-reactant selected to efficiently remove the gas-phase component of interest. The aerosol particles flow through the tube unaffected and are collected on a filter. This enables an examination of the partitioning of volatile components between the gas and condensed phases.

11.3 Determining the Particle Concentration and Size

To characterise the aerosol distribution, it is important to determine the number concentration, the mass concentration and size distribution of particles. The strengths and deficiencies of some of the common methods are described below.

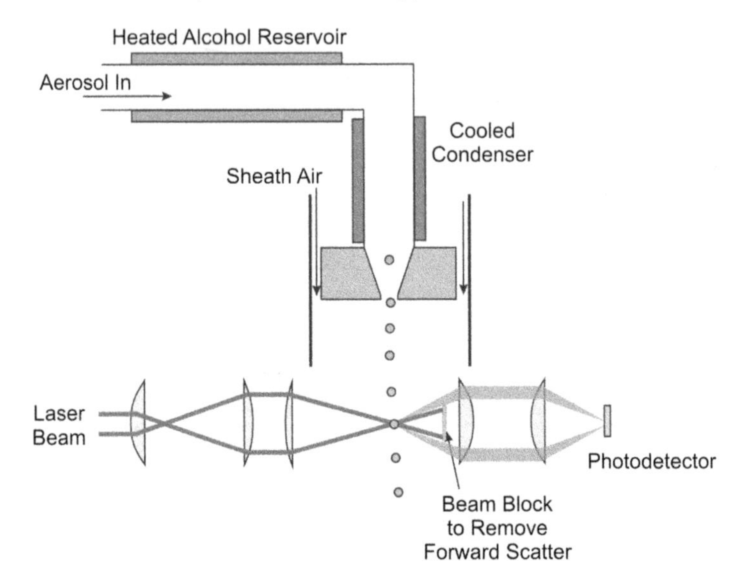

Figure 11.4 *An optical particle counter. Particles first undergo condensational growth prior to counting by optical scattering*

11.3.1 Determining the Number Concentration

The number concentration provides a measure of the number of particles per unit volume (i.e. particles cm^{-3}). The most commonly applied instrument is a condensation particle counter (CPC), also known as a condensation nucleus counter (CNC), shown schematically in Figure 11.4. The operation of the most common commercially available instrument is divided into three parts: the saturator, condenser and particle detection region. The saturator contains an alcohol (usually *n*-butyl alcohol) at elevated temperatures which readily condenses on both hydrophilic or hydrophobic particles. Supersaturated conditions arise as the heated flow leaves the saturator and enters the condenser. Particles grow rapidly to a size at which they can be detected by light scattering, either permitting the counting of individual light scattering events or by indirectly monitoring light attenuation. Single particle counting can be achieved at particle concentrations lower than $10^4\,cm^{-3}$.

Individual particles greater than 10 nm in diameter can be readily detected using a CPC. To detect particles < 10 nm in diameter, supersaturations of several hundred percent may be required to achieve sufficient particle growth for detection. A variant of the CPC is the cloud condensation nuclei counter (CCNC). This instrument assesses the concentration of aerosol particles that can act as cloud condensation nuclei, activating to form cloud droplets in a water supersaturated environment between 0.01 and 1%.

11.3.2 Determining the Mass Concentration

Mass concentration (typically $\mu g\,cm^{-3}$) provides an important measure of aerosol loading, particularly for enforcing regulatory standards for air quality. The most common method involves recording the mass loading of filters, under conditions of controlled temperature and relative humidity, prior to and after sampling a known volume of air over a set period of time. Particles larger than a specific size are removed at the inlet stage. Cascade impactors can enable a determination of the cumulative mass distribution within a number of sampling size ranges. Uncertainties in mass distribution can arise from water absorption/desorption from the filter, evaporative losses of semi-volatile compounds and particle loss during handling. The lower limit on the mass concentration that can be detected is about 2 μg m^{-3}.

Piezoelectric crystals are a sensitive method for determining mass concentrations. The resonant frequency of vibration of the crystal varies with material and thickness. If the crystal increases in mass, the change in resonant frequency can be used to measure the mass deposited from a known volume of air. A typical sensitivity of 1000 Hz μg^{-1} can be achieved for a resonant frequency of 10 MHz. Such instruments are used for mass concentrations in the 10 μg m^{-3} to 10 mg m^{-3} range.

11.3.3 Determining Particle Size

The reported particle size depends on the technique employed and may differ from its geometric size. Indeed, the interconversion of sizes reported by different measurement techniques is nontrivial and measurements of particle shape, refractive index and density are crucial for performing such a conversion. The aerodynamic diameter of a particle is the diameter of the equivalent unit density (1 g cm^{-3}) sphere that has the same settling velocity as the particle (11). This exceeds the geometric diameter for particles with densities larger than 1 g cm^{-3} and depends on particle shape, density and size. For non-spherical particles, it is common to define the Stokes diameter, the diameter of a sphere that has the same density and settling velocity as the particle. More generally for a non-spherical particle, an equivalent diameter is used; the equivalent diameter is the diameter of the sphere that would exhibit the same aerodynamic size as that of the non-spherical particle.

The cascade impactor, described above, and the aerodynamic particle sizer (APS) classify particles according to their aerodynamic diameter. An APS operates by exploiting the varying inertia of particles of varying size in a gas flow that has been accelerated through a nozzle. The acceleration of a particle through the nozzle increases with decreasing particle size and density. The velocity of a single particle is inferred by determining the time-of-flight of the particle between two probe laser beams, with the scattered light acting to define the timing cycle, as illustrated in Figure 11.5. Although the smallest size that can be probed by the APS is 0.2 μm, high resolution information on particle size can be achieved in real-time. For both the impaction technique and the APS, the pressure drop necessary to classify the particle size can lead to changes in relative humidity and particle size, as well as deformations in particle shape.

Electrostatic methods for determining particle size yield a measure of the electrical mobility of a particle. The electrical mobility size is dependent on particle shape and size, but not density and is commonly measured with a differential mobility particle sizer (DMPS), which consists of a differential mobility analyser (DMA, also referred to as an electrostatic classifier) and a particle detector, usually a CPC. The sampled aerosol is exposed to a cloud of positive and negative ions in a bipolar charger, charging the particles to ± 1, ± 2, ... charge units with an aerosol mean charge close to zero. The particles flow from the charger into the electrostatic classifier (Figure 11.6), which consists of two concentric metal cylinders, passing into a laminar flow of clean air flowing in the

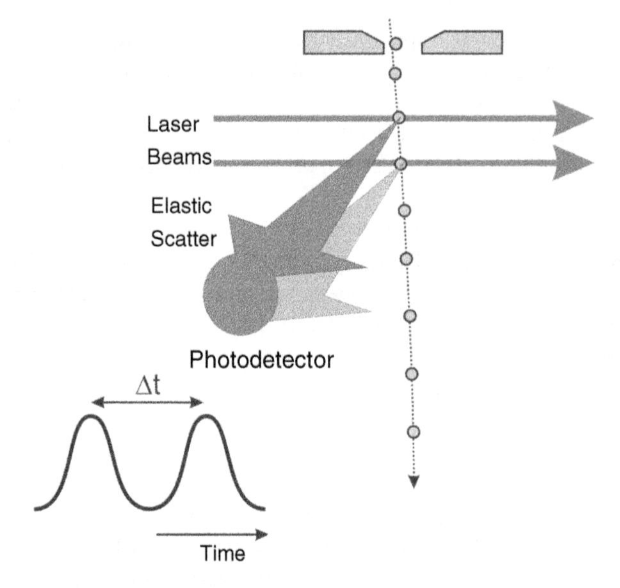

Figure 11.5 *An aerodynamic particle sizer that determines aerodynamic size from the time required to travel between two probe laser beams*

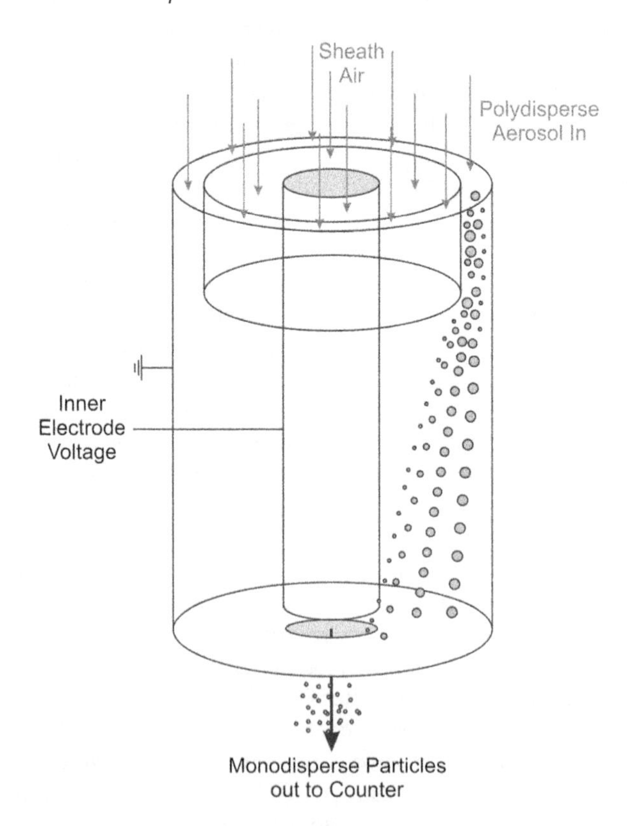

Figure 11.6 *A schematic of a differential mobility analyser (DMA)*

annular space between the two cylinders. In such a coaxial design, the inner cylinder acts as the collection rod and is maintained at a negative voltage (0–10 kV) while the outer cylinder is maintained at ground. Particles travel through the electric field between the two cylinders with a well-defined trajectory determined by their electrical mobility. Only particles within a narrow mobility range are transmitted through a small slit at the exit of the DMA. The size distribution can be determined by monitoring the number of particles exiting the DMA as the classifier voltage is varied. The classifier voltage can also be scanned continuously to measure a size distribution as a scanning mobility particle sizer (SMPS).

Particle concentrations in the diameter range 3 nm to 1 μm can be determined by DMPS and SMPS. The upper limit imposed on the operating particle concentration depends on particle size but can be as high as $2 \times 10^8\,cm^{-3}$. At the lower end of the size range, particle diffusion is significant leading to diffusional losses and broadening of the size distributions. In addition, approximately 1% of the particles at a size of 3 nm may be charged and this can lead to poor detection sensitivity.

The intensity of the light scattered from a particle can be used to measure the optical size, which is dependent on the refractive index, shape and size of the particle (see Chapter 13). An optical particle counter (OPC) records the intensity of scattered light from individual particles integrated over a known solid angle. The particle size can be estimated from a calibration curve, determined from monodisperse spherical particles of known size and refractive index, which is a monotonic function of particle size (e.g. Figure 11.7) (12). The light source can be a monochromatic laser or an incandescent white light source with minimum detectable sizes of ∼50 nm and ∼200 nm, respectively. Instruments capable of measuring particles in the size range 0.1–10 μm are common with size distributions recorded by determining concentrations in up to 50 size-resolved ranges or channels.

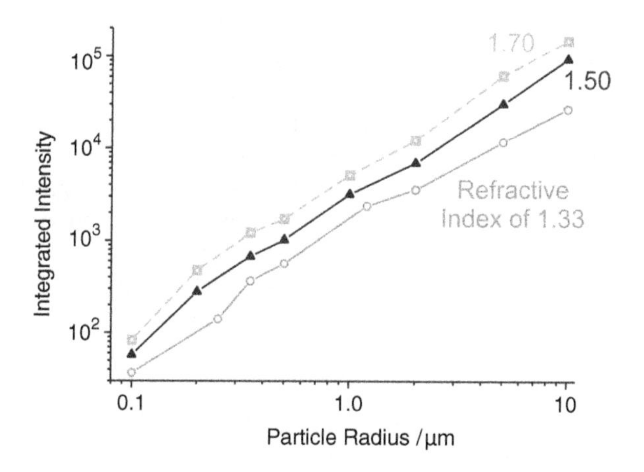

Figure 11.7 *An example of the variation in the scattered light intensity from a spherical particle with change in particle size and refractive index. Mie scattering calculations were performed at sizes shown by the symbols and the lines are included to guide the eye. The illuminating laser wavelength is 650 nm and the light scattering is integrated between 80° and 100° from the forward scattering direction. If the refractive index is unknown or the particles are large, significant errors may result*

Coincidence errors may occur if the aerosol concentration rises above 1000 cm^{-3}. OPCs can be combined with a CPC for examining particle sizes down to and less than 10 nm.

11.4 Determining Particle Composition

Aerosol composition can be determined by the post-analysis of aerosol samples collected on filters using a wide range of conventional techniques that identify and quantify chemical species and composition. Such measurements are susceptible to artefacts that may occur during sampling, transport or storage. Negative artefacts arise through loss (i.e. evaporation or chemical reactions) of particulate species. Although real-time *in situ* analysis is preferred, there are currently very few techniques that allow direct on-line determination of aerosol composition.

11.4.1 Off-line Analysis

A wide range of analytical techniques can be used to analyse aerosol particles collected on filters or impactor substrates. Analytical techniques must be sensitive enough to examine the small amount of material that is collected. The time required for sample collection is dependent on the aerosol mass loading and sampling rates. Typically, sampling durations may extend beyond a day. In addition, the type of filter or impactor substrate used depends on the nature of the samples to be analysed and the analytical technique that will be used. For example, the total carbon loading present in atmospheric aerosols is typically divided into organic and elemental carbon fractions (OC and EC, respectively) and aluminium foil substrates must be used for sampling instead of the filter substrates discussed in Section 11.2.2. A carbon-free substrate is essential as the total carbon loading is determined by measuring the amount of carbon dioxide released on combustion (8). If samples are to be analysed by ion chromatography, precleaned Teflon or Mylar substrates are used.

Teflon membrane filters are also used for non-destructive techniques such as X-ray fluorescence (XRF) analysis or proton-induced X-ray emission (PIXE). Both analytical techniques are rapid, non-destructive and sensitive, providing a powerful approach for the elemental analysis of aerosols. A major advantage of PIXE is the extremely high spatial resolution that can be achieved by the highly focused nature of the proton beam, focused to a spot size of tens of micrometres. Typically, up to 20 elements may be quantified from a single PIXE spectrum. Elemental concentrations as low as 1 ng m^{-3} can be analysed by PIXE (6). XRF can be used to analyse up to 45 elements; however, light elements may be lost from heating by X-rays or the vacuum conditions required for the technique (13). X-ray photoelectron spectroscopy (XPS) is a non-destructive technique that probes the atomic composition, structure and oxidation state of the elements in an aerosol sample and is particularly useful for surface analysis (14).

Atomic absorption spectroscopy (AAS) and atomic emission spectroscopy (AES) can also provide an elemental analysis of aerosol particles. One of the disadvantages of AAS and AES is that samples must be in solution form for analysis and chemical and spectral interferences must then be considered when choosing appropriate solvents and reagents (14). For higher sensitivity trace analysis of elements, inductively coupled plasma coupled with mass spectrometry (ICP-MS) can be used. ICP-MS has better detection limits compared to AAS and is capable of isotope determination (13). Gas chromatography with either a flame ionisation detector (FID) or mass spectrometer (MS) can be used to analyse the volatile organic components of an aerosol. For organics of lower volatility high-performance liquid chromatography (HPLC) may be coupled with absorption and fluorescence detectors or with mass spectrometers.

Laser-induced breakdown spectroscopy (LIBS) is a real-time technique for acquiring an elemental fingerprint of aerosol samples, particularly heavy metals. A laser beam of sufficient energy is used to create a plasma that excites the elemental constituents that are released from the aerosol and the resolved emission spectrum from the elements is recorded. No sample preparation is required and LIBS can be used for multi-element detection and single particle analysis (15). Laser microprobe mass spectrometry (LMMS) uses a pulsed laser beam to generate elemental and molecular ions by laser ablation, providing a signature of the molecular, as well as elemental, constituents. Laser illumination can be constrained to a single particle and, thus, this is a single particle technique, with the ions detected by mass spectrometry. Some discrimination between particle surface and bulk composition can be attained and trace OC compounds can be detected (6).

Raman and infrared (IR) spectroscopy are useful techniques for characterising the molecular composition of an aerosol. Infrared spectroscopy can be used to probe the chemical composition and phase of an aerosol population by passing an IR beam through a suspended aerosol population. Analysis by infrared spectroscopy can be performed at atmospheric pressure and, thus, the concentration of volatile species may be determined. Raman spectroscopy is a non-destructive technique that can provide information about the chemical bonds and the functional groups present in micrometre-sized aerosols.

Information on morphology and elemental composition of individual particles can be obtained by transmission electron microscopy (TEM) or scanning electron microscopy (SEM) coupled with energy-dispersive spectrometry (EDS). Environmental scanning electron microscopy (ESEM) can be carried out at pressures of $0.1–1 \times 10^3$ Pa (16), much higher than the 10^{-4} Pa required for conventional SEM, thereby reducing some of the problems associated with losing volatile components. Additional advantages of ESEM are that very little sample preparation is required, images can be obtained in environments that closely mimic the particles' natural states, and dynamic experiments such as hydration/dehydration cycles can be performed *in situ* (17). For any microscopy technique, collecting data from a statistically significant number of particles can be time intensive.

11.4.2 Real-time Analysis

Despite the advantages of on-line analysis in real-time, such as avoiding sampling artefacts and the need for expensive laboratory procedures, there remain few techniques for determining aerosol composition directly *in situ* with high time resolution. For example, EC in excess of $0.5\,\mu g\,m^{-3}$ can be quantified with a photoacoustic soot sensor. Carbon black particles absorb energy from a laser beam and transfer the heat to the surrounding gas phase. Modulation of the laser beam leads to modulation in the heat transfer which can be detected with a microphone. Polycyclic aromatic hydrocarbons (PAHs) can be quantified using a laser-induced fluorescence analyser. A sample is irradiated with a N_2 laser and PAH concentrations can be determined by fluorescence with a sensitivity of typically $<1\,\mu g\,m^{-3}$.

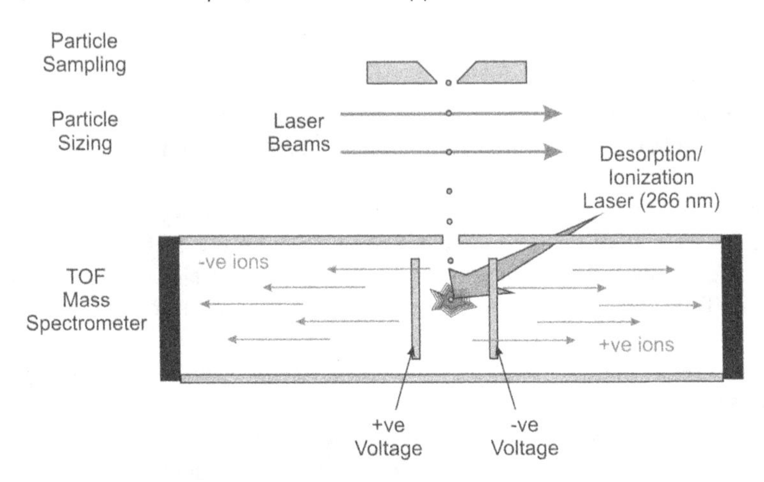

Figure 11.8 *A schematic of a single particle time of flight mass spectrometer showing the sampling, sizing and mass spectrometer stages*

One of the disadvantages of analysing ensembles of particles is that it is impossible to simultaneously determine single particle size and composition. To probe the mixing state of different aerosol components, size-resolved measurements of composition are essential. Recent advances in the versatility of mass spectrometry to detect both inorganic and organic species have been exploited in the development of single particle instruments that now enable simultaneous analysis of particle size, elemental composition and speciation in real-time (Figure 11.8) (18–20). Particle size is determined by an optical technique, either by integrated scattering intensity or by the time required to travel between two probe lasers. Laser desorption ionisation using a high-energy pulsed laser is then used to desorb and ionise components of the aerosol. Alternatively, compounds may be desorbed by thermal methods followed by either electron or chemical ionisation. Chemical ionisation is often used as it is a softer ionisation technique that can allow for determination of the parent molecular weight. Detection is usually by time-of-flight (TOF) mass spectrometry. Dual polarity TOF spectrometers provide simultaneous information on the positive and negative ions formed during the ionisation step, yielding important information on chemical speciation and composition. The identification of key chemical tracers is particularly useful for source apportionment. Most real-time mass spectrometers are able to detect particles between 0.2 and 5.0 μm in size (18), although particles as small as 50 nm may now be analysed (21, 22). Information can be gained on the presence of elemental carbon, organic carbon, sulfates, nitrates, sea salt, dust and metals.

Although qualitative analysis by mass spectrometry is now routine, quantification can be difficult. This is particularly true for organic compounds for which extensive fragmentation can occur at the laser desorption/ionisation stage. Incomplete vaporisation of the aerosol particle can lead to a greater sensitivity for species on the surface of the particle than in the particle bulk. As an alternative, desorption and ionisation can be achieved in two steps with separate lasers performing each step. Lower irradiances are required and this leads to less fragmentation of the desorbed compounds.

Fluorescence detection is often used to discriminate between bacterial and other aerosols. Irradiation of single particles with a UV laser can allow real-time *in situ* detection of fluorescence signatures from flavins, and amino acids such as tryptophan and tyrosine, used as markers of bacterial material (23). Elastic light scattering can be used in parallel to determine the particle shape, refractive index and size, allowing detailed classification of biological aerosols.

A number of approaches have been developed for sampling single particles non-intrusively. Single particle measurements are particularly powerful for performing controlled laboratory investigations of aerosol properties. The sampling of single solid or liquid particles can be achieved using electrostatic traps, acoustic traps or with a focussed laser beam. Electrostatic levitation was first employed by Millikan in the early 20th century to measure the charge of an individual electron and has become a powerful tool for trapping and characterising single aerosol particles. Contemporary instruments not only balance the gravitational force by varying the potential difference between top and bottom electrodes, but also confine the particle in the horizontal plane with AC frequencies (Figure 11.9) (24). Measurements can be routinely made of electric charge, mass and size.

Optical levitation can be achieved with a single focussed laser beam (25). The optical radiation pressure force acts on the particle in the direction of propagation of the laser beam, balancing the gravitational force. To achieve a stable trap, the laser power is controlled by an active feedback mechanism to balance any changes in particle size and the trap well-depth is equal to the weight of the particle. More recently, single transparent particles have been trapped in a three-dimensional single-beam gradient force optical trap, commonly referred to as optical tweezers (see Chapter 14). A tightly focussed laser beam, formed by focussing a laser beam through a high numerical aperture microscope objective, is characterised by a strong gradient in electric field near the focal waist, which acts to pull a particle towards the region of highest light intensity. Under these conditions, the gradient force can dominate the scattering force by many orders of magnitude and a stable gradient trap is formed, strongly confining the particle in three dimensions. Such a trap can be used to study and manipulate

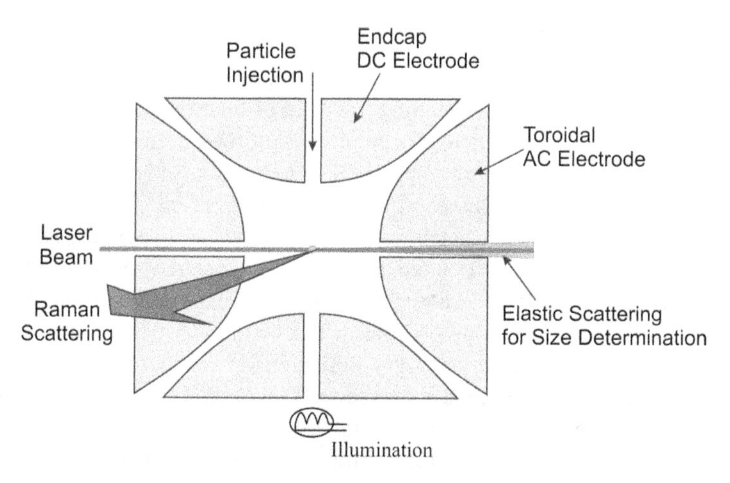

Figure 11.9 *A schematic of an electrodynamic balance. A single charged particle is trapped allowing single particle spectroscopy and dynamics to be studied*

multiple particles held in parallel optical traps, allowing comparative measurements of aerosol particle properties and dynamics, and studies of particle coagulation.

For particles trapped electrostatically, optically or acoustically, a wide range of non-intrusive spectroscopic techniques can be used to probe the particle *in situ*. Raman, fluorescence and infrared spectroscopy are perhaps the most generally applied, providing unambiguous optical signatures of droplet composition. Typical sensitivities for trace analysis *in situ* are 10^{-3} M for Raman scattering. Stimulated Raman scattering from a droplet, observed at wavelengths commensurate with morphology dependent resonances (also referred to as whispering gallery modes), can provide a method for monitoring the size of a spherical liquid droplet with nanometre accuracy (25). The combination of this wide array of trapping and spectroscopic techniques can enable single particle dynamics to be interrogated over extended periods of time, with spatial and temporal resolution of composition and accurate determination of particle size and temperature.

11.5 The Equilibrium State of Aerosols

We now turn our attention to consider the thermodynamic and kinetic factors that determine the size and composition of aerosol particles, considering first the equilibrium state of the aerosol. In particular, we consider the partitioning of water between the gas and condensed phase as this provides an excellent and relevant model system for interpreting the phase behaviour of aerosol. Water is an important component of the atmospheric aerosol and changes in relative humidity (RH) can lead to changes in particle size and composition, which influence the radiative impact of aerosols and clouds on climate, the role of aerosols in mediating heterogeneous chemistry, and the impact of airborne particles on human health. This section describes the thermodynamic principles behind the interaction of particles with water vapour by first examining the phase behaviour of aerosol particles and the dependence of wet particle size on RH through Köhler theory, which determines hygroscopic growth.

11.5.1 Deliquescence and Efflorescence

Knowledge of the phase of an aerosol is of crucial importance for determining the optical properties of the aerosol and for predicting the rates of heterogeneous reactions (26, 27). Although both surface and bulk chemistry are important for a liquid aerosol due to the free diffusion of reactants throughout the droplet bulk, mechanisms of heterogeneous chemistry are largely limited to surface processes when solid aerosol particles are involved. The state of a soluble solid inorganic salt is dependent on the RH of the surrounding environment, as shown in Figure 11.10. The RH is defined as the ratio of the partial pressure of water to the saturation water vapour pressure at a given temperature, expressed as a percentage.

At low RHs, the salt exists as solid particles with a certain dry particle diameter, D_{dry}. As the RH increases, the particle remains dry until the RH reaches the deliquescence RH (DRH), at which point the dry particle spontaneously takes up water and grows to form a salt solution droplet. At the DRH, a solute-saturated solution droplet is formed with a salt concentration determined by the solubility limit of the solute in water. Thus, the DRH is specific to each inorganic salt or organic compound and the DRH of a mixed particle tends to be lower than the DRH of the individual components. For solid multicomponent aerosols,

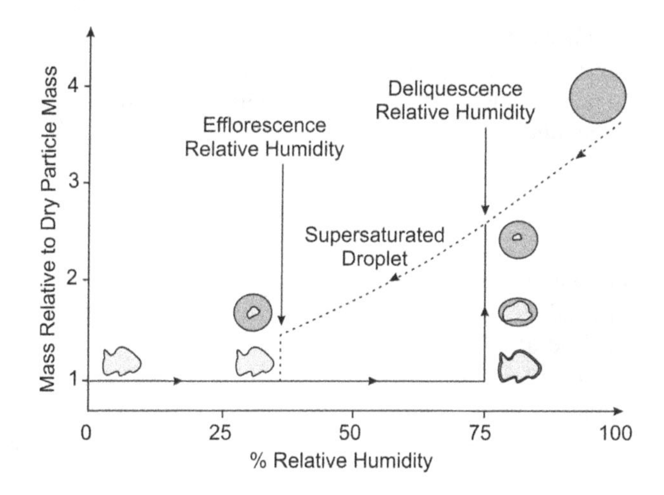

Figure 11.10 *The variation in relative particle mass of a soluble inorganic salt particle (sodium chloride) with relative humidity illustrating the RHs at which the particle changes phase Reprinted with permission from Chem. Soc. Rev., Heterogeneous atmospheric aerosol chemistry: laboratory studies of chemistry on water droplets by Jonathan P. Reid and Robert M. Sayer, 32, 2, 70-79 Copyright (2003) Royal Society of Chemistry*

the solubility of each component must be known to predict the partitioning of the chemical species between the solid and aqueous phases (28).

The solution droplet continues to grow with further increase in RH, with an equilibrium size determined by the dry particle diameter and the RH. The variation in equilibrium wet particle diameter with RH, $D_{wet}(RH)$, is described by Köhler theory and is often reported as a RH-dependent growth factor, i.e. the ratio $D_{wet}(RH)/D_{dry}$. The solution droplet loses water by evaporation with decrease in RH, remaining as a supersaturated solution at RHs below the deliquescence point. Crystallisation is dependent on the slow kinetics for forming salt nuclei leading to a phase change, which cannot be predicted from thermodynamic principles. Thus, the liquid droplet remains in a supersaturated metastable state until a critical supersaturation is achieved at an RH much lower than the DRH, at which point efflorescence occurs. The efflorescence of multicomponent droplets is considerably more difficult to predict than deliquescence behaviour. For example, the particle may exhibit more than one efflorescence point, one for each component of the particle.

It should be noted that not all aerosol species undergo deliquescence or efflorescence behaviour. Many organic aerosol components are observed to take up water continuously across the entire range of RH from dry to wet conditions, behaviour that is similar to a sulfuric acid aerosol. Some water-insoluble organics may not take up water at all, even at RHs approaching 100%, remaining as solid particles across the whole range of RH.

11.5.2 Köhler Theory

Köhler theory is used to describe how the equilibrium wet particle size varies with RH, recognising that the vapour pressure of the droplet must equal the surrounding RH for the droplet to remain in equilibrium with the vapour phase. There are two competing effects that must be considered when determining the vapour pressure of a solution droplet: the Kelvin effect and the solute effect. The Köhler equation (Equation (11.1)) is used to describe how the Kelvin effect and the solute effect act in combination to influence the equilibrium particle size at a particular RH.

$$\mathrm{RH} = a_{\mathrm{w}} \exp\left(\frac{2V_{\mathrm{w}}\gamma}{RTr}\right) \tag{11.1}$$

a_{w} is the activity of water in an equivalent bulk solution with the same solute molality as the droplet; V_{w} is the partial molar volume of water in the solution; γ is the surface tension of the solution, r is the radius of the liquid droplet, R is the ideal gas constant and T is the temperature. The left-hand side defines the RH of the surrounding environment that must be balanced by the droplet vapour pressure, the right-hand side. The first term on the right-hand side describes the influence of the solute effect on vapour pressure, while the second term describes the impact of surface curvature through the Kelvin effect. The contributions from the two terms are shown in Figure 11.11, along with the combined effect.

The solute effect describes how the presence of a solute can lower the vapour pressure of the droplet and is proportional to the solute concentration and inversely related to particle size. As the RH of the environment decreases, the vapour pressure of the solution droplet must decrease for the droplet to remain in equilibrium. Evaporation of water leads to a reduction in droplet size and an increase in solute mole fraction, reducing the vapour pressure of the aqueous component. At high RH, when the solutes are dilute, this decrease in vapour pressure can be estimated from the Raoult equation for ideal solutions, which relates the vapour pressure of the solution to the vapour pressure of pure water and the mole fraction of water in the solution. However, as the RH decreases and the solute concentration increases, the ideality of the solution can no longer be assumed and the differences between

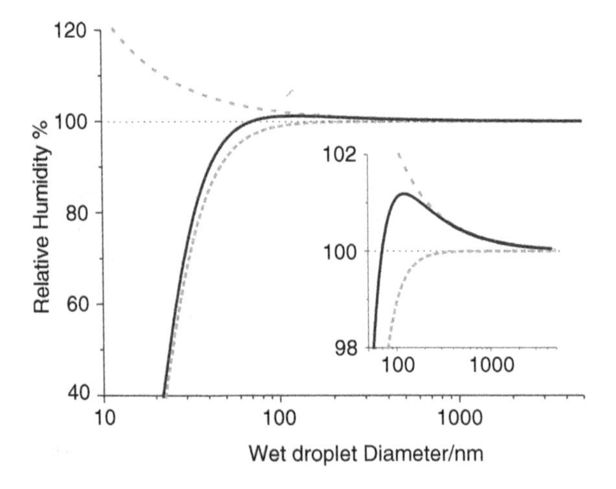

Figure 11.11 *A Köhler plot of the dependence of the equilibrium size of an aqueous sodium chloride droplet on RH (solid black line). The dry particle size generating the liquid droplet is a 10 nm diameter sodium chloride particle. Contributions to the vapour pressure from the Kelvin effect and solute effect are shown separately (upper and lower dashed curves, respectively). An expanded view of the variation near 100% RH is shown in the inset figure*

solute–solute and solute–solvent interactions must be accounted for. The activity coefficient of water must be included in the relation of the solvent activity to the solvent mole fraction, behaviour characteristic of real solutions.

$$a_w = \gamma_w \times x_w \tag{11.2}$$

In the limit of large droplet size ($r \rightarrow \infty$), the equilibrium vapour pressure of the droplet is simply determined by the solute term.

The Kelvin effect describes the influence of surface curvature on the droplet vapour pressure and is increasingly significant with diminishing particle size, becoming relatively insignificant for particles greater than 1 μm in diameter (28). The Kelvin effect leads to an elevation of the vapour pressure of a droplet over that expected for a flat surface and has profound consequences for the hygroscopic behaviour of aerosol particles. When the contributions from the solute and Kelvin effects are both considered, the RH required for an equilibrium size to be established can be in excess of 100%, as evident in Figure 11.11. If an aerosol particle grows from small size with increasing RH, the RH must exceed a critical supersaturation for the particle to become activated and grow larger than the activation diameter. If this critical supersaturation is not achieved, the aerosol particle cannot grow spontaneously to large size and remains unactivated. The critical supersaturation is directly dependent on the aerosol composition through the influence of surface tension and also through the contribution of the solute term. Quantifying the supersaturation required for the activation of cloud condensation nuclei is important for understanding the influence of aerosol on the properties and lifetimes of clouds in the atmosphere.

11.5.3 Measurements of Hygroscopic Growth

Hygroscopic growth studies are used to provide information on the water uptake behaviour and mixing state of wet aerosol particles, in effect leading to direct measurements and parameterisations of the Köhler equation for a specific aerosol component. Particle composition and relative humidity dictate the water content of an aerosol. By recording the dependence of the size of a single particle on RH, or the distribution of sizes in an ensemble measurement, the variation in growth factor with RH can be determined. Inorganic compounds are mostly hygroscopic and, although organics are known in general to influence the water content of particles (29), the hygroscopic properties of organics are much less well characterised (30). Hygroscopic growth measurements are valuable in investigating how particles interact with water vapour under sub- and super-saturation conditions. A microbalance may be used to obtain hygroscopic growth information, based on the sensitivity of the balance to changes in the mass of particles on filters or impactor substrates with varying RH. The long equilibration times required for the deposited microparticles to equilibrate with the surrounding vapour have limited the application of this method.

The standard and more universal method for performing hygroscopic growth measurements on submicrometre particles is to use a humidified tandem DMA (HTDMA) set-up (31). Such an approach is frequently used to study the influence of organic components on the water uptake of inorganic salts and is advantageous because it can be used in both ambient and laboratory studies. Aerosols are generated using an atomiser and dried by passing the flow through a diffusion dryer, a region kept at a RH below the efflorescence point of the particle's components. A DMA is used to select a monodisperse aerosol population from the polydisperse sample. The monodisperse aerosol flows into an equilibration region of variable RH. The size distribution at the outflow from the equilibration region is determined by a second DMA and a CPC. The growth in particle size distribution is assumed to be due to the condensation of water on the particles. The dependence of the particle growth factor on RH can then be determined through measurements over a range of RH. The size changes that are observed are based on measurements of the electrical mobility diameter and, thus, provide information on the volume change associated with water uptake.

Single particle measurements on electrostatically or optically trapped particles have also been used extensively to investigate the hygroscopic properties of mixed component aerosols, particularly in examining aerosols in supersaturated solute states. However, these measurements have been universally made on particles considerably larger than 1 μm in diameter and are, as a consequence, insensitive to the influence of surface curvature on the equilibrium size, providing a detailed analysis of the solute effect alone.

11.5.4 Other Phases

An aerosol may contain immiscible hydrophobic and hydrophilic phases and these may be internally mixed within the same particle or externally mixed as separate aerosol components with distinct size distributions. In polluted urban environments, an organic aerosol component is often observed that is externally mixed from the dominant mass of inorganic-containing aerosol, which is peaked at larger size within the accumulation mode. The externally mixed organic phase arises from combustion sources and has a particle size distribution that peaks at a diameter of ~100 nm. The partitioning of components between internally mixed hydrophobic and hydrophilic phases can be dependent on RH. Many inorganic solutes, such as sodium chloride, 'salt out' any organic components dissolved in an aqueous phase as the RH is lowered, increasing the mass of the organic partitioned to the hydrophobic phase.

Pure and mixed organic particles may exist as supercooled droplets, which is a non-equilibrium state (27). Supercooled droplets remain liquid even though the temperature is below the freezing point of the droplet components. Small particle volumes and lack of contact surfaces allow particles to remain supercooled for wider temperature ranges than comparable bulk studies (32). Such conditions reduce the rate of nucleation preventing crystallisation from occurring, similar to the formation of supersaturated solute droplets below the DRH. Although the existence of supercooled particles in the atmosphere remains to be confirmed, results of recent laboratory studies suggest that supercooled aerosols may change the reactivity and lifetime of aerosol particles (27, 32).

11.6 The Kinetics of Aerosol Transformation

Defining the equilibrium state of an aerosol is central to understanding the properties of aerosol in many complex environments. However, it is also imperative to quantify the kinetics of particle transformations and the rate at which the equilibrium state is attained. In many scenarios, such as in the delivery of active pharmaceutical ingredients or in industrial processes such as spray drying in which the aerosol components are dispersed in a volatile solvent, the initial aerosol system may start far from an equilibrium state. Rapid changes in particle size, temperature and composition may occur once the aerosol is generated and it is essential to understand the intimate coupling between heat and mass transfer during particle evaporation or growth.

11.6.1 Steady and Unsteady Mass and Heat Transfer

It is useful to define a relative physical length scale, known as the Knudsen number (Kn), which is equal to the ratio of the mean free path of diffusing gas-phase molecules to the particle radius. At low Knudsen numbers ($\ll 1$), referred to as the continuum regime, the mean free path is much smaller than the size of the particle and the gas phase behaves as a continuous fluid surrounding a large particle. Under conditions representative of this limit, diffusion in the gas phase is slow compared to processes at the aerosol particle surface or within the particle bulk, and gas phase diffusion limits mass transfer between the phases. If the Knudsen number is large ($\gg 1$), referred to as the free-molecule regime, the particle is much smaller than the mean free path and the gas-phase molecules move discretely around it. Under conditions representative of this limit, molecular processes at the particle surface (accommodation and adsorption) can limit the rate of mass transfer. A transition between these limits is observed at Knudsen numbers of ~ 1, referred to as the transition regime.

An immediate distinction can be drawn between the limiting conditions relevant for mass and heat transfer for small accumulation mode particles ($\ll 1\,\mu m$) and large coarse particles ($\gg 10\,\mu m$). In the former case, the Knudsen number is greater than 1 for particles at pressures of 100 kPa or lower. Thus, for the dominant distribution of particles in the atmosphere, we must understand the dynamics of molecular processes occurring at particle surfaces to quantify the kinetics of particle transformation. By contrast, for cloud droplets and coarse particles in many other disciplines, an understanding of gas-diffusion limited mass transfer is required.

For droplets containing components of low volatility, the mass flux from a particle is slow and evaporation occurs isothermally, with the latent heat lost during evaporation returned to the droplet by conduction from the gas phase. The mass flux is dependent on the gradient in partial pressure of the evaporating component and can be calculated from Equation (11.3), which is attributed to Maxwell.

$$\frac{dm_i}{dt} = -\frac{4\pi r D_{ij} M_i}{R}\left[\frac{p_i^0(T_r)}{T_r} - \frac{p_{i,\infty}}{T_\infty}\right] \quad (11.3)$$

m_i is the mass of component i in the droplet, r is the droplet radius, D_{ij} is diffusion constant of component i in gas j, M_i is the molecular mass of component i, $p_i^0(T_r)$ is the vapour pressure of i at the temperature of the surface, T_r, and $p_{i,\infty}$ is the partial pressure at infinite distance from the droplet surface, where the temperature is specified by T_∞.

From Equation (11.3), the time-dependence of the change in particle radius can be derived and is given by Equation (11.4) when the heat flux is in balance to and from the droplet and the droplet and surroundings remain at a temperature T.

$$r^2 = r_0^2 - S_{ij}(t - t_0) \quad \text{where} \quad S_{ij} = \frac{2D_{ij}M_i p_i^0(T)}{\rho_i RT} \quad (11.4)$$

ρ_i is the density of the droplet and r_0 is the initial droplet radius. Evaporation occurs at a steady rate, with a time dependence that is characterised by a steady decline in the square of the radius. Indeed, single-particle measurements of evolving particle mass or size have been used to determine the vapour pressures of semi-volatile and low-volatility compounds that cannot be routinely probed by standard techniques.

If the loss of latent heat from an evaporating droplet is fast, either because the vapour pressure is high or the diffusion constant is large (e.g. at low pressure), the droplet surface and bulk decrease in temperature. This is accompanied by a decrease in the vapour pressures of the evaporating components, leading to a lowering of the mass flux from the droplet. This is referred to as the period of unsteady evaporation. Eventually the droplet attains the constant wet-bulb temperature, at which the rate of latent heat loss from the droplet is balanced by the rate of heat transfer to the droplet from the surrounding gas. The mass-flux from the droplet then remains constant and the droplet remains at a steady temperature that is depressed below that of the surrounding environment. Such a process requires that the coupling of the mass and heat transfer be considered when examining the dynamics of evaporation or growth for aerosols initially far removed from an equilibrium state.

11.6.2 Uptake of Trace Species and Heterogeneous Chemistry

On many occasions, a particle may be at equilibrium with the surrounding vapour, with the vapour pressures of volatile components equal to their surrounding partial pressures. However, it is important to be able to quantify the rate of equilibration when an aerosol is exposed to a trace concentration of a gas-phase component. Uptake of the trace component may lead to heterogeneous chemistry and the chemical ageing of the aerosol phase. Heterogeneous chemistry often competes effectively with chemistry that occurs in the gas phase as a result of the lower activation barriers that reactants can encounter in the condensed phase due to the solvation of ionic intermediates, reactants and products (33). For example, the oxidation of sulfur dioxide to sulfuric acid occurs rapidly in humid air containing aerosol particles leading to the formation of acid rain.

$$\mathrm{SO_2(ads) + H_2O \rightleftharpoons H^+(aq) + HSO_3^-(aq)}$$

$$\mathrm{HSO_3^-(aq) \rightleftharpoons H^+(aq) + SO_3^{2-}(aq)}$$

$$\mathrm{HSO_3^-(aq) + OH^- + O_3 \rightarrow H_2O + SO_4^{2-}(aq) + O_2}$$

At the microscopic level, reactant molecules must first diffuse through the gas phase (shown schematically in Figure 11.12). On collision with the surface of the particle, the reactant gas molecule must accommodate on the surface, a process that is quantified by the mass accommodation coefficient, α, which is defined as the ratio of the number of molecules adsorbed by the particle to the number of molecular collisions with the surface. Reaction can occur either on the surface of the particle or in the bulk of a liquid aerosol.

Equation (11.5), which follows directly from Equation (11.3), can be used to describe the mass flux per unit area of a trace species accommodating on an aerosol particle that initially does not contain the trace species,

$$J_c = 4\pi r D_g (c_\infty - c_s) \quad \text{or} \quad \frac{J_c}{4\pi r^2} = c_\infty \frac{D_g}{r} \tag{11.5}$$

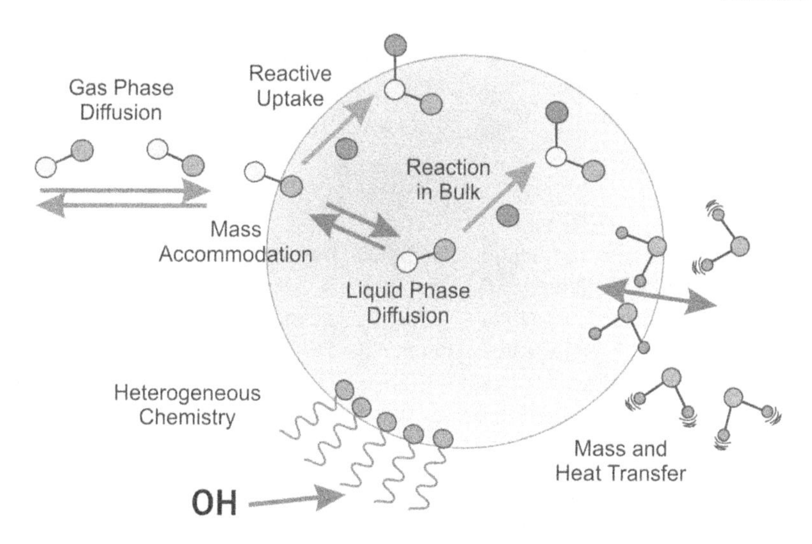

Figure 11.12 *Schematic of the fundamental processes involved in the partitioning of a gas phase species to the aerosol condensed phase*

where the partial pressures have been replaced by number concentrations for the trace species at the particle surface (c_s) and at an infinite distance from the particle (c_∞) and the diffusion constant of the trace species in the gas phase is denoted by D_g. From the relationship of the diffusion constant to the mean free path and the definition of the Knudsen number, it can be shown that:

$$\frac{J_c}{4\pi r^2} = c_\infty \bar{c} \frac{\Gamma_{diff}}{4} \quad \text{where} \quad \Gamma_{diff} \sim Kn \frac{4}{3} \tag{11.6}$$

and Γ_{diff} is commonly referred to as the diffusional correction, accounting for the slowing of the mass flux or uptake due to gas diffusion, and $\bar{c}$ is the root mean square speed of molecules in the gas phase. Although Equation (11.6) is appropriate for considering the diffusion limited mass flux of a trace species, the mass flux may be limited by mass accommodation or by Henry's law saturation (Γ_{sol}) when the solute reaches the saturation concentration in the condensed aerosol phase. This can be reflected by adapting Equation (11.6) to incorporate an uptake coefficient, γ_{mea}, in place of the diffusional correction, thereby allowing the estimation of the mass flux of the trace adsorber while recognising that gas diffusion may not be the rate limiting process.

$$\frac{J_c}{4\pi r^2} = c_\infty \bar{c} \frac{\gamma_{mea}}{4} \quad \text{where} \quad \frac{1}{\gamma_{mea}} = \frac{1}{\Gamma_{diff}} + \frac{1}{\alpha} + \frac{1}{\Gamma_{sol}} \tag{11.7}$$

This formalism is commonly referred to as the resistance model and has been widely used to interpret and predict the mass flux of trace constituents from the gas to the particle phase. Gas diffusion, mass accommodation and solubility (Γ_{diff}, α, Γ_{sol}, respectively) can each be considered to provide a certain conductance (or resistance, i.e. $1/\Gamma_{diff}$, etc.). When any one of the three processes for non-reactive uptake is the rate-limiting step in mass transfer, the conductance value is small thereby contributing a large resistance in the determination of

γ_{mea}. Thus, if the gas diffusion rate is fast and the solubility of the trace component in the condensed phase is large, mass accommodation may limit the mass transfer rate and Equation (11.7) can be written as:

$$\frac{J_c}{4\pi r^2} = c_\infty \bar{c} \frac{\alpha}{4} \tag{11.8}$$

Although the mass accommodation coefficient for water accommodating on a water surface is considered to be large (>0.1), coefficients for many trace molecules have been measured to be considerably less than 1. Under these conditions, the mass accommodation of the trace species may limit the rates of chemical reactions in the condensed aerosol phase and the rate of aerosol ageing. For example, the mass accommodation coefficients of ozone and the hydroxyl radical on water surfaces have been determined to have a lower limit of 0.002 and 0.004, respectively.

11.7 Concluding Remarks

In this chapter we have illustrated the full breadth of information that can be gained in the analysis of aerosols – size, morphology, concentration and composition. This array of information allows one to better understand the thermodynamics and kinetics of processes involving aerosols in a myriad of disciplines. In recent decades, the possibility of analysing single particles has become a reality and has enabled the microscopic factors that control the composition and reactivity of the aerosol to be studied with unprecedented detail. There still remain many challenges in aerosol science and we have highlighted some of them in Section 11.1, particularly in regard to atmospheric science. The chemical and physical properties of multicomponent and multiphase inorganic/organic/aqueous aerosol remain poorly characterised and understood. The development of new analytical techniques for field measurements and laboratory studies are set to play a crucial role in unravelling the chemistry that occurs in such a complex environment.

References

(1) IPCC (2007) Climate change 2007: The physical science basis – contribution of working group to the fourth assessment report of the intergovernmental panel on climate change. Cambridge University Press, Cambridge.
(2) Anderson, H. R. (2009) Atmos. Environ., 43: 142–152.
(3) Pope, C. A. (2000) Environ. Health Perspect., 108: 713–723.
(4) Hinds, W. C. (1982) Aerosol Technology: Properties, Behaviour, and Measurements of Airborne Particles. John Wiley & Sons, Ltd., New York.
(5) Shaw, D. T. (ed.) (1978) Recent Developments in Aerosol Science. John Wiley & Sons, Ltd., New York.
(6) Spurny, K. R. (ed.) (1999) Analytical Chemistry of Aerosols. Lewis Publishers, Washington, D.C.
(7) Finlayson-Pitts, B. J., Pitts, Jr. J. N. (2000) Chemistry of the Upper and Lower Atmosphere. Academic Press, Toronto.
(8) McMurry, P. H. (2000) Atmos. Environ., 34: 1959–1999.
(9) Liu, P., Ziemann, P. J., Kittelson, D. B., *et al.* (1995) Aerosol Sci. Technol., 22: 293–313.
(10) Chow, J. C. (1995) J. Air Waste Manage. Assoc., 45: 320–382.

(11) Seinfeld, J. H., Pandis, S. N. (2006) Atmospheric Chemistry and Physics: From Air Pollution to Climate Change. John Wiley & Sons, Ltd., New York.
(12) Bohren, C. F., Huffman, D. R. (1983) Absorption and Scattering of Light by Small Particles. John Wiley & Sons, Ltd., New York.
(13) Solomon, P. A., Norris, G., Landis, M., *et al.* (2005) Chemical analysis methods for atmospheric aerosol components, in Aerosol Measurements: Principles, Techniques and Applications. Baron P. A., Willeke K. (eds.) John Wiley & Sons, Ltd., Hoboken, New Jersey.
(14) Skoog, D. A., Holler, F. J., Nieman, T. A. (1998) Principles of Instrumental Analysis. 5th edn. Brooks/Cole (eds.). Thomson Learning, London.
(15) Martin, M. Z., Cheng, M. D., Martin, R. C. (1999) Aerosol Sci. Technol., 31: 409–421.
(16) Stokes, D. J. (2003) Philos. Trans. R. Soc. Lond. Ser. A – Math. Phys. Eng. Sci., 361: 2771–2787.
(17) Livio Muscariello, F. R., Marino, G., Giordano, A., Barbarisi, M., Cafiero, G., Barbarisi, A. (2005) J. Cell. Physiol., 205: 328–334.
(18) Sullivan, R. C., Prather, K. A. (2005) Anal. Chem., 77: 3861–3885.
(19) Sipin, M. F., Guazzotti, S. A., Prather, K. A. (2003) Anal. Chem., 75: 2929–2940.
(20) Noble, C. A., Nordmeyer, T., Salt, K., *et al.* (1994) Trends Anal. Chem., 13: 218–222.
(21) Lake, D. A., Tolocka, M. P., Johnston, M. V., *et al.* (2003) Environ. Sci. Technol., 37: 3268–3274.
(22) Su, Y. X., Sipin, M. F., Furutani, H., *et al.* (2004) Anal. Chem., 76: 712–719.
(23) Kaye, P. H., Stanley, W. R., Hirst, E., *et al.* (2005) Opt. Express, 13: 3583–3593.
(24) Davis, E. J. (1997) Aerosol Sci. Tech., 26: 212–254.
(25) Mitchem, L., Reid, J. P. (2008) Chem. Soc. Rev., 37: 756–769.
(26) Moise, T., Rudich, Y. (2002) J. Phys. Chem. A, 106: 6469–6476.
(27) Hearn, J. D., Smith, G. D. (2005) Phys. Chem. Chem. Phys., 7: 2549–2551.
(28) Martin, S. T. (2000) Chem. Rev., 100: 3403–3453.
(29) Saxena, P., Hildemann, L. M., McMurry, P. H., *et al.* (1995) J. Geophys. Res. -Atmos., 100: 18755–18770.
(30) Swietlicki, E., Hansson, H. C., Hameri, K., *et al.* (2008) Tellus Ser. B-Chem. Phys. Meteorol., 60: 432–469.
(31) Rader, D. J., McMurry, P. H. (1986) J. Aerosol. Sci., 17: 771–787.
(32) Hearn, J. D., Smith, G. A. (2007) J. Phys. Chem. A, 111: 11059–11065.
(33) Ravishankara, A. R. (1997) Science, 276: 1058–1065.

12

Practical Rheology

Roy Hughes

Bristol Colloid Centre, University of Bristol, UK

12.1 Introduction

In this chapter we are going to adopt a hands-on approach to the study of rheology. This subject area is concerned with the science of the deformation and flow of matter. We will introduce the concepts with the minimum of mathematics and consider how to make measurements that are free from artefacts and relevant to many applications. We will investigate the relationship between the rheology of a material and its storage stability. Finally, we will consider ways in which we may impart the flow behaviour we desire by the addition of colloid and polymer species to our formulation.

12.2 Making Measurements

It is easy to become daunted by the range of measurements that can be performed by modern instruments (1–3). The unfamiliar terms and units that are used to enumerate the results can be intimidating to the first-time user. The best way to overcome this difficulty is to grapple with some basic definitions.

In rest of this chapter we are only going to consider measurements in shearing flows. This type of flow is that which is most commonly encountered in instrument laboratories. You should be aware that the processing of many materials results in complex flow patterns. These may involve stretching flows or forces normal to the flow direction and even

Colloid Science: Principles, methods and applications, Second Edition Edited by Terence Cosgrove

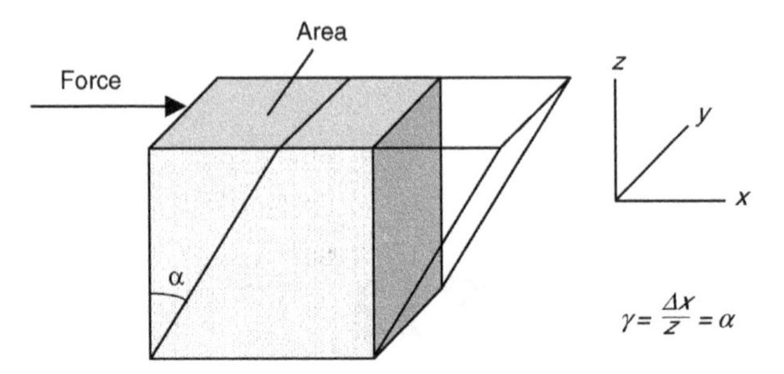

Figure 12.1 *The shear stress and shear strain on a cube*

turbulence. You should always bear this in mind when comparing instrumental data with an application for your material or formulation.

12.2.1 Definitions

Imagine you have a small deformable cube of material. Suppose we apply a force to the upper plane of the cube whilst holding it steady. If the force is only gentle the cube will deform only slightly. The vertical edge will move through an angle α. This is shown in Figure 12.1. There is a net stretch on the cube due to the applied force. For a homogeneous material the relative deformation is affine, which means it is essentially continuous throughout the body of the cube. The relationship between the force and the deformation is due to the physical properties of the cube and it is this that we are interested in. In principle we could measure the stretch of the cube and record this value for an applied force. This would be different for different sizes of cube and is not a very convenient measure. The *relative deformation* of the cube is termed the (shear) *strain*, γ, and removes the dependence on the size of the body. It is the increase in the dimension in the x direction of the cube, Δx, relative to the height of the cube z. This equates for small deformations to the angular displacement. It is a dimensionless quantity. The force required to cause the deformation also depends on the area of the cube on which it acts. If we double the area we have to double the force to cause the same deformation. So we use (shear) *stress* as a convenient tool. This is the force divided by the area of the face of the cube. It has the units Pa or N m^{-2} and is denoted by the symbol σ. The stress divided by the strain gives us a fundamental property of our material, G, the *shear modulus*, as shown in Equation (12.1):

$$G = \frac{\sigma}{\gamma} \tag{12.1}$$

For a material which is described as *linear* the shear modulus maintains a constant value regardless of the stress or strain applied. Such a material is called a Hookean solid.

Whilst a simple elastic modulus might be an appropriate measure for rigid materials or perhaps some gels, many colloid, polymer and surfactant systems appear essentially fluid. It is difficult to practically arrange a simple cube to define our terms for these materials! We need to visualise a slightly more complex scenario. Imagine a fluid contained in a vessel with

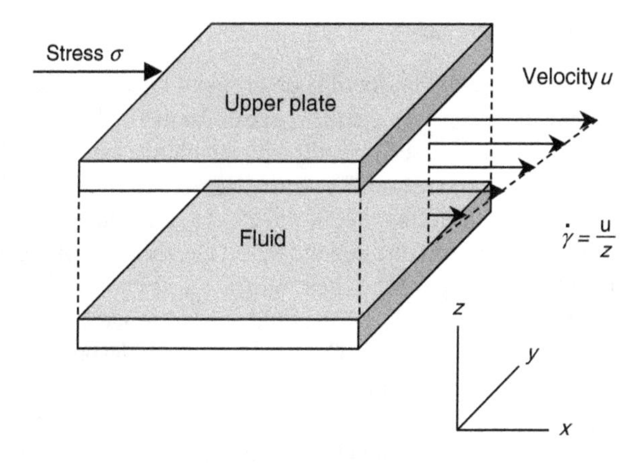

Figure 12.2 *The shear stress and shear rate on two plates*

planar walls parallel to one another. We can consider just one small element of the vessel, as shown in Figure 12.2.

We will now apply a stress to the upper surface whilst maintaining the lower surface static. If the molecules forming the fluid are firmly attracted to the upper and lower plates a velocity gradient will develop with the distance z across the gap. As the material is a liquid the displacement in the x direction is continuous. The strain is given by Equation (12.2) provided the gap is small and the velocity change is linear. The strain increases constantly with time t. At any time since the start of the experiment, the strain is given by the ratio of the distance moved x relative to z.

$$\gamma(t) = \frac{x}{z} \tag{12.2}$$

The velocity u is the rate of change of distance x with time. We can write this as in Equation (12.3).

$$\dot{\gamma} = \frac{\mathrm{d}\gamma(t)}{\mathrm{d}t} = \frac{1}{z}\frac{\mathrm{d}x}{\mathrm{d}t} = \frac{u}{z} \quad \left(= \frac{\mathrm{d}u}{\mathrm{d}z} \text{ for large gaps}\right) \tag{12.3}$$

The term $\dot{\gamma}$ is the shear rate or strain rate and for a constant applied velocity this is invariant with time. Thus the shear rate, which is constant with time, is a much more convenient term to use than strain, which is varying with time. The dot on the strain is called 'Newton's dot' and represents the time derivative of the strain. Shear rate has the units of reciprocal seconds, s^{-1}. In practise you set the shear rate by adjusting the conditions on the instrument you are going to use. The most important relationship to emerge is that between stress and shear rate. The stress divided by the shear rate gives us a fundamental property of our material, η, the *shear viscosity*, as shown in Equation (12.4):

$$\eta = \frac{\sigma}{\dot{\gamma}} \tag{12.4}$$

For a material which is described as *linear* the shear viscosity maintains a constant value regardless of the stress or shear rate applied. Such a material is a Newtonian liquid.

12.2.2 Designing an Experiment

Designing the correct experimental protocol is determined by what it is you would like to know about your sample, the sensitivity and the range of the method you are using. There are numerous measurement methods based on industrial standards. For example, viscous flow can be monitored using a flow cup (ISO or Ford cup). Here a cup with a small hole formed in the base is filled with a liquid. The flow out of the cup through the hole is monitored with time. The stress acting on the fluid is due to gravity and the viscosity of the fluid determines the speed at which the liquid escapes. The shear rate is not single valued, but is determined by the complex flow patterns that develop as the fluid approaches and exits the nozzle. Some examples of cups are shown in Figure 12.3. Cups have found widespread and successful use in the paint industry for assessing coatings, although their use is not restricted to just this application. Drilling fluids are often characterised in terms of a pour point. If you take a viscous liquid in a beaker the rate at which it pours depends upon the gravitational stresses on the fluid and its viscosity. If we now reduce the temperature the viscosity of simple liquids increases. At some point it will no longer 'apparently' pour. We can define a pour point temperature for this material. Under these conditions it will prove hard to pump and filter. As a quality control tool and in difficult environments such methods have an important place.

These approaches become more difficult to use when we want to understand how a particular formulation performs under a range of flow conditions: particularly when we have no previous knowledge of the relationship between measurement and performance. This is partly because they impose complex flow patterns on the material and partly because they may not be operating on timescales appropriate to our needs. Viscosity and elasticity tend to be functions of stress and shear rate once a certain value is exceeded (i.e. non-linear). The aforementioned measures will do only a little to explore these phenomena and so tend to be more limited as tools for developing materials or understanding processes.

The best method to determine a rheological property for such a purpose is to control the force or deformation applied to the sample. Most of the colloidal materials we are interested

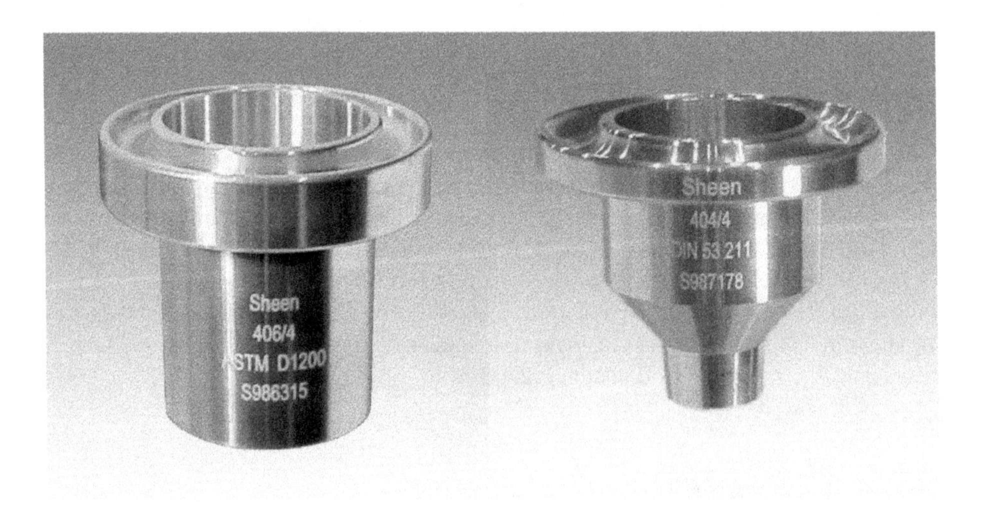

Figure 12.3 *Flow cups (pictures courtesy of Sheen Instruments Ltd).*

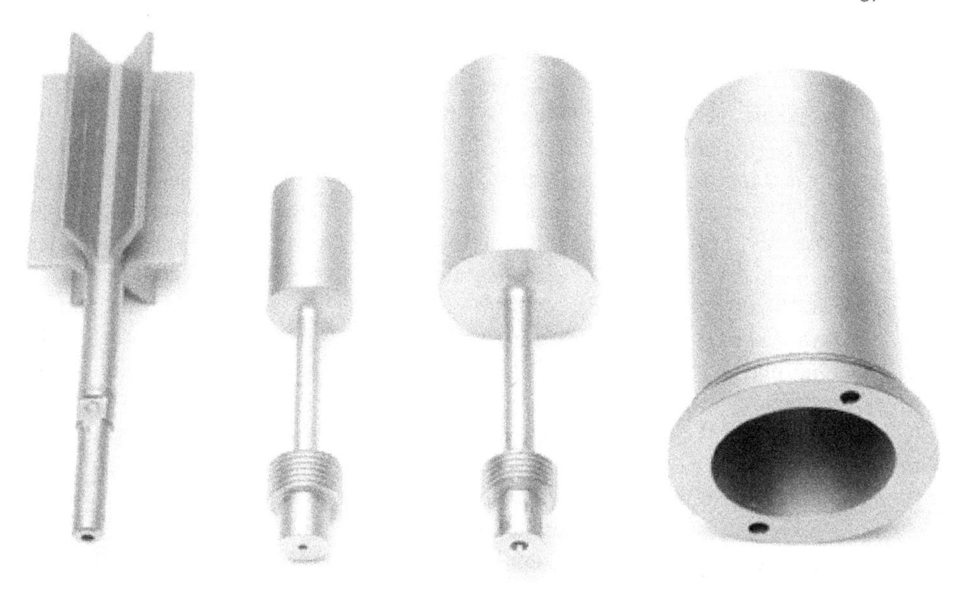

Figure 12.4 *A range of instrument geometries from Bohlin Instruments.*

in have a fluid or near fluid like character. This means that the materials are not self constraining and need to be introduced to a sample cell in order to hold their shape. Such sample cells have organically acquired the name *measuring geometries* or simply *geometries*. Figure 12.4 shows some designs of geometries. There are many variations. Typically they contain sample volumes from about 0.5 mL to 50 mL; perhaps a little more. The choice of geometry can be critical and the wrong selection can result in meaningless data. Most viscometers or rheometers operate by using an electric motor to apply a known torque or known velocity. The stress, strain or shear rate that is applied to the sample depends upon the shape of the geometry attached to the motor and the performance of the motor.

Suppose you are investigating spraying or blade coating and you know that the sample will experience very high deformation rates. In order to explore how a new formulation may respond in the laboratory you would want a high shear rate. Consider Equation (12.3), this shows that for a given velocity u (supplied by the motor) in order to obtain a high shear rate you need a small gap z (determined by the geometry). A typical high shear geometry is the Mooney Ewart which consists of a bob placed concentrically in a cup with a small gap between the two. One might achieve $10\,000\ s^{-1}$ with such a design. As the gap becomes smaller the effect of any large contaminating particles becomes increasingly significant. Clearly particles can bridge the gap between the geometries, an extreme case, which would totally disrupt the measurements. There is a less extreme condition when particles have a diameter of about 1/10th of the gap size. The shear rate and the strain would not be affine and these effects alone begin to change the stress you would achieve in a larger gap at the same rate. It is equally more prone to particle jamming and local flow effects causing unusual and irreversible phenomena such as particle aggregation. This might be desirable if you are trying to mimic a process but it would not provide a well-controlled measure of the rheological properties of a sample.

An overlooked problem with making a good measurement is that of wall slip. In order to get the shear rate we believe our motor and geometry selection has determined we need the assumption that the molecules would be firmly associated with the walls to be true (see Section 12.2.1). If this is not the case, then we will form a slip plane between the geometry surface and the liquid next to it. Thus when you rotate the geometry it will move at a speed much faster than the neighbouring liquid and the shear again will be non-affine. The data will prove difficult to interpret. The extent to which a material might slip as a function of the velocity of the shearing surface is surprisingly poorly understood. Emulsions and foams are particularly prone to this problem, as are high concentration particulate systems or skin-forming materials such as doughs. The wetting characteristics of the material play a role. So if you roughen the shearing surface for example by sand blasting or attaching double sided sticky tape, you can reduce the energy of contact between the fluid and the surface. This increases the 'wettability of the surface' and can eliminate slip planes occurring. The local flow patterns can change, altering the hydrodynamics close to the surface. You can also change the materials of the geometries as a whole, switching between stainless steel and plastics or glass. This is also important for achieving a desired chemical resistance of your geometries. You should not always consider the adsorption of species as a positive benefit to your measurements. An example of this is the addition of high molecular weight polymers in order to reduce the drag between the walls of a pipe and the liquid it contains. This reduces pumping energy and costs in long pipelines. The extent to which this effects polymer rheological measurements particularly in mixed systems in the lab is not well understood and almost always ignored.

A related phenomenon to wall slip is fracture. You see examples of this in very viscous and slightly elastic slurries. Generally this occurs as the shear rate is increased during an experiment. The velocity reduces across the gap between the moving and the stationary surface. At some point after a given time and at some position across the gap a critical strain is achieved and the sample breaks creating a plane of slip between a rapidly moving and a slowly moving surface. The reason for the breakage depends upon the rheological properties of the material and so it is difficult to generalise when you would expect to see this.

Other considerations include the volume of the sample compared with the exposed area. Large areas and small volumes accelerate the effects of evaporation. This will of course lead to concentration changes but could also lead to films forming and bridging the moving and static surfaces, significantly increasing the apparent viscosity of the material.

High concentrations of polymer solutions can produce forces normal to the plane of rotation of the geometries. It is not uncommon to see materials 'climb out' of the geometry at moderate shear rates.

12.2.3 Geometries

Geometries are not always constructed to achieve a near perfect shear rate. For example, the DIN standard cup and bob (Figure 12.5) has a well-defined shear rate that is achieved in the narrow gap between the walls. The conical base of the bob does not result in a constant shear rate across the geometry, thus leading to systematic errors in the data collected. A constant rate can be achieved using a narrow gap and a shallow cone angle. This is the basis of the

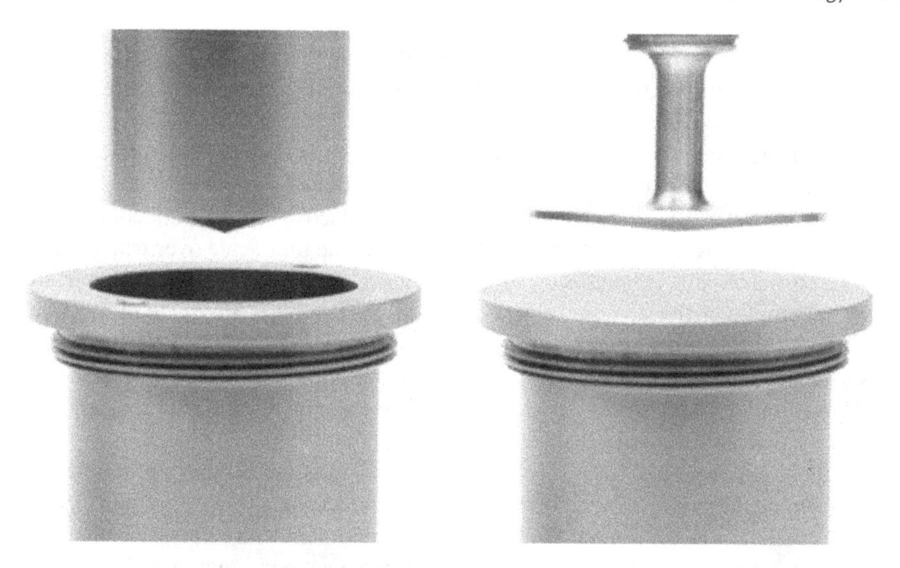

Figure 12.5 *A comparison of a cone and plate, and, a cup and bob geometry*

cone and plate geometry. As the cone rotates the outer edge of the cone moves faster than the centre of the cone. Thus the velocity is not constant across the geometry. However, by angling the cone such that the increase in speed corresponds to an increase in the gap the shear rate can be made constant. The point of the cone must of course be truncated, to prevent a frictional drag between cone and plate. This truncation is usually by a distance of about 50 μm to 200 μm and clearly this is an area where large particles could become trapped.

We need to account for another factor which can impinge on the quality of the rheological measurement. When we shear the fluid in a geometry, lines of flow develop. If we consider a small cubic element there will be a velocity gradient across the element. If we imagine the cube was a rigid block, one side of the block would be travelling faster than the other. This results in a twisting motion or a vorticity. At low shear rates the viscous forces tend to dampen out this tendency but as the shear rate increases there is a tendency towards forming small vortices. These secondary flow patterns occur at a critical rate dependent upon the geometry, the density and the viscosity of the sample. The lower the viscosity, the greater the tendency to form these flows. The shear rate where we see these secondary flows can be estimated using the Taylor number. Once a critical value is exceeded, the shear stress at a particular rate is no longer a measure of viscous processes alone; it includes inertial properties.

When a very high shear rate is applied to the sample the vorticity becomes the dominant feature of the flow. At this point complex turbulent flow patterns develop and mixing occurs. The onset of this is determined by the ratio of the inertial to viscous forces. We term this ratio the Reynolds number and when we achieve high Reynolds numbers the flow is dominated by inertial rather than viscous processes (1, 3).

The lesson to learn here is that even the most wisely selected geometry can be used in regions where it produces data of very little value.

12.2.4 Viscometry

We can measure a viscosity by applying a shear rate to a sample and recording the shear stress we obtain. Suppose for a moment that we have the perfect device so that it was able to apply an instantaneous velocity to the moving part of a geometry. A velocity gradient would develop with time across the sample and a *steady state* stress would be recorded. Simple liquids such as water or oils are Newtonian; that is the viscosity is independent of the shear rate. The molecules respond rapidly to the gradient and the time scale over which the stress reaches a constant value is practically instantaneous on the timescales we are interested in. However, complex materials, the subject of our discussion here, are made of much larger species than single atoms or small molecules. As a result the timescale for the system to reach a steady state is comparatively long. This is an important observation since if you wished to record a steady state viscosity at *any* shear rate you need to wait long enough for the larger species to respond.

The rate at which steady state is achieved depends upon the diffusive properties of the species forming the system. The time required to diffuse a given distance is controlled by the diffusion coefficient of the sample. If we look at an experiment with a timescale much longer than some characteristic diffusion time we certainly expect our structure to have achieved steady state. We usually define a characteristic time, τ, for a movement relative to the dimensions of the species forming the system. So, for example, for a dispersion of monodisperse particles with a radius a, we would write (3, 4),

$$\tau = \frac{6\pi\eta(0)a^3}{k_B T} \tag{12.5}$$

where $k_B T$ is the thermal energy of the particles and $\eta(0)$ is the viscosity at low stresses or rates where the material is relatively unperturbed. As we suggested earlier the molecules forming the dispersion medium respond instantly. We could test Equation (12.5) above and show that atoms are so small, of the order of Angstroms, that the characteristic time would be tiny compared to that of say a 1 micrometre silica particle. We have not made an arbitrary choice in selecting the particle length to characterise our system. Many concentrated particulate systems show a short range order akin to a true molecular liquid. The order typically occurs over a particle diameter or so. Thus when a particle moves a distance of the order of a particle dimension substantial local order is lost. We are characterising the size and life time of the arrangement.

The practical outcome of this for our experimental design is that we must wait long enough for the structure to respond to our applied stress in order to get steady state data. Unfortunately Equation (12.5) is only a guide and also depends on our knowledge of the viscosity at a low stress or shear rate. We can illustrate this using some experimental data shown in Figure 12.6. This shows the stress that is achieved with time when four different fixed shear rates are applied to a dispersion. The first thing to notice is that as the shear rate is increased the shear stress increases. In order to see the data more clearly a logarithmic scale is used. At low shear rates the stress gradually increases to reach a steady state response, seen after about a minute. As the rate is increased, a steady state is achieved more rapidly and at $1000\ s^{-1}$ it has occurred faster than the instrument was able to record readily.

In order to achieve a steady state response it is important to allow enough time for the instrument to measure at each shear rate. It is possible to use feedback mechanisms in

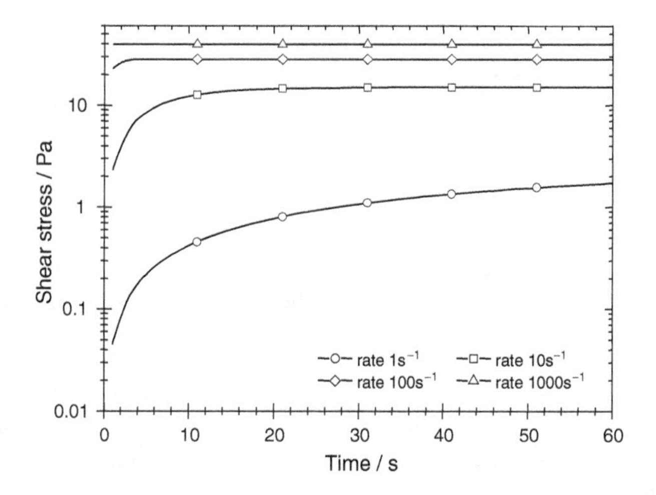

Figure 12.6 *A plot of shear stress versus time*

some instruments to sense when the viscosity reaches a plateau value. This can be convenient but should be used with caution. An example of why you should take care is shown by the response from a heavily entangled polymer solution (5, 6). Figure 12.7 shows the stress response at a low shear rate (less than a few reciprocal seconds) and at high rates (tens of reciprocal seconds). At low shear rates the viscosity, determined by the shear rate divided into the stress at any instant, increases to a plateau value. The polymer chains are taking time to rearrange to reach their steady state configuration. However, when a high rate is applied the chains still attempt to follow the flow but they are unable to change their mutual entanglements fast enough to respond to the higher flow rate. They begin to store energy in highly distorted chains and much more stress is required to achieve a similar rate.

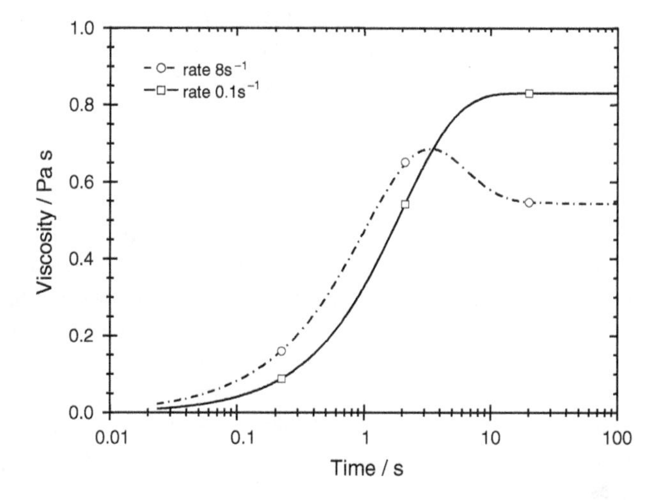

Figure 12.7 *An example of 'stress overshoot' illustrating difficulties in measuring time dependent rheology*

This leads to a peak in the viscosity or a stress overshoot. Eventually enough shear energy is input to the system to enforce a new steady state conformation and the viscosity falls to a plateau value. A material which looses its viscosity with *time* when a single shear rate is applied and which recovers the viscosity once the shearing is stopped is termed *thixotropic* (the reverse being *anti-thixotropic*). You can see from our polymer example (Figure 12.7) that this simple classification is not always adequate. It is important to clearly distinguish the difference between time-dependent responses at any one shear rate separate from the steady state behaviour you observe at different shear rates. It is very common for *shear thinning* and *thixotropy* to be confused. This can lead to very serious formulation design problems.

Although steady state behaviour is often the most desirable form of characterisation, sometimes the time required to achieve this state can be prohibitively long. It is quite feasible for an hour or more to pass before an equilibrium response is achieved at low shear rates. Under such circumstances a *creep* test (Section 12.3.3) is much more appropriate for obtaining information.

It is worth considering that there are some circumstances where you may not wish to achieve a steady state response. For example, you may be imitating a process where your sample is transported from a trough to a blade coater and the fluid is rapidly deformed. A steady state response is probably not a relevant measure here.

One experimental method, which is commonly available on viscometers, is a sweep of shear rates. These experiments are designed to increase the shear rate from say $0\ s^{-1}$ to $500\ s^{-1}$ then back to $0\ s^{-1}$ in a known time regardless of the steady state response. This can result in very complex looking curves, optimistically called thixotropic loops. Although the experiment can be performed in a very controlled manner the data is very difficult to interpret. Not only is it unclear at which rates, if any, a steady state response is achieved but also the material has been exposed to a complicated *shear history*.

The rheological properties depend on the shear history of the sample. This means that the response depends upon all the previous deformations and stresses applied to the sample. This also includes loading the sample into the geometry. Often the best we can do is to avoid applying high shear on filling the geometry and leaving the sample a reasonable time to recover. However, what constitutes a 'reasonable time' is a matter of opinion and patience!

12.2.5 Shear Thinning and Thickening Behaviour

By using a range of geometries it is possible to cover a very wide range of shear rates and shear stresses. Accordingly, the best way of representing this data is to use logarithmic scales. Typical example plots for two different silica dispersions are shown in Figure 12.8. They typify two different classes of behaviour. For both samples we see that as the shear rate is reduced the stress reduces. One curve goes through a point of inflection and then continues reducing. At both high and low shear rates the data goes to a slope of unity. This means that the shear stress divided by the shear rate gives a constant viscosity both at low and high shear rates. The material will flow at both at low and high shear. This defines a low shear rate $\eta(0)$ and a high shear rate viscosity $\eta(\infty)$. This type of flow curve characterises a *pseudoplastic* material. The other sample, however, does not display a low shear viscosity but a yield stress σ_y. If a stress is applied which is less than this value no flow occurs. This is a *plastic* material.

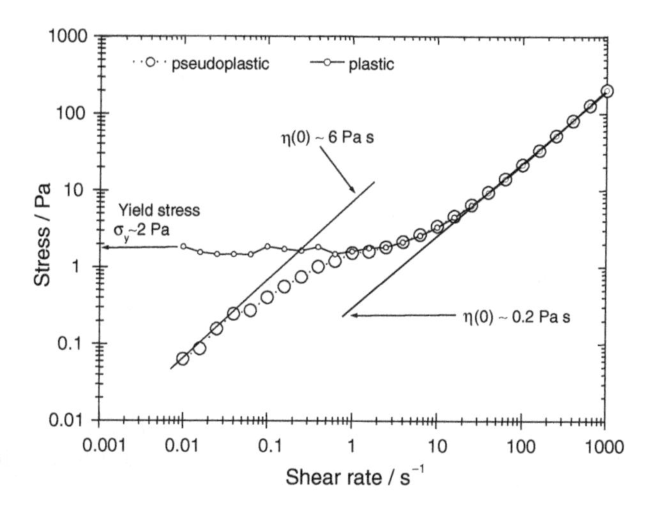

Figure 12.8 *A double logarithmic plot of shear stress against shear rate.*

We can establish the viscosity on a log/log plot by extrapolating to the value on the y-axis when the line fitted to the shear stress shear rate curve intersects with a shear rate of 1 s^{-1}. One feature of these two sets of data is the similarity of the high shear viscosity. If the data is replotted on a linear/linear scale the two samples appear almost identical and it would be hard to distinguish the material with a yield stress. A yield stress essentially implies the material is solid-like at low rates and stresses. The presence of a yield stress is important for storage stability and this is considered further in this chapter.

A fuller characterisation of these curves is usually achieved by fitting the data to a mathematical model. This allows us to compare trends in the data for materials of varying composition using a limited number of constants. Fitting flow curves to a model and extracting the constants as functions of composition also allows systems to be compared more readily than plotting a large number of curves side by side. These models describe the phenomena and are termed, unimaginatively, phenomenological models (3).

12.2.5.1 Plastic Models

$$\text{Bingham}: \quad \sigma = \sigma_{\text{by}} + \eta_{\text{pl}}\dot{\gamma} \tag{12.6}$$

$$\text{Casson}: \quad \sigma = \left(\sqrt{\sigma_{\text{c}}} + \sqrt{\eta_{\text{pl}}\dot{\gamma}}\right)^2 \tag{12.7}$$

$$\text{Herchel-Bulkley}: \quad \sigma = \sigma_{\text{hb}} + (\eta_{\text{pl}}\dot{\gamma})^{-n+1} \tag{12.8}$$

The yield stress is given by σ_{by}, σ_{c}, σ_{hb} and η_{pl} is the plastic (high shear) viscosity.

12.2.5.2 Pseudoplastic Models

$$\text{Cross}: \quad \eta(\dot{\gamma}) = \eta(\infty) + \frac{\eta(0) - \eta(\infty)}{1 + (\beta\dot{\gamma})^n} \tag{12.9}$$

$$\text{Kreiger}: \quad \eta(\sigma) = \eta(\infty) + \frac{\eta(0) - \eta(\infty)}{1 + (\alpha\sigma)^m} \tag{12.10}$$

where α, β, m and n are constants. It is not always possible to measure either the high or low shear rate behaviour. Data can still be fitted using power law behaviour.

$$\text{Ostwald-De Waele}: \quad \sigma = A\dot{\gamma}^{-n+1} \tag{12.11}$$

At very high shear rates or stresses with viscous slurries it is possible to observe a log jamming effect and this can lead to increases in viscosity. Trying to establish the steady state nature of these is extremely difficult.

12.3 Rheometry and Viscoelasticity

In Section 12.2.2 we suggested that most of the materials of interest to us in this discussion had near fluid-like character. This intentionally vague terminology allowed us to dismiss concerns about the physical state of a material. The ‘true state’ of a material, whether it is solid or liquid-like can be of great importance to us. We will now start to explore this in more detail in the next sections and consider its relationship to sedimentation, one form of storage instability.

12.3.1 Viscoelasticity and Deborah Number

Suppose a strain is imposed on a solid sample by the application of a stress. The size of the stress is determined by the shear modulus (Equation 12.1). Now just imagine there is a little fluid character to our solid and the molecules that form the material begin to rearrange and move relative to each other. The energy stored in the structure is dissipated by the flow and the stress begins to reduce. The rate at which the sample relaxes is determined by the balance of elastic and viscous processes. We can define a relaxation time, τ, as the ratio of the viscosity to the elasticity.

$$\tau = \frac{\eta}{G} \tag{12.12}$$

This time is related to that defined in Equation (12.5) which measures a diffusion timescale for particles to begin to lose their local order. If we were to make experimental observations at times much shorter than this we would see a more elastic character in our sample, and at times much longer, we would see a more viscous character. We can define a ratio of the relaxation time to the experimental observation time, t.

$$De = \frac{\tau}{t} \tag{12.13}$$

This ratio is the Deborah number De and this is a measure of the tendency of a material to appear either viscous or elastic. We can classify the material into three classes of behaviour.

$De \gg 1$	$De \sim 1$	$De \ll 1$
solid-like	viscoelastic	liquid-like

So when the observational time of the material approaches the relaxation time it will display both elastic and viscous characteristics and is termed *viscoelastic*.

12.3.2 Oscillation and Linearity

Characterising viscoelastic materials requires very subtle tests (5, 7, 8). It is most convenient to work in a linear regime. This means that the shear modulus and the shear viscosity are independent of the strain or the stress used to measure them. They are only functions of time or frequency. There is a very good reason for doing this, which is that when we apply small deformations or forces we do not disrupt the structure of the sample too greatly. This makes it easier to interpret the data.

Normally the first experiment you would perform is one to test the limits of linearity (3). We would probably need to replace the sample in our geometry after this test to ensure that the structure is not permanently damaged by the experiment. A typical test is a sweep of stresses applied to a sample at a fixed frequency. The material responds by developing an oscillating strain. If the material were purely elastic the strain and stress would be coincident. This means a peak in the oscillating stress is accompanied by a peak in the oscillating strain. The waves are said to be in-phase. Now as we introduce some viscous character to the sample some energy is dissipated and the stress and strain waves begin to mismatch. A phase difference has occurred between the waveforms. We can utilise the phase difference to divide the waveforms into two terms:

- a strain in phase with the stress, an elastic energy stored component
- a strain 90° out of phase with the stress, a viscous energy dissipation component.

The ratio of the peak in the stress to the peak in the strain is the storage modulus G' for the in-phase component and the loss modulus G'' for the out-of-phase component.

The protocol used for the measurements consists of selecting a frequency, say 1 Hz, and applying this to the sample at a given stress. The procedure is then repeated at a higher stress and so on. The storage and loss moduli should be constant (or in a practical sense nearly so) until a critical strain or stress is reached. This sort of behaviour is shown in Figure 12.9 for a concentrated emulsion in the form of a cream. Here you clearly observe the change in the storage modulus as the critical stress is exceeded. Ideally you would like to work at a stress which is in the linear region, so in the example (Figure 12.9) a value below 5 to 10 Pa would be a good choice. The question of how low a stress you should use is a subtle balance. If you pick a value too close to the end of the linear region this can lead to problems since the application of a stress in an apparently 'linear' region over a very long period of time can occasionally result in slow structural breakdown. However, if you select a stress which is too small the strain can reduce to a point where the displacement is small and the signal contains a high level of 'noise'.

Suppose we take the critical stress and divide this by the storage modulus at the point where the breakdown in linearity occurs, we get a strain of 0.05 or 5% (i.e. $G' = 200$ Pa and

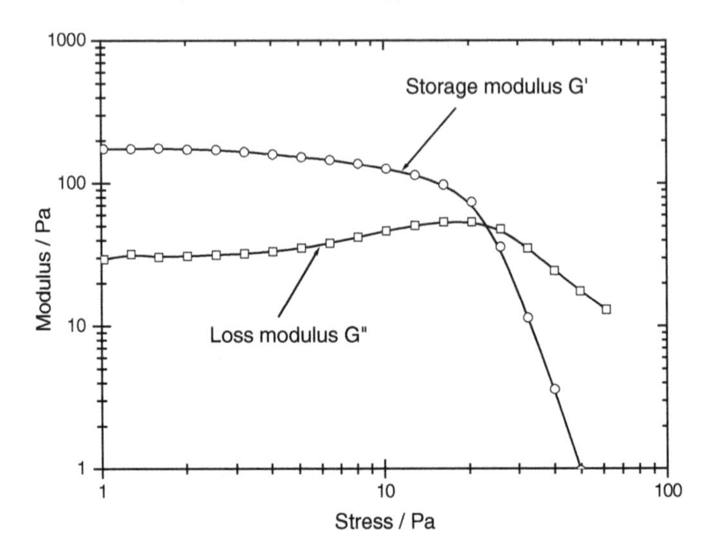

Figure 12.9 *A shear stress sweep for a cream*

$\sigma = 10\,\text{Pa}$). This is a typical value for a dispersion or an emulsion. If the system is very aggregated you might expect a lower strain will lead to a loss in linearity. Many polymers can sustain much higher strains.

We have not considered the choice of the geometry for viscoelastic experiments in any detail and it is important to consider all the aspects mentioned in the previous section. There is an additional concern which is the issue of geometry inertia. Geometries have a finite mass and so they require energy to enforce their motion. You can best aid the efforts of your instrument manufacturer in their designs to allow for the energy dissipated in the motion of the geometry by selecting lightweight geometries. If you want low stresses you need a large sensing/application area and in order to reduce the weight of the geometry you can use less dense but rigid materials such as titanium. This will improve the quality of the low stress and strain data.

12.3.3 Creep Compliance

Sensorial testing tends to show that we have a good intuitive feel for the viscosity of a material. We can compare two materials and providing there is a reasonable difference between their viscosity and they are not too thick or thin we can say which is the most or least viscous by feeling it. We do not have such a good 'feel' for elasticity, the other key property of a viscoelastic material. For example, if one compares a block of metal with a bowl of jelly you will find there is a tendency to say the jelly 'is more springy', 'it is more elastic'. This common usage of the word is entirely at odds with the scientific usage of the word. Whereas a gel may have a shear modulus of 100 Pa say, the metal block has a shear modulus of 100 GPa (1 000 000 000 times larger!). In fact, it is better if we think about the quantity compliance. The shear compliance is denoted by the symbol J. It is the ratio of the shear strain to shear stress, as shown in Equation (12.14) below.

$$J = \frac{\gamma}{\sigma} \tag{12.14}$$

It is reciprocally related to the shear modulus. In a creep compliance experiment we apply a stress to the sample and follow the compliance with time, $J(t)$, after the application of a shear stress. This experiment is a very good method for establishing if a material is essentially viscous or elastic.

12.3.4 Liquid and Solid Behaviour

The application of a step stress to a material results in a displacement of part of the geometry. Initially an instantaneous compliance, $J(\infty)$ is recorded and then the strain grows to show a change in compliance with time. In practice, the instantaneous response can be difficult to observe: in an experiment it depends on how the signal is filtered and how quickly the strain can be applied by the instrument. This test is referred to as a creep test (5) and a typical curve for a fluid is shown in Figure 12.10. The remainder of the creep curve for a viscoelastic material is made of two types of processes. We obtain a curve as the body is deformed because the components which form the material rearrange and take time to do so. The response is not instant but is slowed and is normally termed a retarded response. It has a characteristic retardation time τ_r and an elastic component or compliance J_g. At longer times we generally see one of two processes occurring.

For a material which is pseudoplastic and has a low shear viscosity the strain and the compliance will keep increasing all the time the strain is applied. This is shown in Figure 12.10. At long times the strain increases in direct proportion to the time. The material has attained a constant shear rate across it; the strain divided by the time is the shear rate.

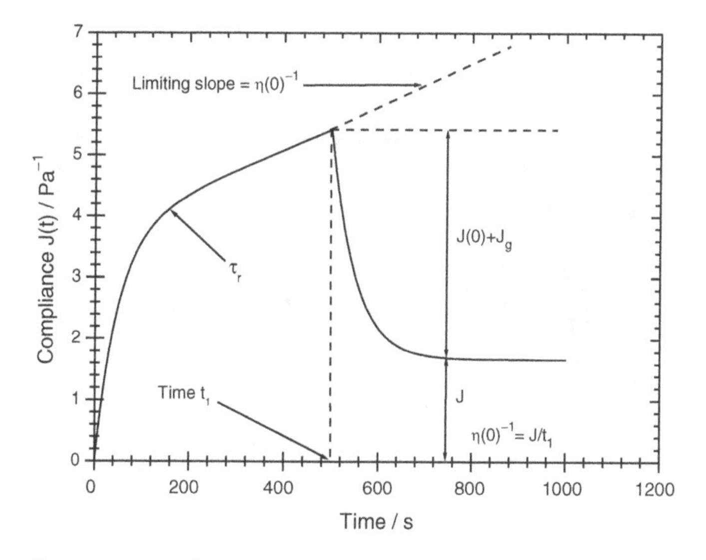

Figure 12.10 *The creep compliance curve of a viscoelastic fluid*

$$\dot{\gamma}_{t\to\infty} = \frac{\gamma}{t} \tag{12.15}$$

The viscosity is the shear stress divided by the shear rate. We may express this in terms of the strain,

$$\eta(0) = \frac{\sigma}{\dot{\gamma}_{t\to\infty}} = \frac{\sigma t_{t\to\infty}}{\gamma} \quad \text{or} \quad \eta(0) = \frac{t}{J(t)_{t\to\infty}} \tag{12.16}$$

As shown above, we can replace the strain divided by the stress with the compliance. This equation shows us that the gradient of the compliance versus time curve gives us the reciprocal of the viscosity. The point being made here is that if you run a creep experiment for an appropriate time and determine the slope you can determine the viscosity.

Suppose we now remove the stress after a time t_1. If all the rearrangements of the species forming the material have occurred by the time the stress is removed the geometry will stop moving and the strain will be constant. If the time t_1 is too short compared to the retardation time τ_r not all the rearrangements of species will have occurred and the material will still possess some elastic properties. When the stress is removed the material will recover some of its original shape. A portion of the original displacement will be returned and the compliance reduced. This portion of the curve is known as creep recovery.

If the material is a viscoelastic liquid you can *never recover all* the strain, some will be lost through the process of flow. If it is a viscoelastic solid it will not possess a zero shear viscosity but will have a yield point. This material will recover all the strain applied to the sample provided the yield stress is not exceeded. Such a response is shown in Figure 12.11.

For very viscous materials it can be difficult to distinguish between the two states. Mathematical analysis of the curve can help the distinction to be made. The plots also show several ways of obtaining viscous and elastic responses. Creep tests do have some direct analogues in terms of applications. For example, the levelling of a film under the action of a gravitational stress has some similarities to the response of a material under a step shearing

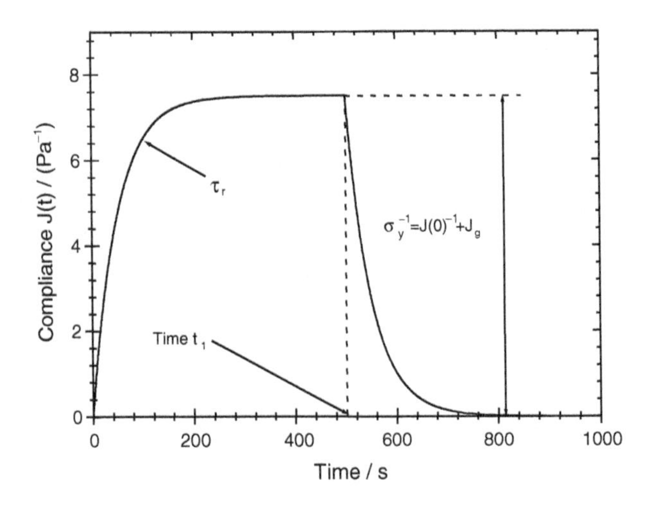

Figure 12.11 *The creep compliance curve of a viscoelastic solid*

stress. Anomalous behaviour in one is reflected in the other, so it can prove a good tool. This rule tends to work less well with high loadings of polymeric materials. Another use of this technique is suggested in the following section.

12.3.5 Sedimentation and Storage Stability

Rheology is a very important discipline when it comes to distinguishing between the states of matter. As we have seen in the previously described creep test we can differentiate between viscoelastic materials that have viscous or elastic properties at low shear stresses. This is a very useful indicator of the ability of a material to support solids (3). Imagine we have a dispersion of particles. Let us suppose that the particles are denser than the medium. The gravitational force g will cause particles of radius a, to sediment. An isolated particle will sediment with a velocity v proportional to its frictional drag in the medium and the density difference $\Delta\rho$ between the particle and the medium.

$$v = \frac{2}{9}\frac{\Delta\rho g a^2}{\eta_0} \quad (12.17)$$

The viscosity in this expression is the solvent viscosity η_0. This equation applies equally to creaming where the particles are less dense than the medium and migrate to the surface of the fluid.

If, in a sedimenting system, we were to increase the concentration of the particles, they will interact with each other, slowing their velocity of descent. In a pseudoplastic material with a zero shear rate viscosity we can assume the gravitational stresses are too small to induce thinning; this gives us an expression which has a zero shear rate viscosity as opposed to the solvent viscosity.

$$v = \frac{2}{9}\frac{\Delta\rho g a^2}{\eta(0)} \quad (12.18)$$

A cursory inspection of the equation shows that the higher the low shear viscosity the more slowly the material will separate. Therefore, storage stability will tend to increase with higher low shear viscosity. We can do a little better with our prediction. Let us set ourselves a storage stability rejection criterion. We will reject our product if it sediments more than 1 mm in a month. This gives us a velocity of nearly $4 \times 10^{-10}\,\mathrm{m\,s^{-1}}$. We may now use Equation (12.18) to determine the viscosity we require to achieve this stability. Figure 12.12 shows a plot of particle radius on the x-axis versus the viscosity on the y-axis required to achieve our desired storage stability. This plot shows that small particles, nanoparticles, require a much lower viscosity than those of larger species. For a particle 2 μm in diameter and with a density difference between the particle and the medium of only $0.2\,\mathrm{g\,cm^{-3}}$, the viscosity required to stabilise the system is already at 1 Pa s, which is 1000 times that of water. For some processes this is unacceptably large. The solution to this difficulty is to reformulate the system to introduce a yield stress.

A plastic material has a yield stress. This has probably arisen through the formation of a network of interconnecting species. Suppose we know that gravity tends to cause the sedimentation of particles in the system. We may calculate the stress on these particles due to the gravitational forces. This is controlled by the particle mass, their density difference

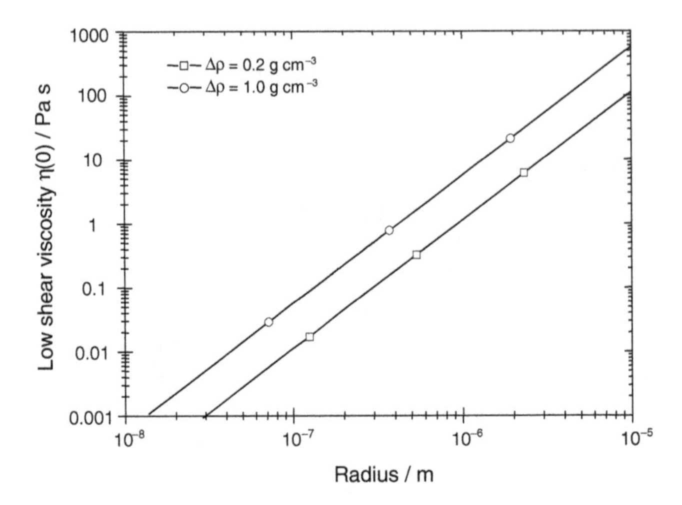

Figure 12.12 *A plot of the low shear viscosity required to achieved stability in a month as a function of particle size and density*

compared to the medium and the area over which the gravitational force is acting. Considering all these elements we obtain,

$$\sigma_y > \frac{\Delta\rho g a}{3} \tag{12.19}$$

So if the sample has a yield stress greater than the value calculated on the right-hand side of the expression we have good reason to expect some stability. We should not treat the results from this calculation as providing us with an exact solution, more as a very good guide to the yield stress you require in order to gain storage stability.

There is an important practical limitation to this approach. Let us consider a typical limit of sensitivity for an instrument as 0.01 Pa. This means we cannot determine yield stresses below this value. Now take a 2 μm diameter particle in water with a density of 1.1 g cm^{-3}. The yield stress to keep this stable should exceed 0.0003 Pa, which is 33 times smaller than we can measure! Even if the density difference was 10 times larger, it is still 3.3 times smaller than we can measure. One lesson to be learnt here is that it can be difficult to be truly sure of the stability of your system from a single measurement. Often it is the balance of evidence from viscoelastic studies at low stresses that provides the clues as to whether the material will store well. Another lesson to be learnt here is the inadequacy of flow curves when it comes to establishing yield stresses for sedimentation control. Attempting to extrapolate from a flow curve, where stresses of say 100 Pa or 1000 Pa have been used, down to a stress less than 0.1 Pa is fraught with errors.

Complex pastes can often give rise to another form of 'phase separation' called syneresis. This can be mistaken for sedimentation. This is understandable as it gives rise to a separated fluid on the upper surface of your paste. You see this sometimes in English mustard. In a container it can have the appearance of sedimentation. The process that is normally responsible for this is a network rearrangement between species in the paste. Following processing, the material begins to relax and rearrange to give a new network structure. The new network occupies a smaller volume than the original mixture, squeezing out liquid as it forms. It can differ from sedimentation visually in that you can sometimes observe the paste dewetting from the container to produce a lubricating layer between the vessel and the solid body of the paste.

Opaque systems present another difficulty. Larger particles separate more readily than smaller ones. So size polydispersity can lead to an uneven distribution of particles that is difficult to observe by eye since the opacity may appear unaltered throughout the system. This can lead to thick gooey sediments developing undetected. One should consider Equations (12.18) and (12.19) for all the species in your system.

Finally, you should also be aware that as the system sediments the concentration of species increases down the vessel. The rheological properties tend to increase in magnitude. This can halt sedimentation, so you can observe over a few days that a clear layer develops, gets a little larger and then becomes static. The system reaches equilibrium. An old trick is to sample this new concentration in the sedimented layer and reproduce it: in most cases your new rheology should prove to be just right to maintain stability.

12.4 Examples of Soft Materials

Many of the systems of practical interest to us consist of complex mixtures of materials. They may carry an active ingredient and require the correct rheological properties to deliver it effectively. Alternatively, the rheological properties may be an intrinsic part of their application, such as in a paint. It would be convenient if these complex mixtures of materials can be blended to give 'exactly the right' properties. So suppose we have a blend of a polyacrylate dispersion and a surfactant solution. If we know the viscosity of the dispersion is 100 mPa s and the viscosity of the surfactant solution was 3 mPa s it would be very convenient if we found the combination was 103 mPa s or 51.5 mPa s or some simple combining rule. Unfortunately, such relationships rarely hold and it is very difficult to predict the viscosity of many blends of materials. The main reason for this is straightforward. Colloidal materials tend to show interfacial activity and when you introduce one material to another new relationships develop between the components. It is no longer a simple mixture of the original components. This leads to complex changes in the rheological fingerprint of the material. The key thing is to understand the interactions between the components that form your material, then you can begin to design a formulation to give you what you want.

We can ask ourselves a question – what are the main features that control the rheology of a system? Clearly it depends in detail on whether we are discussing polymers or particles or surfactants and so forth. We can identify some key embracing rules that determine the magnitude of the properties. These are as follows (3–6):

- the number of interactions between species per unit volume (the number concentration; entities per mL, for example)
- the size of the entities
- the strength of the interactions (for example, in a mixed polymer surfactant system; is the polymer attracted to the surfactant and how strongly)
- the spatial arrangement of the species.

It is quite a tall order to have all this information in sufficient detail even in highly contrived model systems. In general the way we should approach designing or controlling the rheology is to have a good grasp of what we want and the type of colloidal interactions that can achieve it.

12.4.1 Simple Particles and Polymers

Frequently the rheological properties of formulations are adjusted by the addition of thickeners. These come in two forms: particles and polymers. Let us begin with a simple model for the flow of these species. Imagine a system of species that only interact through the flow lines they generate as they move through the medium and the rearrangements they can undergo. These are referred to as non-interacting species since the interactions are not influenced significantly by other types of interaction forces. Particles of this nature are referred to as hard spheres, and polymers are described as ideal.

Polymer chains will change their conformation both internally and with respect to each other. The primary cause for the flow we observe in non-functionalised polymers is the entanglement of the chains (5, 6). The more concentrated the system, the more entangled the chains become. There is a critical molecular weight, hence size, where this happens. The viscosity changes gradually with concentration, roughly linearly. At high concentrations power laws of 3–4 are observed before the systems become too viscous to measure using most practical methods. The onset of the higher power law behaviour is dictated by a critical concentration where polymer entanglements dominate the arrangement of chains. The trend in viscosity is shown in Figure 12.13.

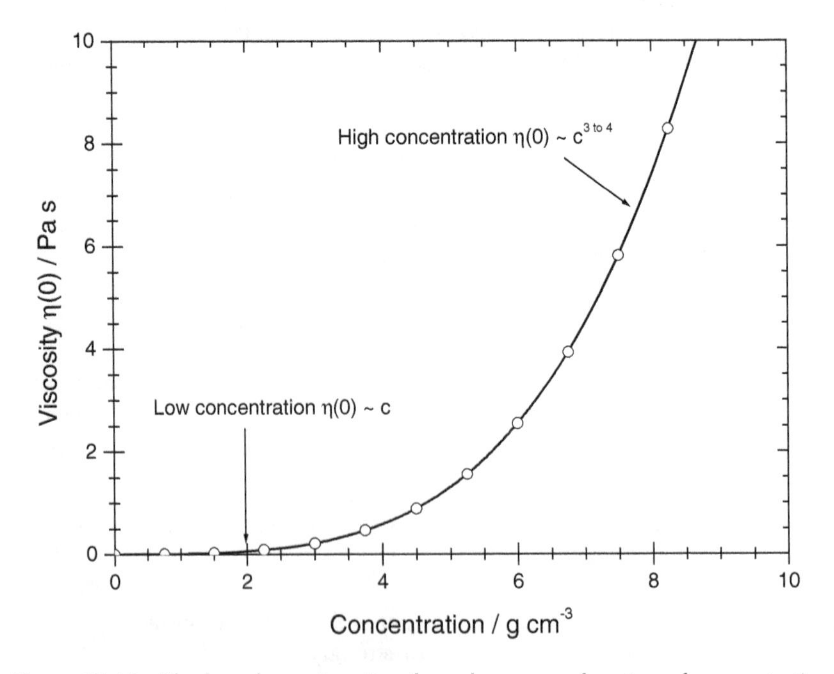

Figure 12.13 *The low shear viscosity of a polymer as a function of concentration*

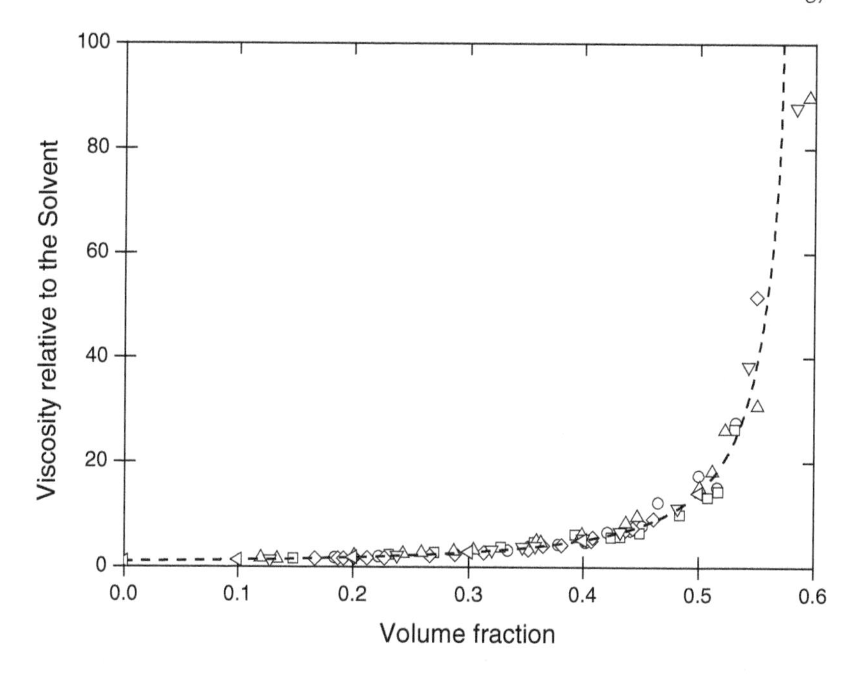

Figure 12.14 *The viscosity of a dispersion at high shear rates as a function of volume fraction*

Particles are restricted to rearranging their spatial distribution since they are too rigid to allow internal rearrangements to occur (3, 4). The concentration of the species determines the viscosity at any point. There are similarities between the change in viscosity with concentration for both particles and with polymers. We usually use the volume fraction to represent our concentration profile. A typical curve is shown in Figure 12.14.

The viscosity increases more rapidly with concentration for dispersions than it does for polymers. The rapid increase occurs near the maximum packing fraction of the dispersion. This is the point where the particles become space filling; it occurs at a volume fraction called the maximum packing fraction; typically in the range $\phi_m = 0.5$–0.7. This applies to both the high and low shear rate limiting viscosity. The packing is always smaller at the low shear rate limit than at the high shear rate limit.

In general the shear thinning profile of the systems have a similar form, as illustrated in Figures 12.15 and 12.16.

The key points illustrated by these flow profiles are as follows:

- that it is easier to achieve a high shear rate limiting viscosity with particles than with polymers
- the rate of change of viscosity with shear rate is more rapid with particles than with polymeric species
- the low shear rate viscosity and the degree of shear thinning are linked in polymers; as the low shear viscosity is increased the degree of shear thinning increases
- significant shear thinning is only achieved with high concentrations of particles and often close to a point where small changes in concentration leads to massive changes in viscosity.

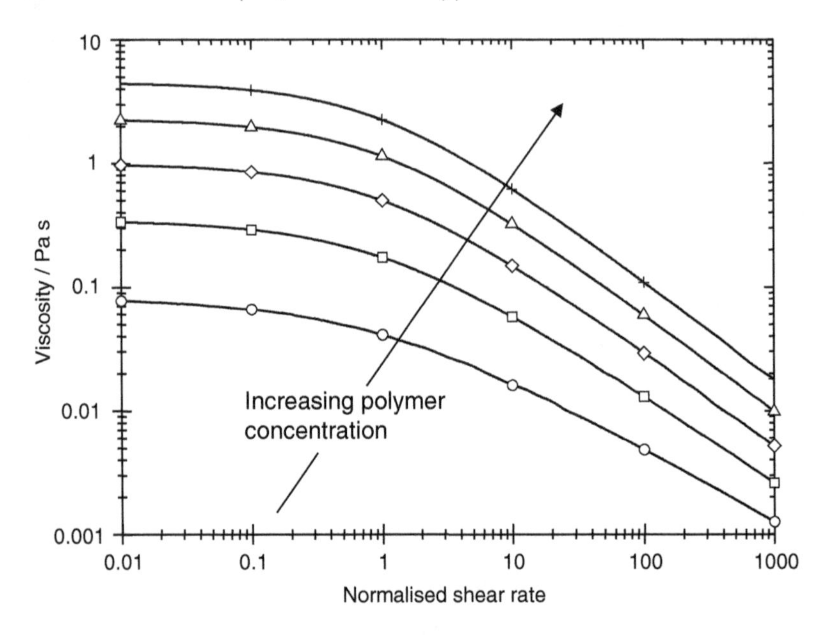

Figure 12.15 *The flow curve for a simple entangled polymer as a function of polymer concentration*

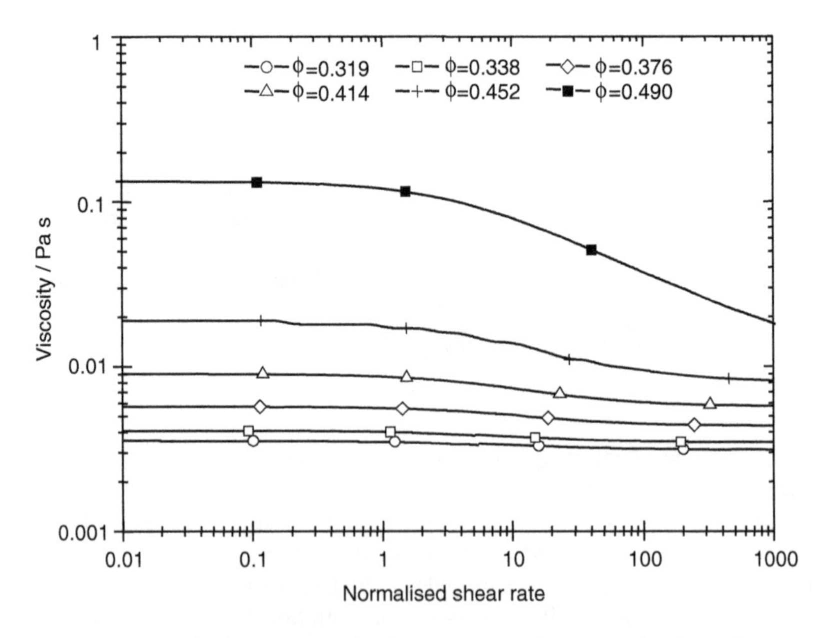

Figure 12.16 *The flow curve of a dispersion as a function of volume fraction*

The latter condition is not usually a good position to find yourself in with a formulated system: small changes in concentration would cause large changes in viscosity. In summary: without building in specific interactions neither system usually offers the ideal flow profile. This has led to the use of highly functionalised systems.

12.4.2 Networks and Functionalisation

Functionalisation of polymer and particulate systems is very important for achieving good rheology modifiers (3, 4). Manufacturers are constantly striving to develop better additives with lower environmental impact and greater tuneability. The basic idea behind the functionalisation of materials is usually two fold:

- to improve the compatability of the additive to a particular class of materials or commercial sector
- to improve the 'tuneability' of the rheology through particle–particle or polymer–polymer interactions.

You will find that it is almost universally the case that the rheological properties are determined by controlling the connectivity between the polymers or particles or whatever species we might be using, as illustrated in Figure 12.17. In order to get a more viscous or elastic structure we need to have a stronger network. The functionalisation controls how this network is assembled, its strength and what conditions will cause its collapse.

The range of this functionality is vast and it is rare that one additive will uniquely solve the questions you ask of your formulation. The most typical form of functionalisation is the introduction of charge interactions between species. This is most commonly, although not exclusively, used in aqueous systems.

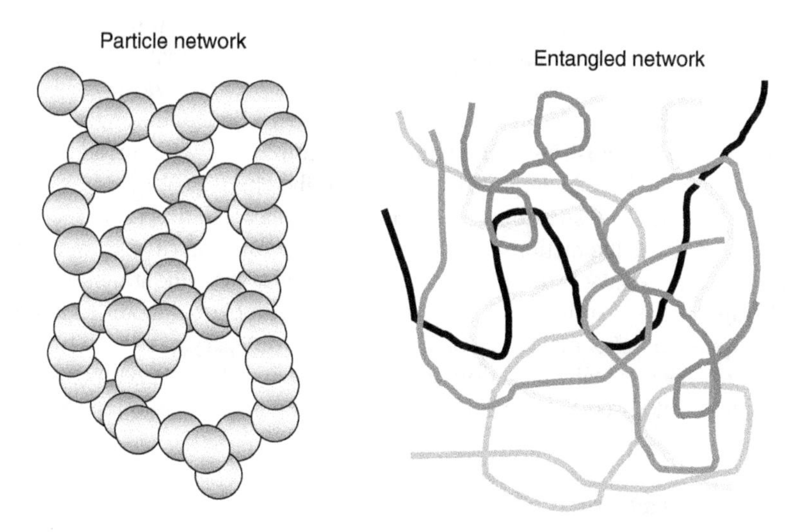

Figure 12.17 *A schematic of the different ways in which connectivity is achieved and networks can be formed*

12.4.3 Polymeric Additives

Copolymerisation of polymers can produce block copolymers with charged groups down their backbone. The introduction of acrylic acid, for example, produces carboxylic acid groups down the backbone of the polymer. This causes elements of the chain to repel one another and the chain to expand. This polymer achieves a higher viscosity at lower concentrations than its non-functionalised cousin. Of course it increases its sensitivity to both pH and electrolyte. This can be both a problem and an opportunity. It allows the chain expansion and entanglements to be tuned to an extent. At low pH the carboxylic acid groups are relatively poorly dissociated leading to lower viscosity systems. One of the difficulties of using such materials is the care with which they must be introduced to your formulation. If exposed to the wrong environment during processing they can precipitate and lose much of their functionality. This can be a 'one way trip' and despite adjustment to the optimum conditions later in the process path, the precipitated material may prove to remain stubbornly in the precipitated state.

It is possible to build network structures by using multivalent cations. For example the introduction of calcium cations to a polymer rich in carboxylic acid groups can result in ion bridging. The divalent ion can link together charged groups on the same chain or on neighbouring chains. This can even lead to precipitation of the polymer or network formation, depending on the relative concentrations and the architecture of the chain. Blends of natural polymers with opposing charges such as acacia and gelatine are good examples where networks can form and modify the rheology but in the correct ratio and at the correct pH they will precipitate to form a coacervate: a solid precipitate. It is possible to produce ionomers where a largely non-aqueous backbone has grafted charge groups. These sites can be used to bridge chains together in low dielectric media since they prefer to associate with each other than expose their charges to media where their opportunity for dissociation is highly limited.

One of the characteristics of particles that is hard to reproduce in polymers is their rapid shear thinning properties. However, these characteristics can be achieved using hydrophobic interactions. You can modify an aqueous polymer with hydrophobic groups. For example, you can place small chain length alkyl groups as terminating species at either end of a polyethylene oxide chain or as side groups on an hydroxycellulose backbone. These groups prefer to associate with each other rather than the medium or the chain. A schematic of this type of interaction is shown in Figure 12.18, in this case with micelles associating with network sites and the hydrophobic branches of the chain.

The forces holding these sites together are relatively weak. The alkyl chains are attracted by weak van der Waals interactions and the water molecules form a hydrogen bonded cage around them. When we place such a networking point under shear the structure is relatively easy to disrupt to give 'particle like' shear thinning. Hydrophobic modification has been achieved in quite complex molecules such as branched polyacrylamides to give tuneable flow behaviour. The alkyl tails are sensitive to surfactants and alcohols, which can be used to tune the strength of associations between groups. They provide an interesting range of possibilities.

12.4.4 Particle Additives

Particles almost cannot help being functionalised. In general a great deal of care must be taken to produce a dispersion of particles which do not either mutually attract or repel. In

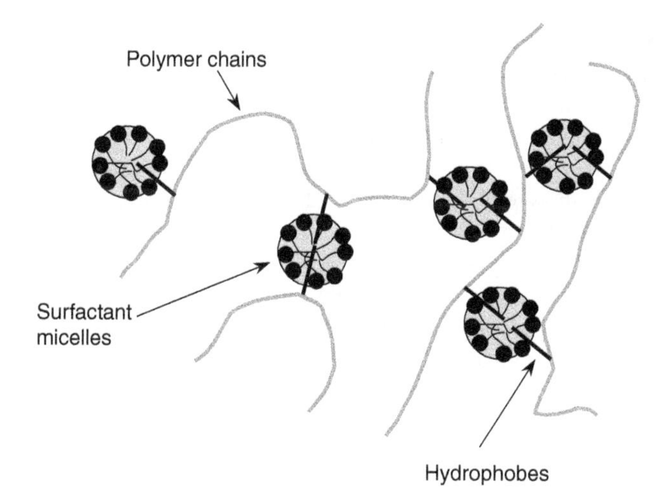

Figure 12.18 A schematic of the role played by hydrophobic modification in forming polymer networks. In this case surfactant is also shown as present in the system

polymer latex systems fragments of the initiator can result in particle surfaces possessing charged groups. This enhances the interactions between the particles producing more viscous systems at lower concentrations than systems interacting through hydrodynamic forces alone.

Colloidal particles are in constant Brownian motion in quiescent conditions. The onset of shear thinning in pseudoplastic dispersions is caused by the shear field that is applied overcoming the Brownian motion of the particles. The particles are forced to follow the flow lines induced by the movement of the geometry. If we could predict where this occurred we could design materials to flow at a rate appropriate to our needs. In Section 12.2.4 we defined a relaxation time characteristic for the structure. Suppose we consider this time relative to our applied shear rate. The dimensions of shear rate are s^{-1}. So the reciprocal of the rate can be thought of as a characteristic time for the shearing process. If the product of the shear rate and the relaxation time is greater than 1 the convective forces will be stronger than the Brownian forces. If it is less than 1 then Brownian forces win out. This ratio is termed the Peclet number, *Pe* and it is given by the expression below (3).

$$Pe = \dot{\gamma}\tau = \frac{6\pi\eta(0)\dot{\gamma}a^3}{kT} = \frac{6\pi\sigma a^3}{kT} \qquad (12.20)$$

The Peclet number defines the stress where we would start to see shear thinning. So if we set the Peclet number equal to 1 we can define a relationship between the critical stress for shear thinning σ_c, the particle radius and temperature.

$$\sigma_c = \frac{kT}{6\pi a^3} \qquad (12.21)$$

From this we can see that the smaller the particle the higher the stress you require to cause it to undergo shear thinning. In order to disrupt this relationship we need to impart additional forces to the particles.

Attractive forces between particles can be controlled to provide weak interactions and the tendency to weakly aggregate. This can aid storage stability. At first it seems counterintuitive to use aggregation to control this type of stability. You might suppose as particles aggregate they become larger and thus more prone to separate. However, if the concentration is high enough it can lead to fairly stable networks. These become space filling at much lower concentrations than 60%. They can develop self-supporting network structures that resist sedimentation and that are highly shear thinning. As we saw in a previous section (Section 12.3.5), frequently you only need a weak yield stress to oppose the effects of gravitational forces. This can be achieved with relatively weak attractions between the particles. The application of even a small shearing stress readily disrupts this network and causes substantial flow (and hence shear thinning) to occur. Thus systems which aggregate weakly can be used to build networks stable to sedimentation but fragile at greater shear stresses.

So far we have restricted our thinking to spherical colloidal particles. The most catastrophic shear thinning is achieved with particles with strong shape anisotropy. Rod-like or disk-like particles can show tremendous changes in rheology. For example the synthetic hectorite clay, Laponite, is a small disc-like entity. The edges and the faces can take different charges depending on the pH and ionic strength of the solution. They can form clusters, they can mutually repel or perhaps form edge-face 'house of cards'-like structures, as shown in Figure 12.19.

At high pH both the faces and edges carry the same sign of electrical charge and mutual repulsion results at low levels of ionic strength. This forms a gel-like material at high particle concentrations and a pseudoplastic fluid at low particle concentrations. As the electrolyte is increased the system begins to flocculate, leading to weak yield stresses. This material is highly tuneable in the appropriate formulation. The phase diagram for transitions at high pH is outlined in Figure 12.20.

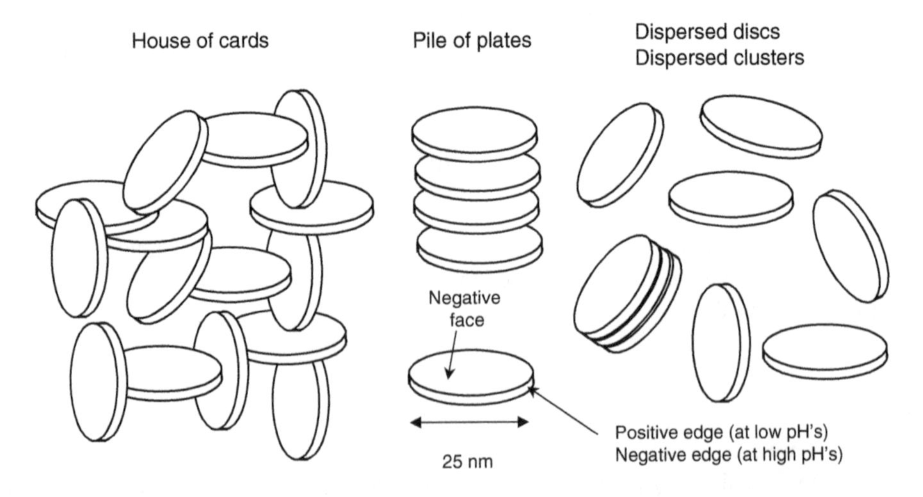

Figure 12.19 *An example of some of the ways in which Laponite discs may associate*

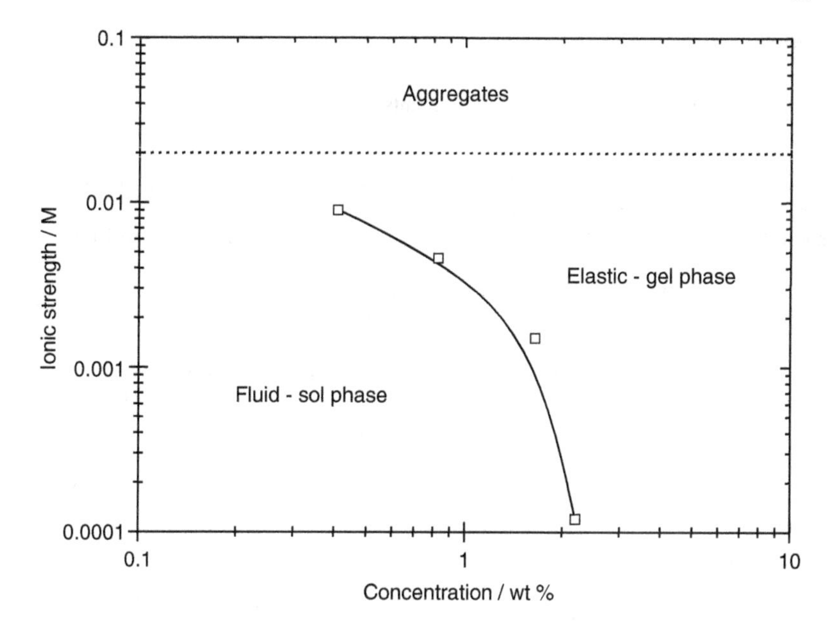

Figure 12.20 *A phase diagram for a Laponite dispersion at a high pH*

Laponite is just one example of a whole range of clays. It is particularly useful in 'clean' applications such as pharmaceutical and personal care products. Organophilic clays possess some of the flexibility of their aqueous relations although they tend to be less tuneable.

A simple trick you can play with particulate systems is to increase their porosity, although the word 'simple' should be viewed advisedly. Porous particles, especially those that are silicaceous, tend to form highly space filling structures when dispersed into a formulation. The maximum packing fraction occurs at very low mass concentrations and a small amount of additive can result in large rheological changes. These systems are guilty of having large surface areas. They can both adsorb species and leach ions, leading to temporal drifts in properties in the wrong formulation. However, there are a wide variety of choices of materials to select from. They can often be tailored to produce the required behaviour.

12.5 Summary

In this chapter we have examined some of the important features of experimental design, methods and the chemical influences on the rheological properties of colloidal systems. This treatise has indicated problems to be aware of when making measurements in order to avoid artefact-laden data. Your choice of experiment should be such that it is 'fit for purpose' and is either related to your process or can be used to give an interpretation of the structural properties of your system. Finally, we have touched upon the behaviour of polymers and particles and how their properties relate to their rheology. It is difficult to do full justice to the field of experimental rheology in this chapter. The reader is urged to examine the referenced texts and be rewarded by greater insights into specific aspects of the subject.

References

(1) Macosko, C.W. (1994) Principles, Measurements and Applications. Wiley-VCH, New York.
(2) Bird, R.B., Stewart, W.E., Lightfoot, E.N. (1960) Transport Phenomena. John Wiley & Sons, Ltd, New York.
(3) Goodwin, J.W., Hughes, R.W. (2008) Rheology for Chemists, An Introduction. 2nd edition. The Royal Society of Chemistry, Cambridge.
(4) Larson, R.G. (1999) The Structure and Rheology of Complex Fluids. Oxford University Press, USA.
(5) Ferry, J.D. (1989) Viscoelastic Properties of Polymers. John Wiley & Sons, Ltd, New York.
(6) Doi, M., Edwards, S.F. (1986) The Theory of Polymer Dynamics. Oxford University Press, Oxford.
(7) Gross, B. (1968) Mathematical Structures of the Theories of Viscoelasticity. Hermann, Paris.
(8) Tschoegl, N.W. (1989) The Phenomenological Theory of Linear Viscoelastic Behaviour. Springer-Verlag, Berlin.

13

Scattering and Reflection Techniques

Robert Richardson

Department of Physics, University of Bristol, UK

13.1 Introduction

Scattering of radiation is an essential tool for the colloid scientist. This chapter aims to introduce the use of scattering techniques in colloid science. For more detailed information, one can refer to several books that are based on the science that can be done using particular types of radiation (1–3). There are also several books that describe the structure determination for crystals and glasses where the focus is the arrangement of atoms rather than the larger scale structures of interest to colloid scientists. Nevertheless, these are useful resources for information of relevance to colloidal systems (4). The most useful books for a colloid scientist seeking to consolidate this introduction are those that describe the scattering methods applied to colloids and polymers (5).

Colloids are generally dispersions of particles with dimensions between 1 nm and 1 μm, i.e. between 10 and 10 000 Å (since 10 Å = 1 nm). There are many experimental techniques for characterising particles in this size range and their approximate ranges of sensitivity are shown in Figure 13.1 and the strengths and weaknesses of the different techniques are discussed below.

Electron microscopy (discussed in Chapter 15) can cover the whole range of sizes and gives very detailed results. The only limitation is that in general the sample is not in its natural equilibrium state. A vacuum is generally necessary so a colloidal dispersion would be ‘frozen’ or its particles extracted and dried for ‘ex situ’ electron microscopy.

Colloid Science: Principles, methods and applications, Second Edition Edited by Terence Cosgrove

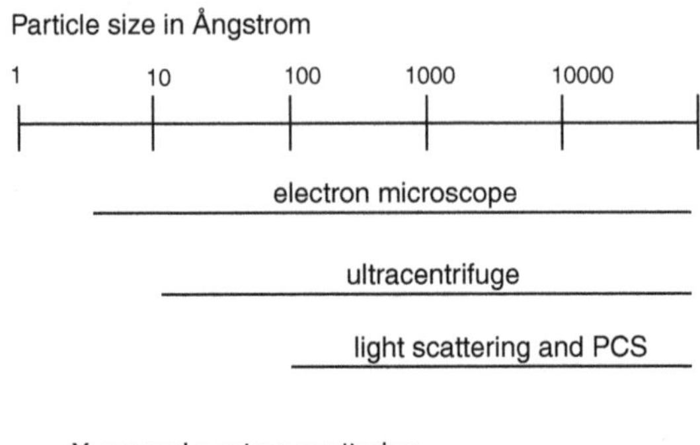

Figure 13.1 *Size ranges covered by different sizing techniques*

To overcome these limitations, there are continuing developments in this field and environmental scanning electron microscopy is capable of operating in a liquid vapour while cryo-transmission electron microscopy seeks to freeze in an equilibrium structure by very rapid cooling.

There are several techniques for particles size determination based on sedimentation. These give useful but rather limited information on particles in an equilibrium state. An example is the ultracentrifuge that can cover a wide range of particle size but experiments are quite demanding. Simple measurements of solution viscosity can also be used to estimate particle size.

Light scattering and small angle X-ray and neutron scattering by the particles in a dispersion are the main topic of this chapter. These scattering techniques cover the entire size range of interest and are capable of measuring the dimensions of particles 'in situ'. These methods may also give more detailed information on the internal structure of particles and their interactions in a dispersion. It is this ability to determine the equilibrium structure of colloids and surfaces in detail using a non-invasive probe that gives scattering methods importance in colloid and surface science.

13.2 The Principle of a Scattering Experiment

The basic scattering experiment is very simple. As shown in Figure 13.2, a monochromatic (i.e. single wavelength) beam is brought onto a sample. The intensity of the scattered radiation is measured as a function of the scattering angle which we will label as θ but note there are other conventions. The important variable, however, is the scattering vector, **Q**, whose magnitude is related to the scattering angle and wavelength.

$$Q = \frac{4\pi \sin \theta/2}{\lambda} \tag{13.1}$$

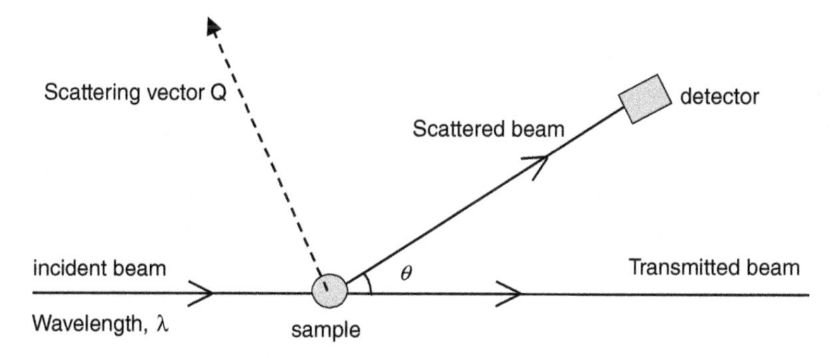

Figure 13.2 *Schematic diagram of a scattering experiment*

In principle, one could measure some scattering with two different wavelengths and plot the intensity against Q and get the same features of the curves at the same Q values. The distances probed in an experiment are inversely proportional to Q (i.e. distance $\sim 2\pi/Q$). That means for large scale structures (e.g. 100 Å or 10 nm) we need a small Q (i.e. $Q \sim 0.06$ Å^{-1}). To achieve small Q in a scattering experiment we need a suitable combination of large wavelength and low scattering angle. For light scattering, a wavelength that is comparable to, or larger than, the size of the scattering particles is generally used. For X-ray and neutron scattering a low scattering angle is generally used.

13.3 Radiation for Scattering Experiments

Table 13.1 summarises the properties of visible light, X-rays and neutrons for scattering experiments from colloidal dispersions.

Visible light has a wavelength of 400 nm to 600 nm and is suited to particles sizes above 0.01 μm although smaller particles can be detected. It is scattered by particles that have a different refractive index from the surrounding solvent.

X-rays (like visible light) are electromagnetic radiation. The useful wavelength range is shorter than about 0.2 nm because longer wavelengths tend to be adsorbed strongly. Small angle X-ray scattering is suited to probing distances in the 1 nm to 1 μm range. For X-rays, it is the difference in mean electron density between a particle and the solvent that scatters the radiation. They are therefore excellent for dispersions of high atomic number materials in low atomic number solvents (e.g. metals or oxides in water). They are less good for

Table 13.1 *Properties of radiation for scattering experiments*

Radiation	Visible light	X-rays	Neutrons
Type	Electromagnetic wave	Electromagnetic wave	Particle/wave
Wavelength, λ	400–600 nm	0.01–0.2 nm	0.01–2.0 nm
Distances probed	>0.01 μm	nm to μm	nm to μm
Scattered by variations of	refractive index	electron density	nuclear scattering properties

dispersions of organic materials in aqueous solvents because the electron densities of the two materials are similar.

Neutrons are particles but have an associated wavelength. The useful wavelength range is about 0.1 nm to 2 nm. Longer wavelength neutrons are difficult to produce with sufficient intensity. Small angle neutron scattering is suited to probing distances in the 1 nm to 1 μm range. Neutrons do not generally interact with the electrons in atoms but they are scattered by an interaction with the nuclei. The scattering is not dependent on the atomic number so they often offer a better option than X-rays for scattering from a dispersion of one material in another of similar atomic number. It is also possible to get different scattering from the same element by using isotopes. Hydrogen/deuterium substitution is particularly useful in colloid science. For instance, organics dispersed in heavy water (D_2O) scatter neutrons strongly.

First we will look at the factors that determine the intensity in a classic light scattering experiment.

13.4 Light Scattering

Light scattering has been used for decades to determine particle sizes. The intensity of light scattered by a suspension of 'small' particles (i.e. particle diameter $\ll$ wavelength) is determined by the equation

$$I(Q) = kcM(1 + \cos^2 \theta) \tag{13.2}$$

where the four factors are the concentration, c, the molar mass of the particles, M, a collection of constants, k, and the polarisation factor, $(1 + \cos^2 \theta)$. The polarisation factor results from the physics of the scattering process and is usually removed by an experimental correction. It contains no interesting information about the sample. This equation was derived by Lord Rayleigh in 1871. An interesting sideline is that the factor k contains the inverse fourth power of the light wavelength which means that short wavelengths are scattered much more strongly than long ones. Hence the sky is blue and the sun appears yellow or red.

When light of wavelength 500 nm is incident on small particles (20 nm radius) the anisotropy of the scattering only results from the polarisation factor. Forward and back scattering intensities are therefore equal, as shown in the polar plot of scattered intensity in Figure 13.3. If the constant, k, and concentration, c, are known, the particle mass may be determined.

However, for larger particles (40 nm) the scattered intensity develops an asymmetry between forward and back scattering. This is a particle size effect and is the basis for the determination of particle size from scattering. The origin of this effect is shown in Figure 13.4. There will be a different path length from source to detector for rays scattered by different parts of the particle. Consider two rays scattered from opposite sides of the particle. They have different source to detector distances. If the particle radius is comparable with the wavelength of the radiation, the path difference means that the two rays arrive at the detector somewhat out of phase. They interfere destructively so there is less intensity detected. In fact the degree of destructive interference tends to increase as the scattering angle increases so the intensity tends to decrease with increasing angle or Q.

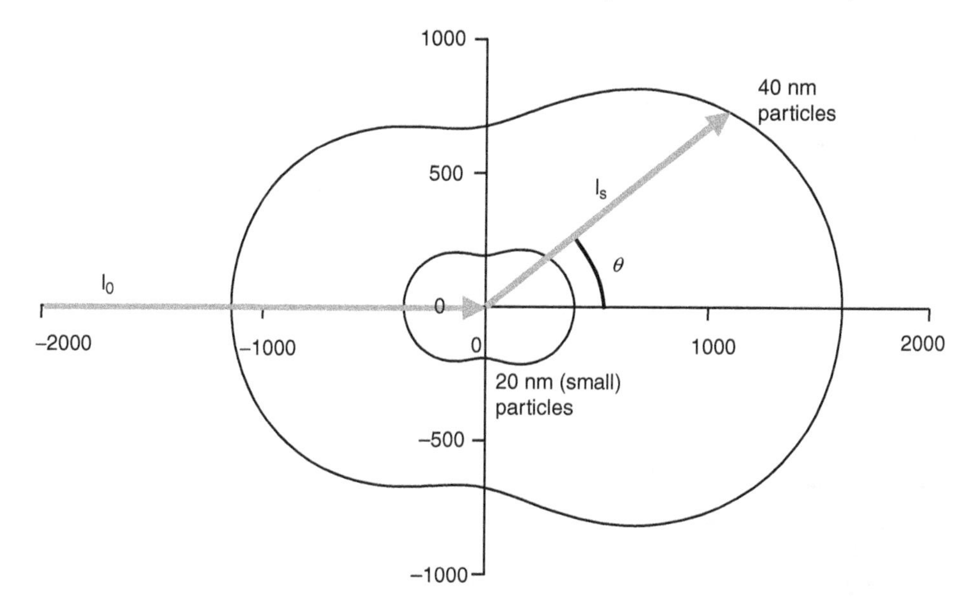

Figure 13.3 *Polar plot of forward and backscattering intensities*

The rate at which the intensity changes with angle (or Q) depends on the size of the particle. However, 'the size of a particle' is a vague concept. More precisely, the intensity depends on a dimension which is called the radius of gyration of the particle because it is roughly analogous to the mechanical quantity.

The radius of gyration, R_G, depends on the size and shape of a particle. In general it may be calculated for any shape by an integral over the volume of a single particle:

$$R_G = \left\langle \sqrt{\frac{1}{V}\int_V r_{//}^2 \, dV} \right\rangle \tag{13.3}$$

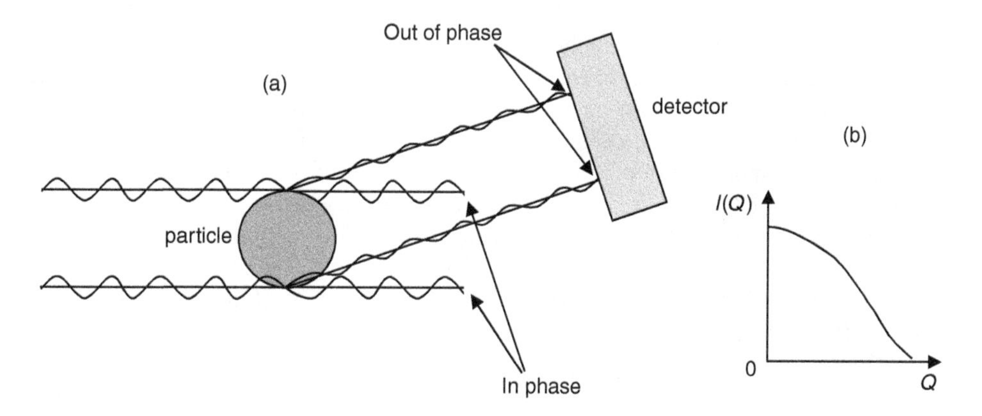

Figure 13.4 *The origin of the particle size effect in scattering*

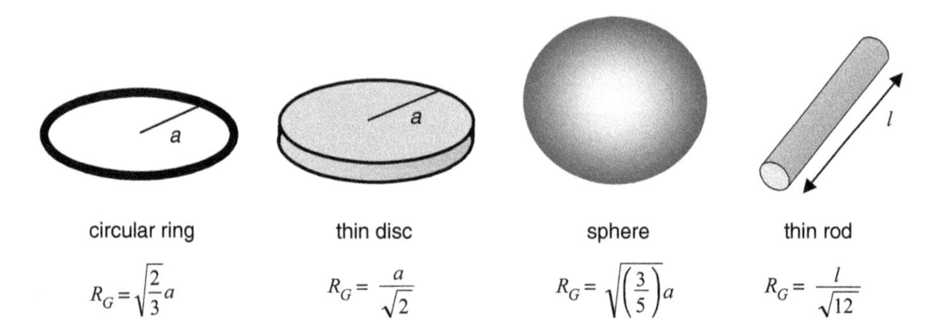

Figure 13.5 *Some simple particle shapes and their radii of gyration*

where $r_{//}$ is the component of distance from the centre of gravity along **Q**, V is the volume of the particle and the brackets indicate an average over all orientations of the particle with respect to **Q**. The relationship of R_G to the dimensions of some simple shapes is given in Figure 13.5.

For large particles, there is an additional factor in the equation that governs the scattered intensity:

$$I(Q) = kcM(1+\cos^2\theta)(1-(QR_G)^2/3+\cdots) \tag{13.4}$$

This new factor depends on the radius of gyration and is approximated by $(1-(QR_G)^2/3)$. Hence the dependence of the scattered intensity on scattering angle or Q may be used to determine the radius of gyration of particles in a dilute dispersion. This theory (known as Rayleigh–Gans–Debye theory) is applicable where particle diameter remains less than the wavelength. For larger particles the scattering pattern becomes extremely complex and Mie theory applies. This is covered in more advanced texts such as (3).

13.5 Dynamic Light Scattering

There is another light scattering technique that operates in a different way. It relies on the fact that particles in a dispersion are moving by diffusion. When a particle scatters a photon of light, there is a small exchange of energy between the photon and the particle. The particle may gain energy from or lose energy to the photon and the photon's energy shifts accordingly. This is the same process (Doppler shifting of frequency) that is used in radar speed traps. The frequency of the radar is changed by reflection from a moving vehicle.

The spectrum of light scattered is measured using the elegant technique of photon correlation spectroscopy (PCS). If the incident spectrum is monochromatic with frequency, ϖ_0, then the spectrum from a colloid generally has a Lorentzian shape and the width of the peak is determined by the diffusion coefficient of the particles, D, multiplied by Q^2, as shown in Figure 13.6(a):

$$I(\omega) \propto \frac{DQ^2}{(\omega-\omega_0)^2 + (DQ^2)^2} \tag{13.5}$$

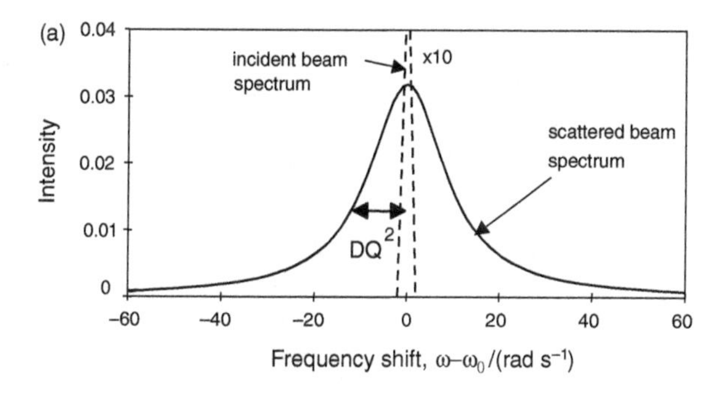

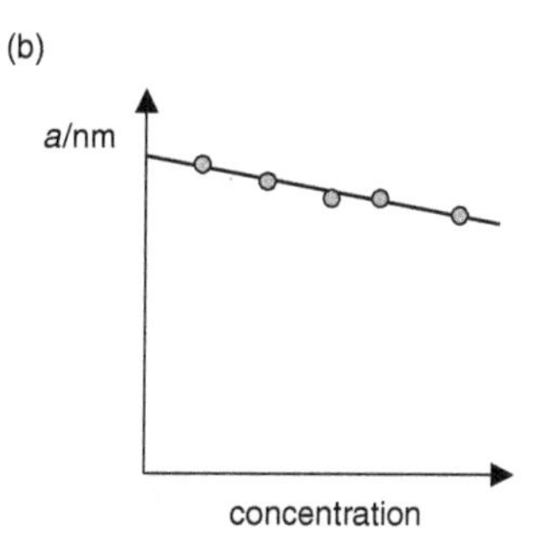

Figure 13.6 *(a) Incident and scattered spectra for dynamic light scattering and (b) extrapolating the hydrodynamic radius to zero concentration*

The hydrodynamic radius, a, may then be calculated from the diffusion coefficient using the Stokes–Einstein equation, provided the solvent viscosity, η, is known. More details may be found in reference (6)

$$a = \frac{kT}{6\pi\eta D} \tag{13.6}$$

Since the diffusion coefficient is influenced by particle–particle interactions as well as viscous drag, it is usually necessary to extrapolate the hydrodynamic radius to zero concentration as shown in Figure 13.6(b). It should be noted that the hydrodynamic radius is often greater than the radius determined by 'static' light scattering because it may include layers of bound solvent.

13.6 Small Angle Scattering

We now consider the techniques of small angle scattering of X-rays and neutrons (SAXS and SANS). A better term would be small Q scattering because the exact combination of scattering angle and wavelength is not important.

Take a gold colloid for example. The scattering from such a sample is shown schematically in Figure 13.7. At small Q, the scattering is sensitive to the size and shape of the particles (as we saw for light scattering) while at large Q the scattering reflects the internal structure of the particles. For particles of a crystalline material the internal structure would give Bragg peaks. The Bragg peaks from colloidal particles are often broader than those from the same material in bulk form because of internal disorder and finite size effects on the diffraction. However, it is the small Q scattering that is generally measured and interpreted in colloid science.

13.7 Sources of Radiation

X-rays may be generated in the laboratory using a sealed tube generator. You can buy a complete SAXS kit from several manufacturers. The intensity from a synchrotron source is

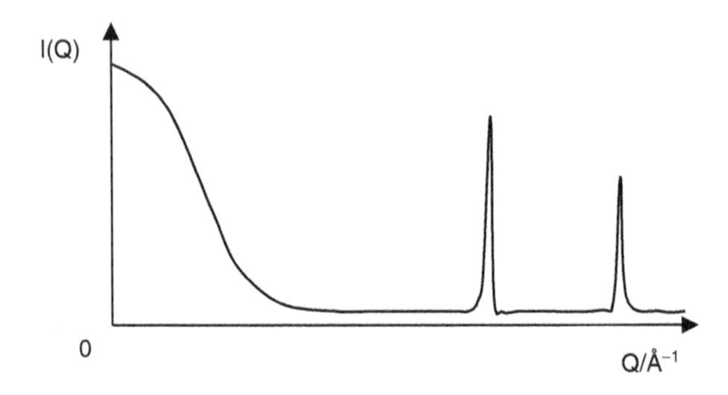

Figure 13.7 Schematic of scattering from colloid of crystalline particles

many orders of magnitude greater. Synchrotron radiation is produced tangentially when a high energy electron beam is deflected by a magnetic field. Facilities exist at the ESRF at Grenoble and Diamond at the Rutherford Appleton Laboratory, Oxfordshire. More details may be found at the facility websites (7, 8). Neutrons in adequate quantities are only available from central facilities. In the UK we have access to the reactor at ILL, Grenoble (9) and the pulsed source at ISIS at the Rutherford Appleton Laboratory (10). A second source (Target Station 2) has recently been constructed at ISIS. Further details and links to neutron sources worldwide may be found at the facility websites. The European Neutron Scattering Association gives information for accessing European facilities and is actively involved in planning future sources (11).

Neutrons produced by the fission process in a reactor have a huge energy ($E \sim 1$ MeV) and by consequence of the de Broglie relationship,

$$\lambda/\text{Å} = 9.04/\sqrt{E/\text{meV}} \tag{13.7}$$

a very short wavelength ($\lambda \sim 0.0003$ Å). This is useless for large scale structures. Fortunately, the energy of the neutrons may be moderated by passing them through material to thermalise. The neutrons adopt the thermal energy of the moderator material. If its temperature is low, the energy of the neutrons becomes low and so their wavelength becomes large. At the institute Laue Langevin, the cold moderator is liquid deuterium at 25 K and it turns out that a 25 K liquid deuterium moderator gives a 'Maxwellian' distribution of wavelengths peaked around 6 Å. Such a cold source is ideal for SANS. Figure 13.8 shows the moderator schematically and its wavelength distribution along with those from ambient and hot moderators. Pulsed neutron sources also use moderators to generate neutrons in the useful wavelength range.

13.8 Small Angle Scattering Apparatus

The basic components of a small angle neutron scattering apparatus are shown in Figure 13.9 (SAXS is similar after the sample). The reactor core is surrounded by D_2O to reflect neutrons back and maximise flux. The cold source is used to maximise the useful flux at ~6 Å. The

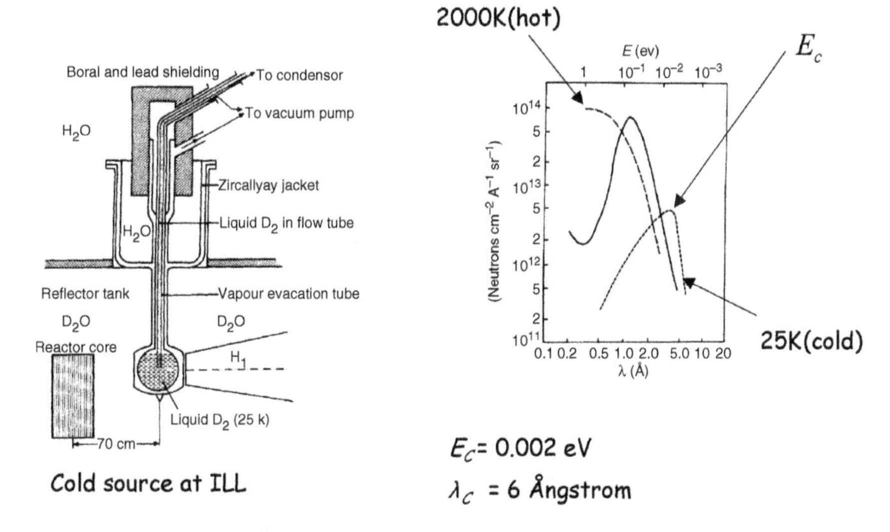

Figure 13.8 *ILL cold moderator and wavelength distribution. Reprinted with permission from Quasielastic Neutron Scattering by M. Bee Copyright (1988) Institute of Physics Publishing Ltd*

velocity selector only passes a narrow(ish) band of neutron velocities. Since velocity, v, and wavelength, λ, are inversely proportional,

$$\lambda \text{ in } \mathring{A} = \frac{3956}{v \text{ in m s}^{-1}} \tag{13.8}$$

a narrow band of wavelengths is passed and so the radiation incident on the sample is nearly monochromatic. The sample is typically 1 cm square and 1 mm thick. A two-dimensional position-sensitive detector collects scattered intensity. Software is often used to regroup the data as intensity against Q. Evacuated tubes reduce background from air scattering.

On a pulsed neutron source, such as ISIS, the velocity selector is unnecessary since the time-of-flight or velocity of the neutrons can be measured. From this the speed and hence wavelength can be calculated. Since a 'white beam' is used the large intensity losses associated with monochromatisation are avoided. Although intrinsically weaker, pulsed sources tend to make efficient use of the neutrons.

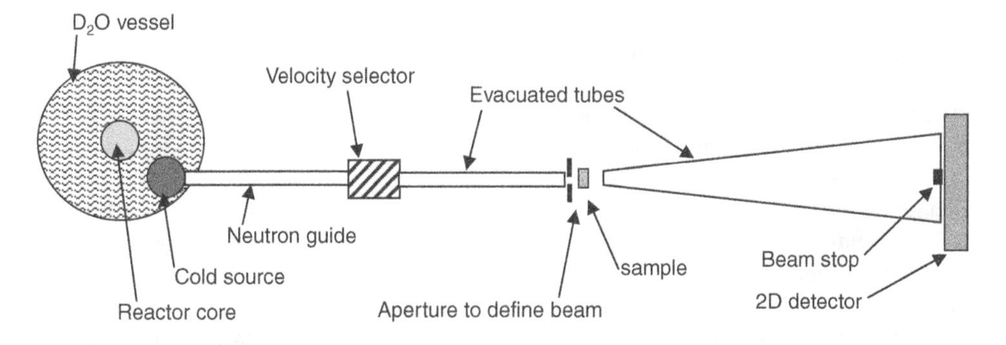

Figure 13.9 *Schematic of SANS Apparatus*

Figure 13.10 The NG3 SANS apparatus at NIST Centre for Neutron Research, Gaithersburg, MD, USA

Figure 13.10 shows the NG3 SANS apparatus at NIST Centre for Neutron Research, Gaithersburg, MD, USA. It shows the shielding around the incident beam in the foreground, the sample position and the large vacuum tank containing the detector in the background. D11 and D22, the SANS instruments at ILL, have a similar layout. Other reactor-based facilities have similar instrumentation. The current SANS apparatus at ISIS is LOQ and a new instrument, SANS-2D, has been built on the second target station is now available.

13.9 Scattering and Absorption by Atoms

The amplitude of the scattering by an atom is characterised by its scattering length, b. For X-rays scattering, b is proportional to the atomic number, z. (Actually $b = z \times a_e$ where $a_e = 2.85 \times 10^{-15}$ m, which is the scattering length for 1 electron.) For neutrons, the scattering length is a nuclear property and it varies irregularly with atomic number and also depends on isotope. Table 13.2 shows the scattering lengths and absorption cross sections for some atoms. We see that hydrogen and deuterium have very different scattering lengths. Physically, a positive or negative scattering length is related to the phase shift of the wave during scattering but we do not need to be concerned over the origin of the sign. We just use it.

Table 13.2 Some scattering lengths and absorption cross sections. Reproduced with permission from Ref. (12). Copyright (2003) US Department of Commerce

Species	$\frac{b_N}{10^{-15}\,\text{m}}$	$\frac{b_X(\text{at } Q=0)}{10^{-15}\,\text{m}}$	$\frac{\sigma_N(abs)}{10^{-28}\,\text{m}^2}$	$\frac{\sigma_X(abs)}{10^{-28}\,\text{m}^2}$
H	–3.74	2.85	0.28	0.73
D	6.67	2.85	0.0	0.73
C	6.65	17.1	0.003	92
O	5.83	22.8	0.0	306
Cd^{2+}	3.7	131.1	$>10^3$	9400

The adsorption cross sections indicate how strongly an element absorbs the radiation. Absorption of X-rays increases very strongly with atomic number so cells for X-ray experiments are made of low-z materials. Neutrons tend to be absorbed less so sample containment is not a problem. There are useful exceptions such as cadmium which can be used for shielding and beam definition apertures.

13.10 Scattering Length Density

In ‘small angle’ experiments (i.e. low Q) the distances probed are generally much greater than inter-atomic spacings so the technique is sensitive to changes in ‘scattering length density’ over distances of up to ~1000 Å rather than the scattering by individual atoms. Scattering length density, ρ, is therefore a very useful concept because it can be used to describe the scattering from a large volume (such as a particle) without having to specify the position of every atom. The scattering length density of a material, ρ, is calculated from the product of the number density of each atom type, N_j, and its scattering length, b_j. The products for different types of atom are then summed.

For neutrons:

$$\rho_N = \sum N_j b_j \tag{13.9}$$

where b_j is the scattering length of an atom for neutrons.

For X-rays:

$$\rho_X = a_e \sum N_j z_j \tag{13.10}$$

where a_e is the scattering length of an electron for X-rays ($a_e = 2.85 \times 10^{-5}$ Å) and z_j is the atomic number.

The scattering length densities for some materials are shown in Table 13.3. It is worth noting that for neutrons the scattering length density depends on the specific isotope. This gives the experimentalist an important tool because it is possible to vary the isotopic content of a sample, for instance by switching from H_2O to D_2O as solvent, in order to emphasise some aspect of the scattering without changing the chemistry of the sample appreciably.

Table 13.3 Scattering length densities of some materials

Material	$\rho_N/(10^{-5}\,Å^{-2})$	$\rho_X/(10^{-5}\,Å^{-2})$
H_2O	–0.05	0.94
D_2O	0.64	0.94
$-(CH_2)_n-$	–0.06	0.65
$-(CD_2)_n-$	0.61	0.65

13.11 Small Angle Scattering from a Dispersion

A simple picture of a dispersion is a number of identical particles suspended in a matrix (the solvent), as shown in Figure 13.11. For a dilute dispersion the inter-particle distance will be able to take almost any value and so inter-particle interference effects are eliminated. The observed scattering intensity, $I(Q)$, then depends only on the four factors in the following equation:

$$I(Q) = (\rho_P - \rho_M)^2 N_P V_P^2 P(Q) \tag{13.11}$$

where $(\rho_P - \rho_M)$ is the *contrast* in scattering length density between a particle and the matrix; N_P is the number of particles in the sample; V_P is the volume of a particle and $P(Q)$ is the particle form factor which is defined by the size and shape of the particle.

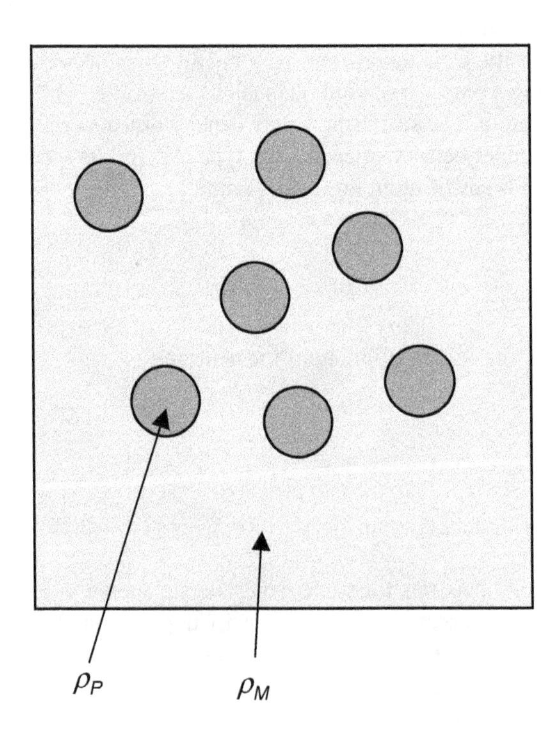

Figure 13.11 Simple picture of a dispersion of homogenous particles in a matrix

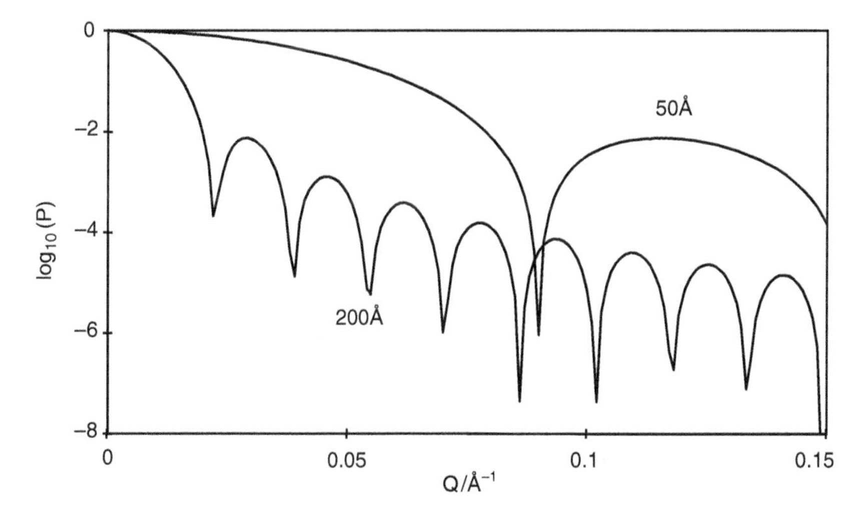

Figure 13.12 Form factor of Spherical Particle

13.12 Form Factor for Spherical Particles

The form factor may be calculated by integration over the volume of a particle of any shape. For a spherical particle, the formula given below is obtained (13):

$$P(Q) = \left\{ \frac{3(\sin QR_s - QR_s \cos QR_s)}{(QR_s)^3} \right\}^2 \tag{13.12}$$

where R_S is the radius of the sphere.

This formula is plotted in Figure 13.12 for two values of the sphere radius. It demonstrates several of the important general features of form factors (note the log scale).

- At $Q = 0$ it has a value of 1
- initially it decreases with increasing Q
- for smaller particles the function is more stretched out in Q and vice versa – in fact it is a function of QR_S
- maxima and minima appear at higher Q.

13.13 Determining Particle Size from SANS and SAXS

There are two complementary approaches to determining the particle dimensions in a dilute dispersion.

The trial and error method involves calculating the scattering from an assumed particle shape (e.g. a sphere) and varying the parameters (e.g. radius, number of particles) until a good agreement is found between measured data and model. If no agreement is found, then try assuming another shape and repeating the fitting process. This is usually done with a least-squares fitting program. It is a useful method but critical interpretation is required because a model that fits is not necessarily unique.

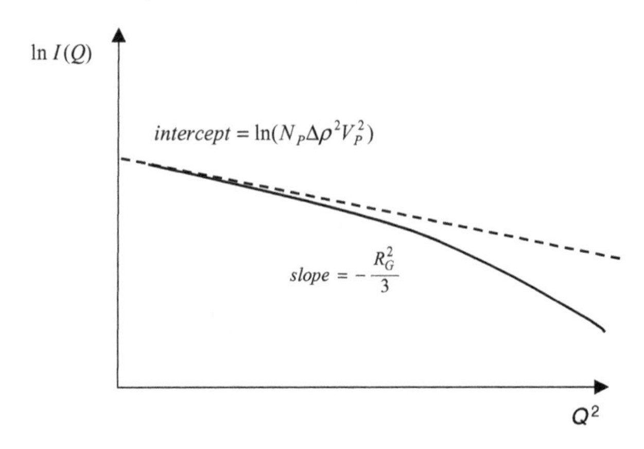

Figure 13.13 *Showing the straight line behaviour on a Guinier plot at low Q*

The Guinier law relates the low Q slope of the scattering to the radius of gyration of the particle and makes no assumptions regarding the particle shape.

13.14 Guinier Plots to Determine Radius of Gyration

It turns out that at low Q (that means $Q < 1/R_G$) the scattering from a dilute dispersion is insensitive to the shape of the particles. The intensity, $I(Q)$, only depends on contrast, number of particles, particle volume and the radius of gyration, as shown in this approximate equation, known as Guinier's law (14).

$$I(Q) = \Delta\rho^2 N_P V_P^2 \exp(-Q^2 R_G^2/3) \tag{13.13}$$

The radius of gyration was introduced in Section 13.4 for light scattering and is a very convenient quantity for characterising the size of a particle. Figure 13.13 shows a Guinier plot of natural log of intensity against Q^2. It has a slope of $-R_G^2/3$ so it is possible to determine R_G without assuming a particle shape.

Note that caution is required because the approximation is only valid for $Q < R_G^{-1}$.

In the next few sections we look at variations and extensions of this basic type of measurement. Much more detail is available in specialised texts (15, 16).

13.15 Determination of Particle Shape

At $Q > 1/R_G$ the shape of the particle does have a major influence on the particle form factor and hence the shape of the scattering from a dilute suspension. This can be seen most clearly on a log/log plot of the particle form factor for a sphere, a thin (i.e. 5 Å thick) disc and a thin (i.e. 5 Å radius) rod, as shown in Figure 13.14. This shows a characteristic region with a slope of -1 for the rod and -2 for the discs. At $Q \gg 1/$(the dimension of the particle), Porod's Law (discussed later) applies. The particle sizes in this example have been chosen to have the same radius of gyration (100 Å) so the form factor is the same for all three in the region below $Q \sim 1/R_G$ where the scattering obeys the Guinier Law.

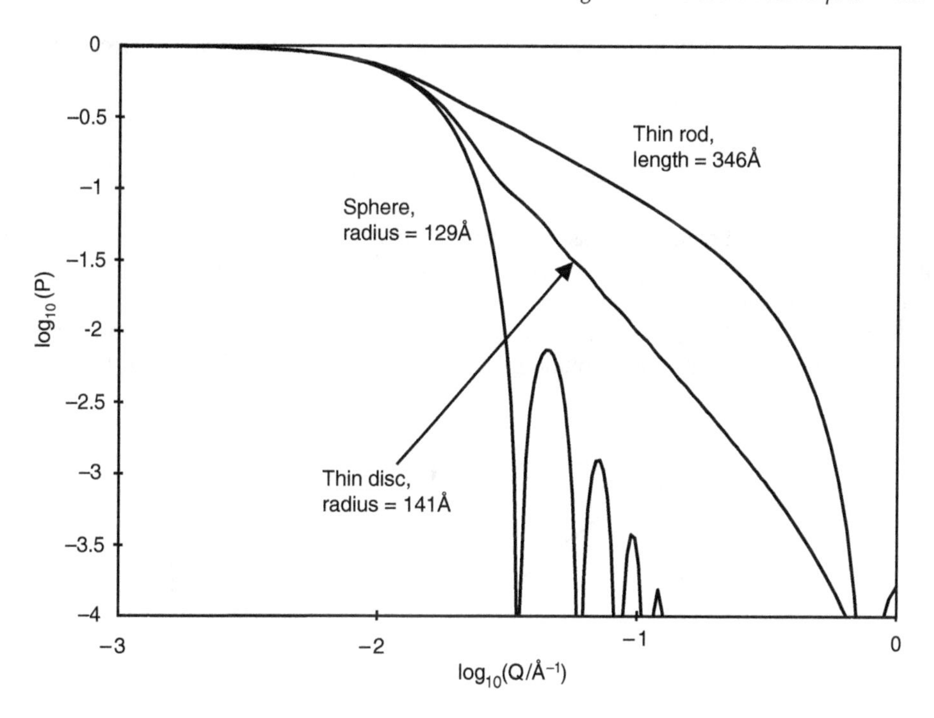

Figure 13.14 Form factors for different particle shapes with same radius of gyration

13.16 Polydispersity

Polydispersity does not greatly affect the low Q slope but it tends to smear out the maxima and minima at higher Q, as shown in Figure 13.15. This can be visualised by averaging form

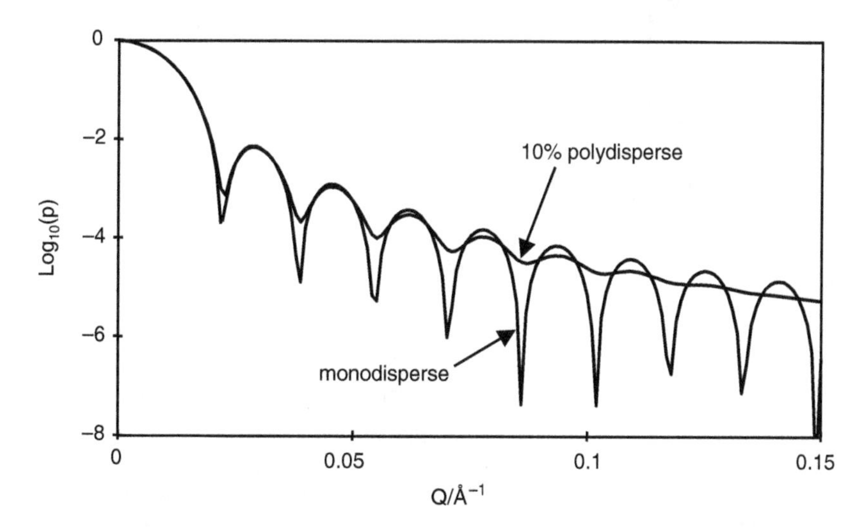

Figure 13.15 Showing effect of polydispersity on form factor

factors with slightly different values of the sphere radius, R_S. The trial and error (fitting) method can be used to deduce the degree of polydispersity.

13.17 Determination of Particle Size Distribution

There are other computer methods for extracting particle size distributions. For instance there is a maximum entropy approach where the smoothest particle size distribution consistent with the scattering curve is determined (17).

Figure 13.16(a) shows an example of SAXS from partly hydrolysed zirconium chloride which forms polynuclear ions in solution. The maximum entropy particle size distribution is

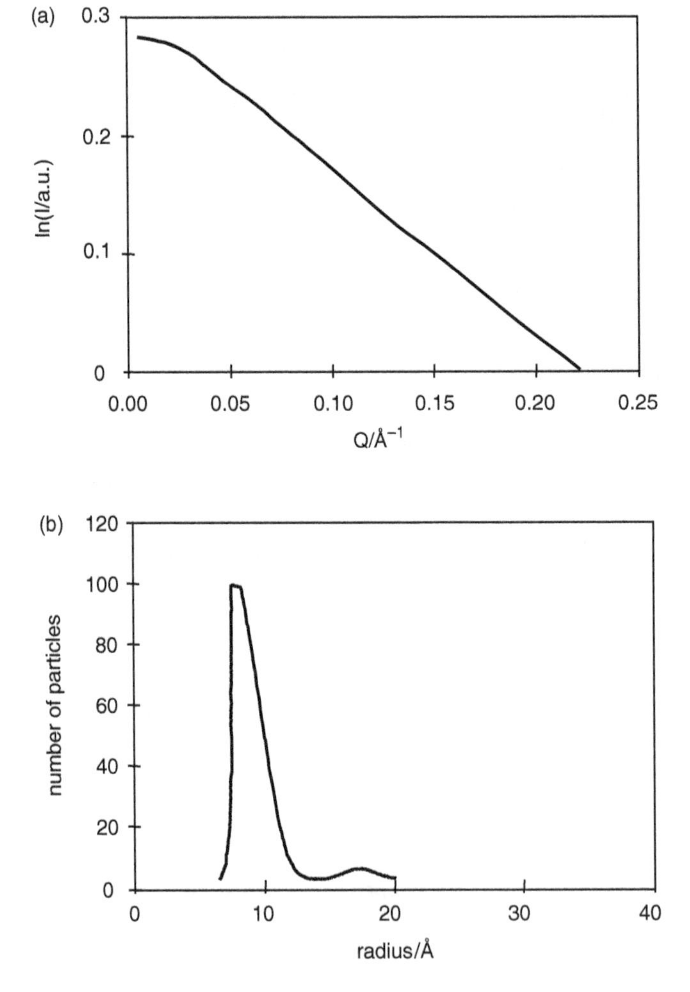

Figure 13.16 (a) SAXS and (b) the particle size distribution determined using maximum entropy method

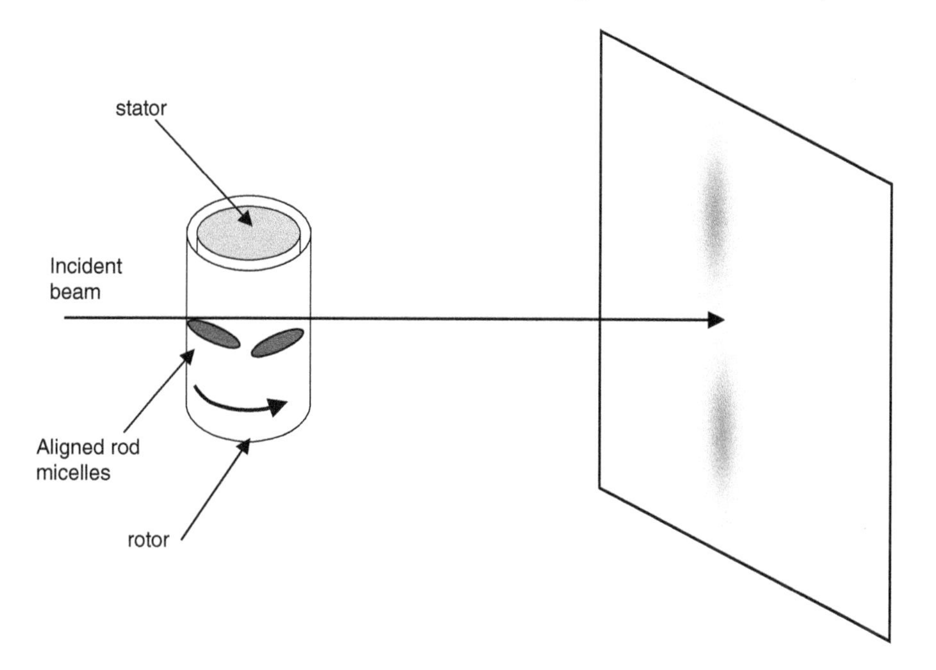

Figure 13.17 Schematic of scattering from shear aligned sample

shown in Figure 13.16(b). The particles appear to have a radius of 10 Å, with a small proportion of larger particles (possibly dimers).

13.18 Alignment of Anisotropic Particles

For non-spherical particles it is advantageous to align the particles for a small angle scattering measurement. The two characteristic dimensions may then be determined by analysing the scattering in the two perpendicular directions on the detector (Figure 13.17). For instance worm-like micelles may be aligned by shearing in a cuvette (18). This is usually made of silica which is transparent to neutrons with a gap of 1 mm or less between the external rotor and the internal stator. Nematic liquid crystals may be aligned by applying a magnetic field.

13.19 Concentrated Dispersions

For concentrated dispersions, rays scattered from different particles will interfere. This inter-particle interference is accounted for by a term called the structure factor, $S(Q)$.

$$I(Q) = (\rho_P - \rho_M)^2 N_P V_P^2 P(Q) S(Q) \qquad (13.14)$$

For dilute dispersions, $S(Q) = 1$. For concentrated dispersions it is an oscillatory function and it can be used to determine how the particles pack together (19). The form of the

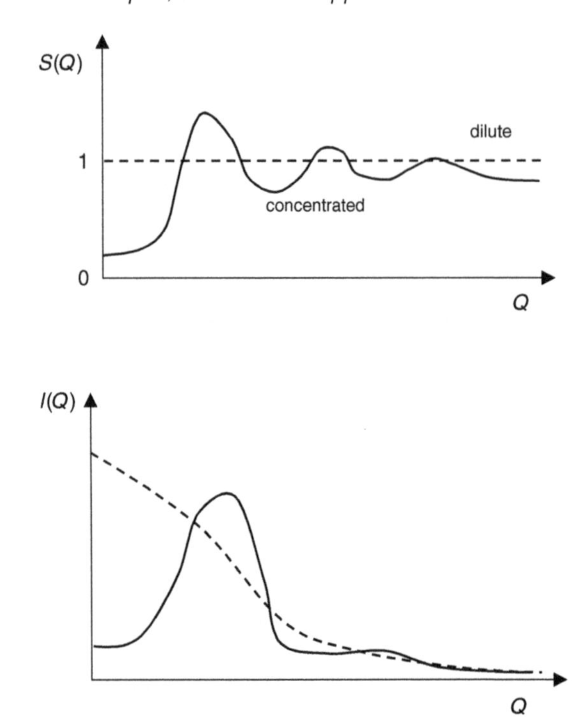

Figure 13.18 S(Q) and I(Q) from concentrated (solid) and dilute (dashed) dispersions

scattering intensity, $I(Q)$ depends on the product of the particle form factor, $P(Q)$, and the structure factor, $S(Q)$, as illustrated in Figure 13.18. Least-squares model fitting is used to determine parameters such as the closest distance of approach of two particles (the hard sphere repulsion radius). For charged particles (e.g. micelles) the surface charge and screening length may be determined by model fitting (20).

One potential pitfall when using the Guinier law to determine radius of gyration is that the slope of the plot is only $-R_G^2/3$ if $S(Q) = 1$. For not very dilute samples this may not be correct and analysis using the Guinier law leads to an incorrect value of R_G.

Figure 13.19 shows SAXS from overbased detergents dispersed in oil. These are used as an engine oil additive. They are calcium carbonate particles stabilised by surfactant. Since the surfactant and the oil have very similar electron densities, which are different from that of the $CaCO_3$ core, the scattering is dominated by the more electron dense core. For the concentrated dispersion, the peak position and shape may be analysed to give the hard sphere radius. On dilution, the peak disappears ($S(Q)$ tends to 1) and a Guinier plot can be used to determine the core radius.

13.20 Contrast Variation using SANS

For SANS the use of contrast variation gives access to more detailed structural information and it is particularly useful for composite particles. Consider such a particle which consists of a core and a relatively thin coating, as shown in Figure 13.20. There will be three

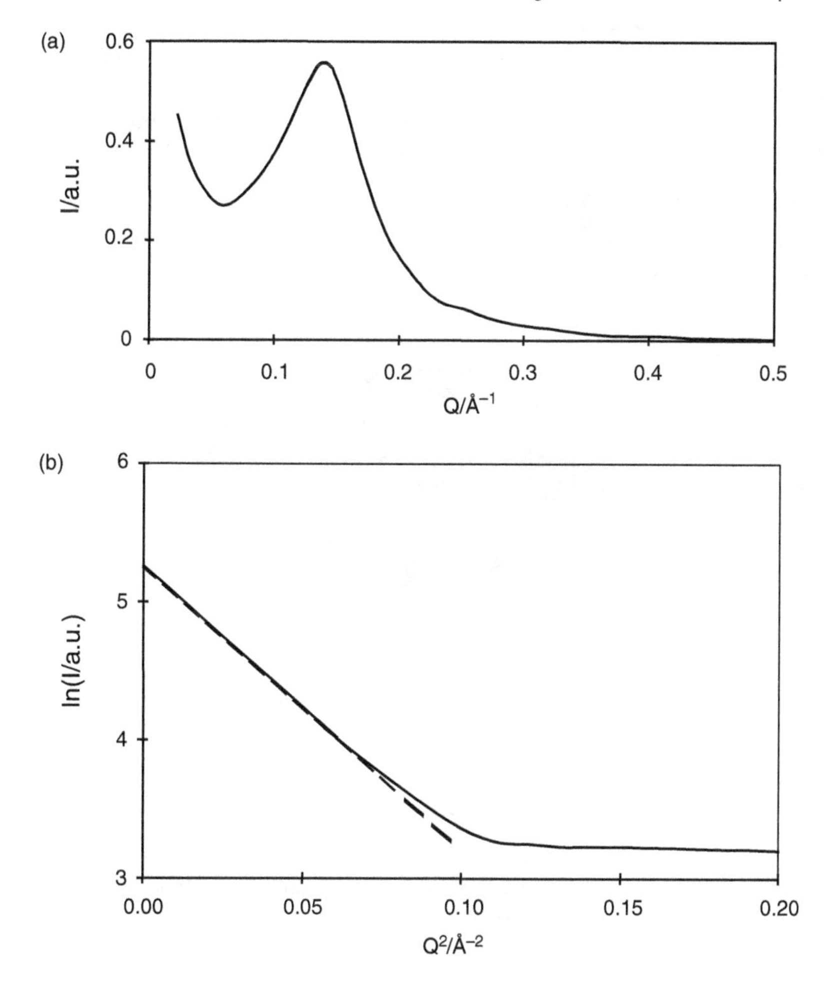

Figure 13.19 *Showing SAXS from concentrated (a) and dilute (b) calcium carbonate dispersions*

contributions: scattering from the core, scattering from the coating and scattering that comes from both. This can be modelled but it is complex and there is a tendency for the core scattering to dominate (because core has more volume than coating) so the coating structure is difficult to extract.

To use contrast variation we first arrange for the solvent to have the same scattering length density as the coating as in Figure 13.20(b). For an aqueous medium, this is done by choosing the correct ratio of H_2O and D_2O. The coating is now 'contrast matched' and the only scattering is from the core so the radius of gyration of the core, R_G, can be determined by a Guinier plot.

Now we arrange for the solvent to have the same scattering length density as the core as in Figure 13.20(c). The core is now 'contrast matched' and the only scattering is from the coating. The thickness of the coating, R_T, can be determined using a version of Guinier's law that applies to the scattering from anisotropic, plate-like objects at $Q > R_{G^{-1}}$ (21).

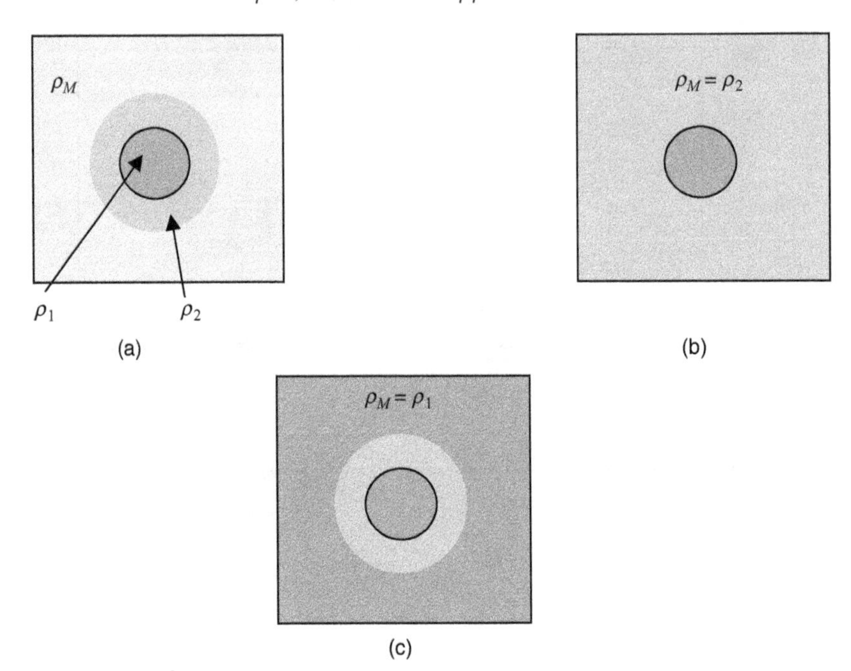

Figure 13.20 Contrast variations from a composite particle

$$I(Q) \propto \frac{1}{Q^2} \exp(-Q^2 R_T^2) \tag{13.15}$$

A plot of $\ln(Q^2 I)$ against Q^2 has a slope of $-R_T^{\ 2}$ so the coating thickness may be determined. Since $R_T = \text{thickness}/\sqrt{12}$.

13.21 High Q Limit: Porod Law

We now consider the form of the scattering at Q well above the Guinier region. Since the distance probed is inversely proportional to Q, very high Q means short distances and the only scattering comes from the step in scattering length density at the surface of the particles in a dispersion, as shown schematically in Figure 13.21.

There are no inter-particle effects because the distances probed are very much shorter than the particle separations. This high Q scattering decays as the fourth power of Q and its strength depends only on the contrast, $\Delta\rho^2$, and the amount of surface area, S, in the sample.

$$I(Q) \approx 2\pi S \Delta\rho^2 Q^{-4} \tag{13.16}$$

This is known as Porod's law (22). The scattering intensity can therefore be used to measure surface area in powders, dispersions, etc.

Porod's Law is modified if the surfaces are not smooth. The nature of a surface can be characterised by its surface fractal dimension, D_S. This concept can be understood as

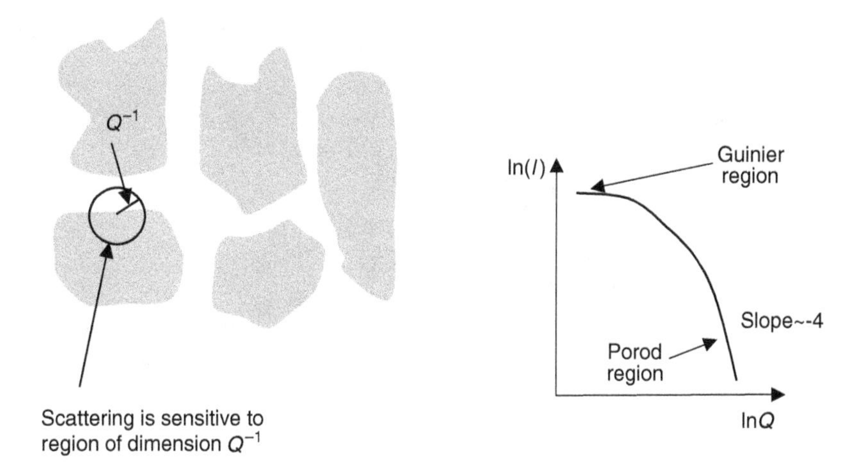

Figure 13.21 *Showing distance probed by high Q and the corresponding Porod region in the scattering curve*

follows. Consider a sphere of radius R on a smooth surface. As the radius of the sphere increases, the area of surface in the sphere increases as the second power of the radius. Hence for a smooth surface, $D_S = 2$. For a very rough, porous surface the surface area inside the sphere will increase as the third power of the radius. Hence for a rough surface, $D_S = 3$. In general the surface fractal dimension will lie between these extremes: $2 < D_S < 3$. Figure 13.22 shows the two extremes schematically.

For a fractal surface, Porod's law is extended by changing the power from 4 to $(6 - D_S)$ and so the fractal dimension may be extracted from the slope of a log/log plot of the high Q scattering, using equation (13.17). Note that for a smooth surface, Porod's law is recovered.

$$\ln(I(Q)) = A - (6 - D_s)\ln(Q) \tag{13.17}$$

where A is a constant

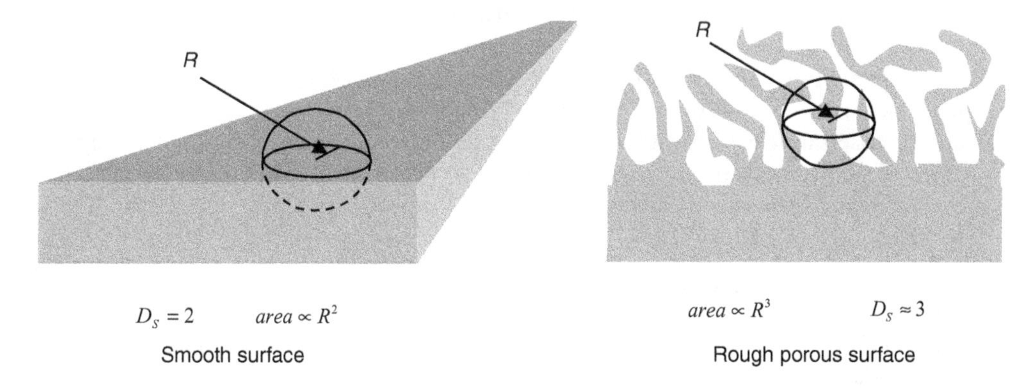

Figure 13.22 *Fractal Surfaces*

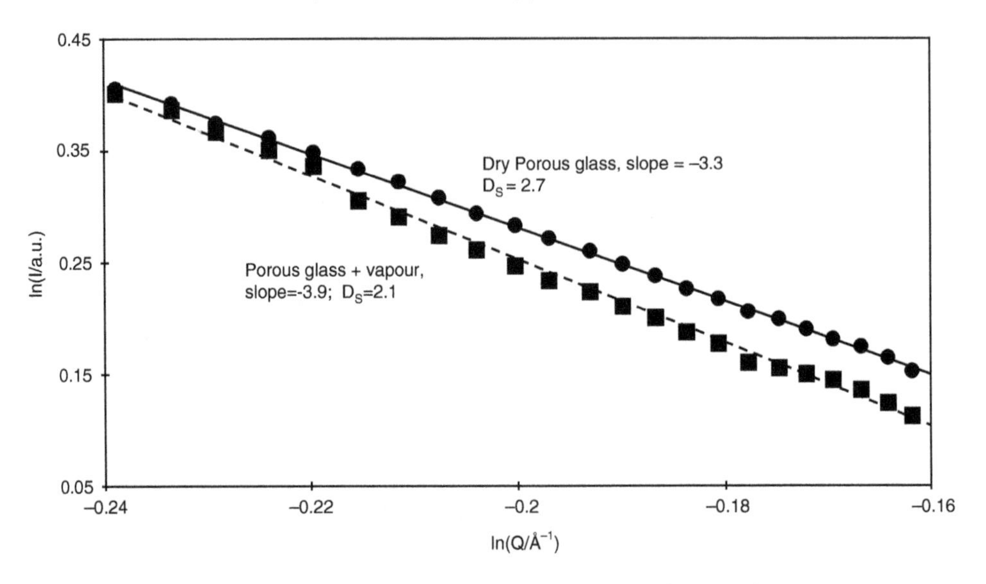

Figure 13.23 *SAXS from porous glass in dry state and exposed to vapour*

Figure 13.23 shows the high Q scattering from a sample of porous glass (Vycor). When it is dry, the slope of the log/log graph is –3.3, indicating a surface fractal dimension of 2.7 (i.e. quite rough). On exposing it to vapour (a halogenated solvent with similar scattering length density to glass) the slope is –3.9 indicating a surface fractal dimension of 2.1 (i.e. nearly perfectly smooth). The conclusion is that the pores have been filled in by capillary condensed vapour of the same scattering length density as the glass so the surface appears smooth.

The concept of fractals has many applications. For instance adsorbed polymer layers and aggregates of particles may be characterised by fractal dimensions. A more detailed discussion of scattering from surface and mass fractals may be found elsewhere (23–25).

13.22 Introduction to X-ray and Neutron Reflection

The reflectivity technique is a recently developed method for studying the structure of macroscopic surfaces (26). We have seen in Section 13.20 above that it is possible to characterise surfaces of particles using small angle scattering by contrast matching the cores of the particles to the solvent. However, reflection from macroscopic surfaces has several advantages as compared to studying surfaces of particles in dispersions. These include:

- reflection is not restricted to stable dispersions
- reflection is not restricted to core contrast matched conditions
- it is more precise because the surface contribution to the scattering is separated out experimentally as a specular reflection
- it is relatively simple to calculate reflectivity from a smooth surface exactly.

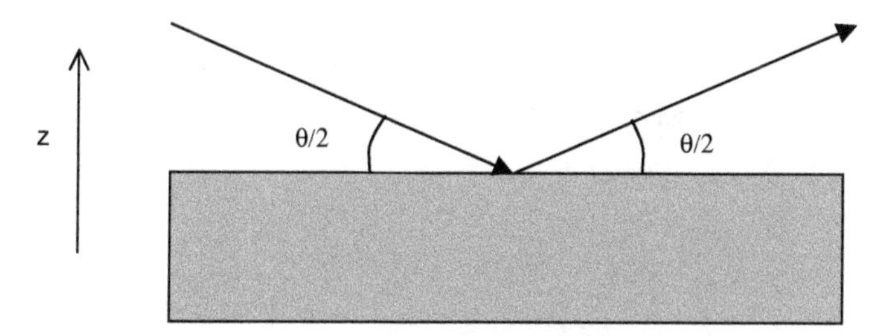

Figure 13.24 Principle of reflection experiment

The big disadvantage is that several square centimetres of flat surface are required. This is simple for the liquid–air interface but more difficult for solid–air, solid–liquid and liquid–liquid interfaces.

13.23 Reflection Experiment

The reflectivity method, shown schematically in Figure 13.24, is very simple in principle. A well-collimated monochromatic beam of X-rays or neutrons is brought onto the surface and the intensity of the reflected beam is measured. The angle of incidence is scanned to vary Q. On a pulsed source, the Q variation can be done by measuring the time of flight at a fixed angle so that Q is varied through the range of wavelengths, λ. The reflectivity as a function of Q is determined and interpreted in terms of the surface structure. It possible to purchase an X-ray reflectometer and X-ray and neutron reflectometers are also available at the central facilities mentioned above.

The reflectivity from a surface $R(Q)$ may be calculated exactly using a method originally developed for the optics of multi-layers (27). However, for the purposes of understanding reflectivity results, the kinematic approximation is very useful (28).

In this approximation there are two factors. The first factor is the reflectivity that would be observed from an ideally smooth sharp interface where the change in scattering length density between the two media is $\Delta\rho$. It is a Q^{-4} decay. The second factor results from any surface structure such as an adsorbed layer or diffuseness of the interface. It is the Fourier transform squared of the scattering length density gradient perpendicular to the surface (i.e. the z direction).

$$R(Q) = \frac{16\pi^2\Delta\rho^2}{Q^4}\left|\frac{1}{\Delta\rho}\int_{-\infty}^{\infty}\frac{\partial\rho(z)}{\partial z}e^{iQz}\mathrm{d}z\right|^2 = \frac{16\pi^2}{Q^4}\left|\int_{-\infty}^{\infty}\frac{\partial\rho(z)}{\partial z}e^{iQz}\mathrm{d}z\right|^2 \qquad (13.18)$$

13.24 A Simple Example of a Reflection Measurement

As an example of neutron reflectivity, consider a monolayer of a deuterated surfactant adsorbed at the surface of water, as shown schematically in Figure 13.25. The water can be made invisible to neutrons by using 8% by volume D_2O so that its scattering length density is zero and it contrast matches to air.

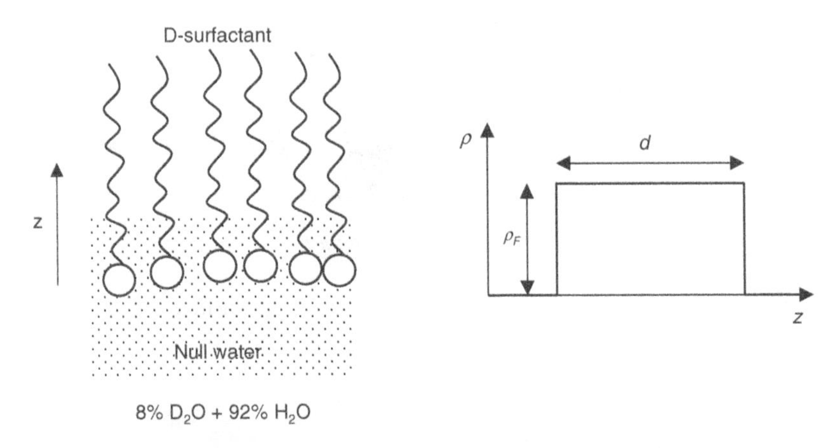

Figure 13.25 *Deuterated surfactant adsorbed at the surface of 'null' water and the corresponding scattering length density profile*

The scattering length density profile is then a simple block shape and it can be shown from standard Fourier transform results (29) that the surface structure factor has a cosine form.

$$R(Q) \approx \frac{16\pi^2}{Q^4}\rho_F^2\, 2(1-\cos Qd) \tag{13.19}$$

where ρ_F is the scattering length density of the surfactant film and d is its thickness.

If the reflectivity from such a system is plotted as RQ^4, the rapid decay is removed from the data. The position of the first maximum of the cosine is easily measured and hence the layer thickness, d, may be determined.

$$d = \pi/Q_{max} \tag{13.20}$$

The amplitude of the cosine oscillations is governed by the scattering length density of the film which can therefore be determined.

$$\rho_F = \sqrt{\frac{(RQ^4)_{max}}{64\pi^2}} \tag{13.21}$$

Since the scattering length density depends on the total scattering length of a surfactant molecule, $\sum_{molecule} b$, and the volume it occupies, the area per molecule, A, may be calculated from the scattering length density of the film.

$$A = \frac{\sum_{molecule} b}{\rho_F\, d} \tag{13.22}$$

where the summation is over the scattering lengths of all the atoms in one molecule.

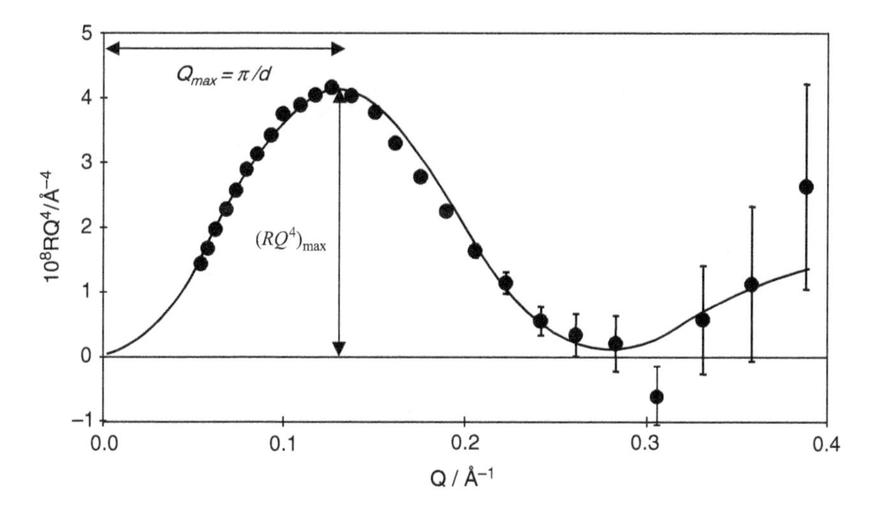

Figure 13.26 *Neutron reflectivity from d-behenic acid spread on water*

Figure 13.26 shows the reflectivity of d-behenic acid spread on water plotted as RQ^4. The data was taken using the CRISP reflectometer at the ISIS neutron source (30).

The position of the maximum indicates that the layer is 24 Å thick and since the total scattering length of the molecule is 441×10^{-5} Å, the area per molecule is determined as 23 Å^2. This simple example shows how the two most important characteristics of the surfactant layer may be determined. The method may be extended to cope with more complex interfaces and to determine more detailed structural information.

13.25 Conclusion

We have discussed how light, X-ray and particularly neutron scattering can give useful information on the structure of colloids and surfaces. Although the techniques do not give 'real space' images, the interpretation of scattering and reflection data in terms of structure is reasonably direct. The data are generally obtained from the samples in an equilibrium state and so are free from artefacts introduced by sample preparation or by the invasive nature of the probe. Hence the methods outlined are widely used in colloid and surface science. Potential users of the methods are encouraged to consult some of the books in the bibliography where more detailed introductions may be found. The web pages of the central facilities are also a useful source of further information, particularly about instrumentation.

References

(1) Kostorz G. (ed.) (1979) Neutron Scattering; Treatise on Materials Science and Technology, Vol. 15. Academic Press, New York.
(2) Als-Nielsen, J., McMorrow, D. (2001) Elements of Modern X-ray Physics. John Wiley & Sons, Ltd., Chichester.

(3) Kerker, M. (1969) The Scattering of Light and other Electromagnetic Radiation. Academic Press, New York.
(4) Guinier, A. (1994) X-ray Diffraction in Crystals, Imperfect Crystals and Amorphous Bodies. Dover, New York.
(5) Lindner, P., Zemb, Th. (eds) (2002) Neutron, X-ray and Light Scattering Methods Applied to Soft Condensed Matter. North-Holland, Amsterdam.
(6) Pusey, P. N., Taugh, R. J. A. (1982) in Dynamic Light Scattering and Velocimetry: Applications of PCS, ed. Pecora, R. Plenum Press.
(7) European Synchrotron Radiation Facility http://www.esrf.fr/
(8) Diamond Light Source http://www.diamond.ac.uk/
(9) Insitut Laue Langevin http://www.ill.fr/
(10) ISIS Pulsed Neutron and Muon Source http://www.isis.rl.ac.uk/
(11) The European Neutron Scattering Association http://neutron.neutron-eu.net/n_ensa/
(12) Sears, V.F., (1992) Neutron News 3: 26 and http://www.ncnr.nist.gov/resources/n-lengths/
(13) Lord Rayleigh (1911) Proc. Roy. Soc. (London), A-84: 25.
(14) Guinier, A. (1939) Ann. Phys., 12: 161.
(15) Feigin, L. A., Svergun, D. I. (1987) Structure Analysis by Small Angle X-ray and Neutron Scattering, Plenum Press.
(16) Glatter, O., Kratky, O. (1982) Small Angle X-ray Scattering, Academic Press, London.
(17) Potton, J. A., Daniell, G. J., Rainford, B. D. (1986) Neutron Scattering Data Analysis; Inst. Phys. Conf. Ser. no 81, IOP Publishing, Ch 3, p. 81.
(18) Hayter, J. B., Penfold, J. (1984) J. Phys. Chem., 88: 4589.
(19) Ottewill, R. H. (1982) in Colloidal Dispersions, Goodwin, J. W. (ed.). Royal Society of Chemistry, London, p. 143.
(20) Hayter, J. B., Penfold, J. (1983) Colloid Polym. Sci., 261: 1022.
(21) Kratky, O., Porod, G. (1948) Acta Physica Austriaca 2: 133.
(22) Porod, G. (1951) Kolloid-Z., 124: 83.
(23) Surface Properties of Silica, Legrand, A. P. (ed.), (1998). John Wiley & Sons, Ltd, Chichester.
(24) Bale, H. D., Schmidt, P. W. (1984) Phys. Rev. Lett., 53: 596.
(25) Allen, A. J., Schofield, P. (1986) Neutron Scattering Data Analysis; Inst. Phys. Conf. Ser. no 81, IOP Publishing, Ch 3, p. 97.
(26) Bucknall, D. G. (1999) in Modern Techniques for Polymer Characterisation, Pethrick, R. A., Dawkins, J. V. (ed.). John Wiley & Sons, Ltd, Chichester.
(27) Penfold, J., Thomas, R. K. (1990) J. Phys. Conden. Matter, 2: 1369.
(28) Als-Nielsen, J. (1885) Z. Phys., B 61: 411.
(29) Champeney, D.C. (1973) Fourier Transforms and Their Physical Applications. Academic Press.
(30) Grundy, M. J., Richardson, R. M., Roser, S. J., Penfold, J., Ward, R.C. (1988) This Solid Films 159: 43.

14

Optical Manipulation

Paul Bartlett

School of Chemistry, University of Bristol, UK

14.1 Introduction

Manipulating, analysing and organising the mesoscopic structure of materials is probably the most challenging problem currently facing Soft Matter science. Soft materials, which include polymers, colloids, microemulsions, micellar systems and their aggregates, are characterised by a wide range of length scales, ranging from tens of nanometers to tens of micrometres, forces from femtonewtons to nanonewtons and timescales which span microseconds to hours. Organising material on these scales has traditionally only been possible through a subtle *chemical* control of interactions and dynamics. However, in the last two decades a new generation of optical techniques has emerged that allow soft matter scientists to *physically* reach down into the microscopic world, grab, move and transform dielectric objects at will, with almost nanometer precision. This chapter summarises the ideas behind these new powerful optical manipulation techniques and highlights a few recent applications in Soft Matter science. More detailed reviews of this area are included in the articles by Grier (1) and Molloy and Padgett (2), while recent developments in optical manipulation techniques have been summarised by Dholokia *et al.* (3).

14.2 Manipulating Matter with Light

Moving objects with light seems, at first sight, the stuff of science fiction stories. Indeed 'tractor beams' play a big role in classic Sci-Fi tales such as Star Trek where the

Colloid Science: Principles, methods and applications, Second Edition Edited by Terence Cosgrove

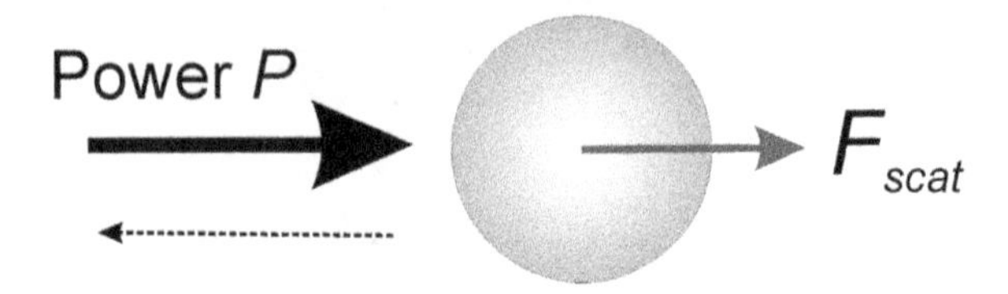

Figure 14.1 *Optical scattering force on a sphere*

U.S.S. Enterprise's laser beam is used to pull in crafts, tow another ship or hinder the escape of an enemy spacecraft (4). The stories may be a little fantastical but the science is sound. Light can move matter because photons carry a momentum. Each photon of wavelength λ has a momentum $p = h/\lambda$, where h is Planck's constant. Illuminating an object leads to a change in the direction of light as a result of refraction, reflection and diffraction. The incident momentum of the beam of photons is changed and so from Newton's laws of motion the object must also experience a force. Of course, the forces are not large enough to move spaceships (unless the beam is phenomenally intense!) but for small objects, such as micrometre-sized particles, the forces are sufficient to allow them to be moved at will.

A very simple calculation, illustrated in Figure 14.1, confirms the magnitude of optical forces. Imagine a light beam of power P incident on a microscopic sphere. As every photon carries an energy $h\nu$ the number of photons incident on the particle per second is $P/h\nu$. If a fraction q of the beam is reflected back then the momentum transferred to the particle leads to a scattering force on the particle of

$$F_{\text{scat}} = 2\left(\frac{P}{h\nu}\right) \times \left(\frac{h}{\lambda}\right) \times q = \frac{2qnP}{c} \tag{14.1}$$

where c is the velocity of light and n is the refractive index of the medium. Inserting typical values gives a crude estimate of the strength of optical forces. Assuming, for instance, a 100 mW laser beam focused on a dielectric sphere of radius λ, for which q is of order 0.05 (5), yields an optical force of about 40 pN. Although this force is obviously far too small to move anything as large as a spacecraft, at the microscopic-level such piconewton-level forces can have a very significant effect. To see this we need to consider the typical magnitudes of forces found in colloidal systems.

Figure 14.2 shows schematically the range of typical forces encountered in soft matter science. Rupture of covalent bonds requires forces of order 1–2 nN while forces of about 20–50 pN are sufficient to unravel polymer chains, convert DNA from a double helix to a single-stranded chain, or to break most van der Waals interactions. Colloidal forces are typically an order of magnitude smaller than the interaction forces between micrometre-sized colloidal particles and are on a scale of a few ten to hundreds of femtonewtons. Probably the weakest forces encountered are those originating from gravity, where the sedimentation force on a colloidal particle is typically a few femtonewtons. Optical techniques, as we shall see below, may be used to apply and measure forces ranging between ~10 fN and ~100 pN, making them ideally suited to the study and manipulation of soft matter.

The first three-dimensional optical traps were built by Arthur Ashkin working at Bell Labs in the early 1970s (6). The trap consisted of two counter-propagating weakly diverging laser beams, as shown schematically in Figure 14.3. At the equilibrium point the axial force

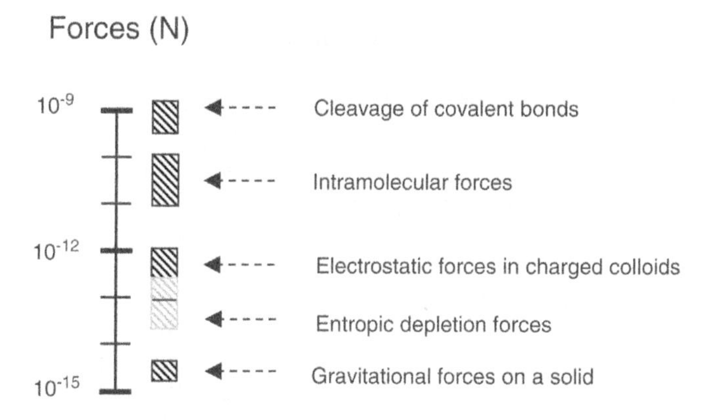

Figure 14.2 *The strengths of forces encountered in soft matter science*

F_{scat}, generated by the *scattering* of photons from both beams, exactly balance and the particle is stably trapped. Any motion along the axis leads to a net scattering force which moves the particle towards the equilibrium point (Figure 14.3a). Although this might be expected from our discussion above Ashkin saw, rather surprisingly, that the particle was also confined radially (Figure 14.3b). This observation, although apparently mundane, is key to the successful development of optical trapping techniques. It was the first demonstration that radiation pressure could also produce a *transverse* force component, which acts perpendicular to the line of the beam. The transverse or *gradient* force F_{grad} acts to move the particle to wherever the laser field is the highest. So in the case of the counter-propagating trap any radial displacement of the particle is opposed by gradient forces generated by both beams. The particle is stable against random displacements in all three dimensions and 'optical trapping' had arrived in the laboratory.

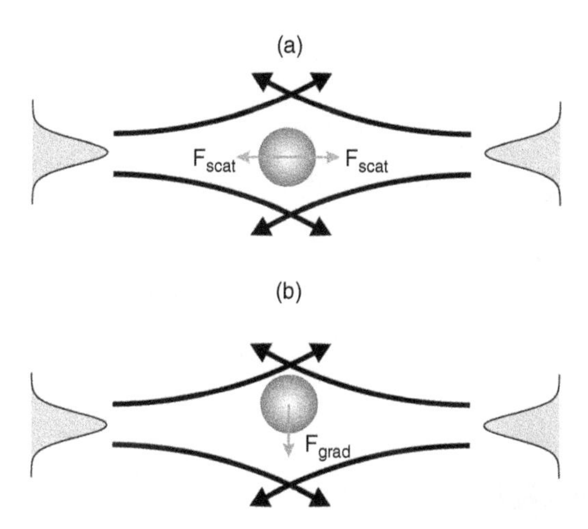

Figure 14.3 *An optical trap generated by two counter-propagating beams. (a) At the equilibrium position the axial scattering forces F_{scat} generated by each beam exactly balance, while (b) when the trapped particle is displaced radially there is an unbalanced lateral gradient force F_{grad}.*

While the counter-propagating trap worked, alignment was rather tedious and the need to get optical access from two sides restricted its use. It was therefore a significant breakthrough when Ashkin showed in 1986 (7) that the gradient forces produced near the focus of a *single* tightly focused laser beam could trap a transparent particle in three dimensions. In the last two decades, this technique, now referred to almost universally as 'optical tweezers', has become a mainstream tool in nanotechnology and biology.

14.3 Force Generation in Optical Tweezers

To understand how a single tightly focused laser trap works, study Figure 14.4.

Figure 14.4 shows the passage of light rays as they travel through a transparent sphere with a high index of refraction. At the surface of the sphere light is refracted and is bent towards the normal on entry and away from the normal on exit, according to Snell's law. Each time the light ray is refracted there is a change in photon momentum and so from Newton's third law the particle will feel an equal and opposite force. Figure 14.4(a) illustrates the case when the particle is displaced to the left of the centre of the beam. More light is refracted to the left than to the right so the particle feels a net force directed to the right and towards the centre of the beam. When the particle is moved up, away from the beam focus, as shown in Figure 14.4(b), then light rays are refracted upwards which generates a reaction force on the sphere which pulls it towards the beam focus. The net effect is that the motion of the particle is constrained in all three dimensions.

This ray optics approach gives remarkably accurate estimates for the strength of optical trapping provided the sphere is significantly larger than the wavelength of the laser (8). For smaller particles it is better to use arguments based on the strength of the electric field at the trapped particle (9). Focusing a laser beam generates an intense electric field at the beam focus, as shown schematically in Figure 14.5. The effect of the electric field is to polarise the

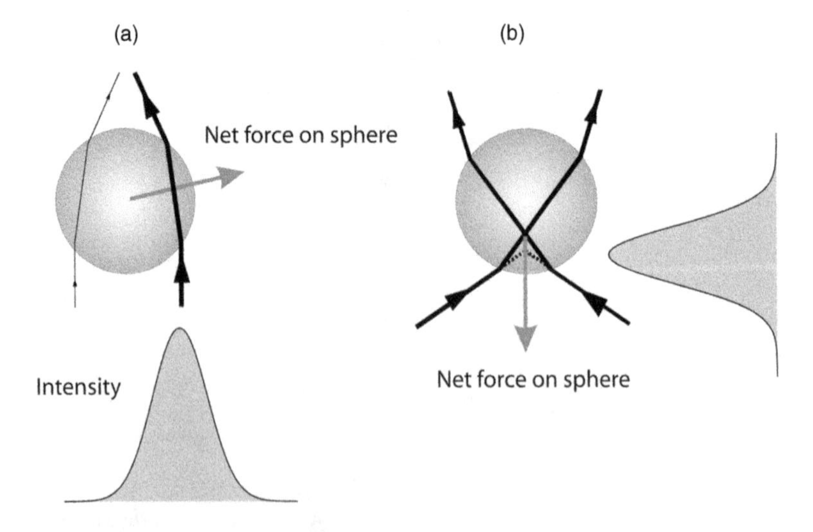

Figure 14.4 *Force generation in a single-beam optical gradient trap*

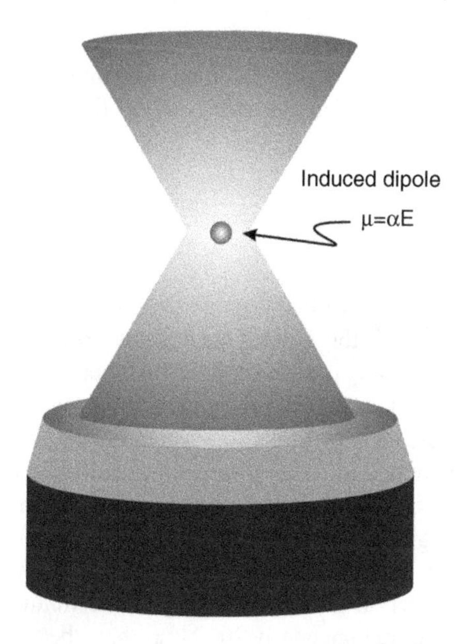

Figure 14.5 *A sphere trapped by the intense electric field at the focus of a single-beam gradient trap*

trapped sphere and generate a time-dependent induced dipole $\mu = \alpha E$, whose size depends upon the polarisibility α of the sphere. To first order, the polarisibility α of a sphere of radius r varies like $\alpha \approx (n_p - n_m)r^3$, where n_p and n_m are the index of refraction of the particle and medium respectively. So a sphere with an index of refraction above that of the medium has a positive polarisibility.

The effect of the oscillating induced dipole is two-fold. First it emits radiation which gives rise to a scattering force on the particle. The intensity of light scattered is proportional to the square of the induced dipole so the strength of the scattering force scales as the square of the polarisibility, $F_{scat} \approx \alpha^2 E^2 \approx (n_p - n_m)^2 r^6 I$ where I is the laser intensity. Although not apparent from this equation, F_{scat} is parallel to the direction of light propagation so the particle is guided along the beam by the scattering force. The second effect arises from the field gradient present near the beam focus. An aligned dipole, which consists of a separated positive and negative centre of charge, feels no net force when placed in a uniform electric field. However, in a spatially varying field the electric fields at the positive and negative centres of charge differ so that there is a net force on the dipole. The strength of this gradient force is $F_{grad} \approx \mu \nabla E$ which using our earlier expression for the induced dipole is easily seen as equivalent to $F_{grad} \approx (\alpha/2)\nabla E^2$ (∇ is the differential operator). The gradient force is linear in the polarisibility and is directed towards the region of highest light intensity in the case where $\alpha > 0$. For stable three-dimensional trapping we must concentrate on increasing the gradient forces so that they exceed the scattering forces. This can not be done by simply adjusting the laser intensity since both the scattering and gradient forces vary linearly with intensity. Instead we need to maximise the intensity gradient near the beam focus. This is most readily achieved by using a high numerical aperture microscope objective which brings light from a wide cone angle to a sharp focus. Finally, we emphasise that the refractive index of the trapped particle must be higher than that of the surrounding medium to ensure that the gradient force is directed towards the maximum intensity region. An air bubble or generally a particle with a low-refractive index is expelled from the beam focus as the gradient forces are reversed.

14.4 Nanofabrication

Optical tweezers offer a highly controlled way of manipulating soft matter systems. There is no direct physical contact with the system so there is no possibility of contamination. Furthermore the ability to remotely position colloids in space means that it is now possible to construct new classes of materials. One of the most exciting developments in optical tweezer technology has been the creation of three-dimensional arrays of optical tweezers. These multiple trap systems are created by using a computer-controlled liquid crystal spatial light modulator to generate a highly controlled phase modulator. When illuminated with a single coherent laser beam the outgoing reflected beam is precisely modulated in phase so that when focused in the tweezer plane there is constructive interference between different parts of the beam and an array of tweezers if generated. Using this technique (known as holographic optical tweezers or HOT), more than nearly 2000 traps have been generated in a plane (10).

Holographic optical tweezers are computer controlled and can be reconfigured rapidly so that the array of traps can be adjusted in space and the structure rotated or modified in real time. Figure 14.6 shows, for instance, a sequence of video images of eight spheres trapped at

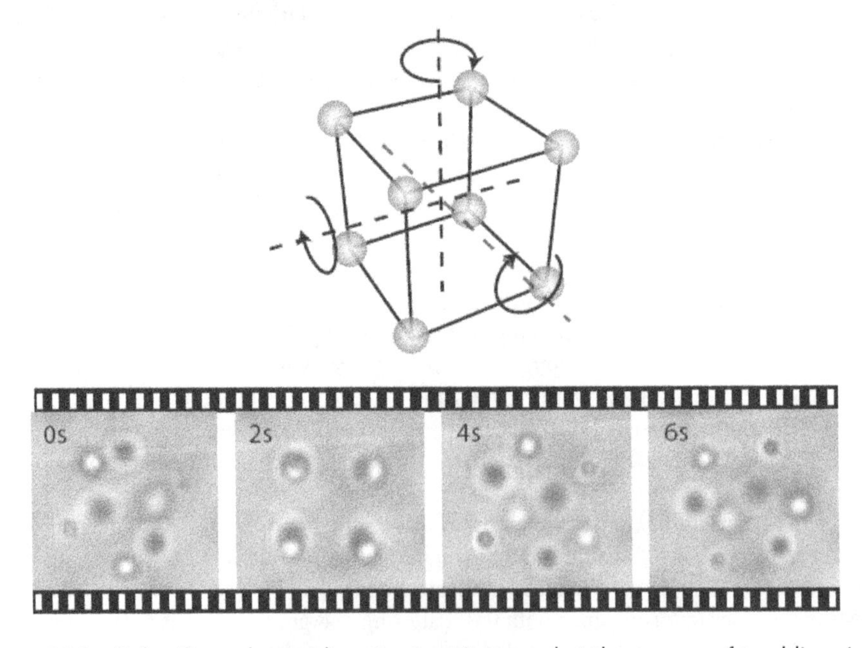

Figure 14.6 *Eight silica spheres (diameter 2 μm) trapped at the corners of tumbling simple cubic lattice. (Reprinted with permission from Ref. (11). Copyright (2004) Optical Society of America)*

positions corresponding to the corners of a 'tumbling' cube where the resolution of the spatial light modulator allows the unit cell size to be set arbitrarily between 4 and 20 μm.

Such controllable three-dimensional patterning is currently under active investigation as a means of organising nanoparticles into photonic band-gap materials.

14.5 Single Particle Dynamics

For the last few decades much of what we know about the dynamics of soft matter has been derived from bulk techniques such as light, X-ray and neutron scattering which measure the properties of a statistically very large ensemble of particles (typically $> 10^{12}$). The averaging inherent in these measurements means that one can not measure the entire distribution of particle properties. Single particle techniques offer a wealth of new data on individual properties which allow much more stringent testing of ideas and also the potential to reveal entirely new behaviour that is not discernible in averaged results particularly from heterogeneous populations. In this section we show how optical tweezer techniques allow us to measure the dynamics of individual colloidal particles with nanometer spatial and microsecond temporal resolution. We illustrate the power of this method by exploring the heterogeneous dynamics of colloidal gels.

14.5.1 Measuring Nanometer Displacements

The three-dimensional position of a sphere held in an optical trap can be measured with a resolution of a few nanometers using the four-quadrant photosensor depicted below.

The sensor relies on interference between the light scattered forward by the particle and the transmitted trapping laser (12). Motion of the particle within the trap changes the direction of the scattered light and so alters the interference pattern. The interference image is projected onto a four-sector quadrant photodetector. The resulting photocurrents are amplified and combined to yield voltage signals which are proportional to the x-, y- and z-coordinate of the trapped sphere. Because of the intense illumination the resolution is very high and using low noise electronics it is possible to achieve nanometer resolution over a bandwidth from about 1 Hz to 10 kHz. Figure 14.7 shows the detector response when

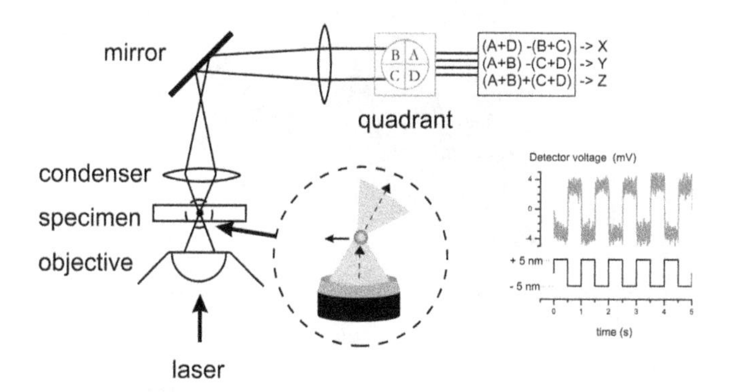

Figure 14.7 *Quadrant photosensor detection of particle displacement*

a sphere is scanned back and forth in 10 nm steps. Clearly the noise level is at the nanometer level.

14.5.2 Brownian Fluctuations in an Optical Trap

A particle held within an optical trap is in fact not fixed but fluctuates in position as a result of a balance between thermal Brownian forces and the optical gradient and scattering forces. The quadrant photosensor provides a very accurate picture of these thermal fluctuations. Figure 14.8 shows an example of the chaotic random trajectory measured for a particle held within an optical trap which can fluctuate about 50 nm around the optical axis.

Analysis of the time-dependent fluctuations of the trapped particle provides a quick and accurate method to characterise the strength of the optical trap. The gradient force provides a restoring force which, over distances of several hundred nanometers, is a linear function of the displacement x of the particle from the centre of the beam focus so that $F_{grad} = -kx$ where k is the stiffness or force constant of the optical trap. The trapped particle is essentially bound by a weak spring to the centre of the trap. The particle, however, does not oscillate (as it would in air or a vacuum) because the motion is heavily damped by the surrounding viscous liquid medium. Figure 14.9(a) shows that the particle fluctuates on a timescale of the order of 10 ms but the motion is very erratic as a consequence of random solvent collisions. The statistics of the thermal fluctuations, however, reveal the nature of the interaction between particle and optical trap. Spectral analysis (which reveals the strength of each Fourier component) of the motion (shown in Figure 14.9b) shows a flat plateau at low frequencies with a high-frequency (ω) Lorentzian decay (proportional to ω^{-2}), characteristic of thermal fluctuations in a harmonic potential. The solid line shows that the measured fluctuations are well described by theory. Similar information is obtained from the

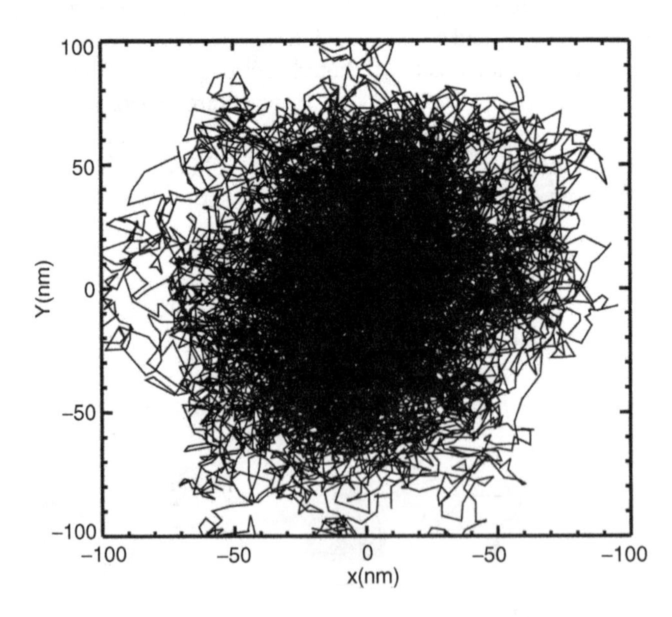

Figure 14.8 The Brownian trajectory (measured for 0.8 s) of a PMMA microsphere of diameter 0.8 μm held in an optical trap

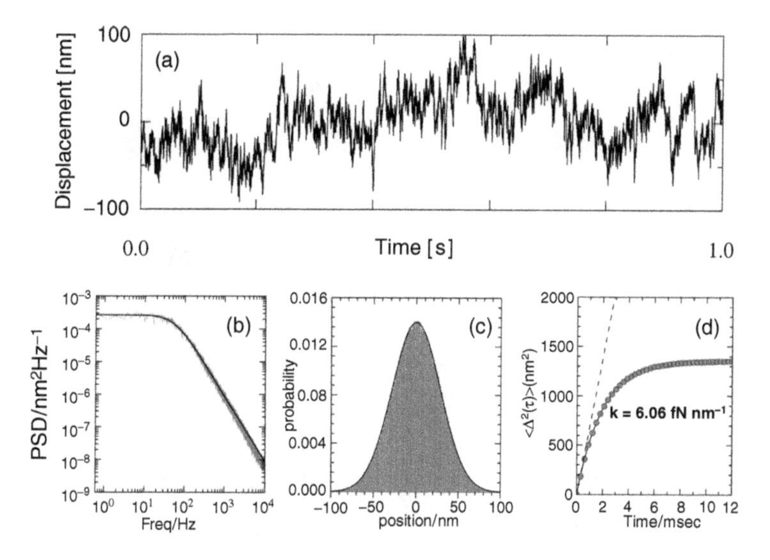

Figure 14.9 *The thermal fluctuations of a particle held within an optical trap*

probability of finding the particle a distance x away from the beam centre, which is readily calculated from the measured trajectory, an example of which is shown in Figure 14.9(c). This distribution yields directly the optical trap stiffness k since from the Boltzmann law $P(\mathrm{x}) = A\ \exp\,(-kx^2/2k_{\mathrm{B}}T)$. Finally, Figure 14.9(d) depicts the mean-square displacement (MSD) $\langle \Delta x^2(t) \rangle = \langle\, |\, x(t) - x(0)\,|\,^2 \rangle$ of the particle as a function of time. Fitting the measured MSD gives the stiffness of the optical trap as $6\,\mathrm{fN\,nm^{-1}}$. Consequently, a typical position resolution of 1 nm equates to a force resolution of about 6 fN.

14.5.3 Dynamical Complexity in Colloidal Gels

Particulate gels, produced by adding a non-adsorbing polymer to a stable colloidal suspension, play a wide role in industry. Gels are intrinsically complex, soft, multiphase systems with an internal organisation or microstructure which varies with length scale. It is this structural complexity which has frustrated previous attempts to link the bulk properties of gels to what is happening on the individual particle scale. An understanding of this local environment is critical for processing and understanding the long-term stability of these materials. Particles with short-ranged attractions show an abrupt change in dynamics as the volume fraction is increased. At the arrest transition, the system transforms from a viscous liquid into a jammed, structurally disordered solid capable of sustaining a shear stress. Since many commonplace materials such as foods, pesticides, coatings and cosmetics consist of colloid or protein gels the molecular mechanism of arrest is a subject of intense scientific debate. The scientific challenge is to explain the rich variety of arrested states seen in protein and colloid gels. Optical tweezers provide a unique method to explore the very different microenvironments present within a gel.

Figure 14.10 shows the trajectory of a single spherical titania particle trapped within a gel, formed from equally sized (1.3 μm), density and index-matched poly(methyl methacrylate) spheres. The different microenvironments present within a gel sample are characterised by

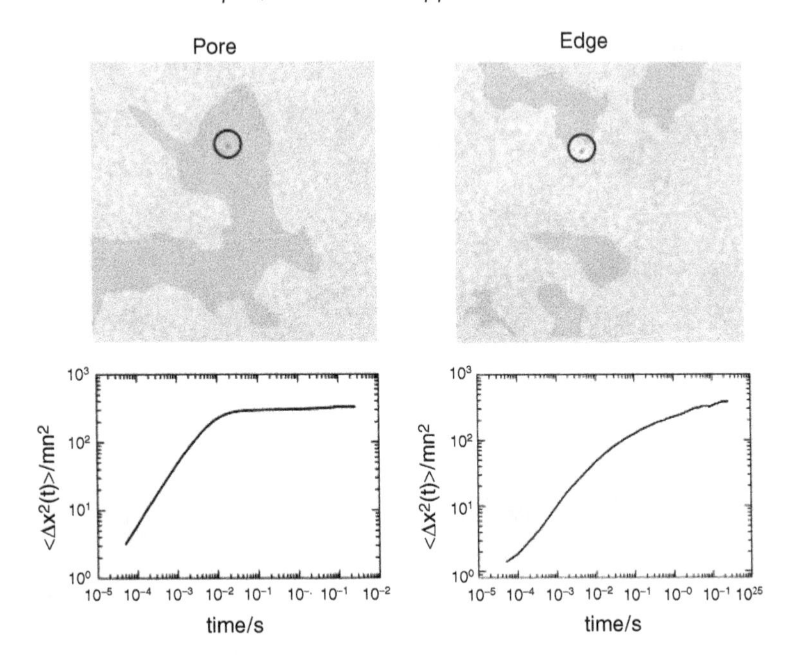

Figure 14.10 *The mean-square displacement of a single colloidal particle located within the pore space and at the edge of a particle chain. Note the significant differences between the two microenvironments*

positioning the probe particle within a pore and at the edge of the particle chains. As is clear from Figure 14.10 the different microenvironments result in very different particle dynamics. A scattering measurement, for instance, would average these different dynamics together and not reveal the true heterogeneous nature of the system.

14.6 Conclusions

This chapter has provided a short introduction to the physics of optical tweezers and has given a few simple applications of the technique to soft matter science. Optical tweezers can now trap and orient large number of particles and measure their properties with high precision. The optical toolkit is now in place. In the next few years, we can expect to see these techniques rapidly developing to the point where they become mainstream, providing researchers with the ability to control the microscopic world with unparalleled precision.

References

(1) Grier, D. G. (2003) A revolution in optical manipulation. Nature, 424(6950): 810–816.
(2) Molloy, J. E., Padgett, M. J. (2002) Lights, action: optical tweezers. Contemp. Phys., 43(4): 241–258.

(3) Dholakia, K., Spalding, G. C., MacDonald, M. P. (2002) Optical tweezers: the next generation. Phys. World, 15(10): 31–35.
(4) Krauss, L. M. (1997) The Physics of Star Trek. Harper Collins, London.
(5) van de Hulst, H. C. (1981) Light Scattering by Small Particles. Dover, New York.
(6) Ashkin, A. (2000) History of optical trapping and manipulation of small-neutral particles, atoms, and molecules. IEEE J. Selected Topics Quantum Electronics, 6(6): 841–856.
(7) Ashkin, A., Dziedzic, J. M., Bjorkholm, J. E., Chu, S. (1986) Observation of a single-beam gradient force optical trap for dielectric particles. Optics Lett., 11: 288–290.
(8) Ashkin, A. (1992) Forces of a single-beam gradient laser trap on a dielectric sphere in the ray optics regime. Biophys. J., 61(2): 569–582.
(9) Harada, Y., Asakura, T. (1996) Radiation forces on a dielectric sphere in the Rayleigh scattering regime. Optics Commun., 124: 529–541.
(10) Curtis, J. E., Koss, B. A., Grier, D. G. (2002) Dynamic holographic optical tweezers. Optics Commun., 207(1–6): 169–175.
(11) Leach, J., Sinclair, G., Jordan, P., Courtial, J., Padgett, M. J., Cooper, J., *et al.* (2004) 3D manipulation of particles into crystal structures using holographic optical tweezers. Optics Express, 12(1): 220–226.
(12) Gittes, F., Schmidt, C. F. (1998) Interference model for back-focal-plane displacement detection in optical tweezers. Optics Lett., 23(1): 7–9.

15

Electron Microscopy

Sean Davis

School of Chemistry, University of Bristol, UK

15.1 General Features of (Electron) Optical Imaging Systems

The aim of any imaging system is to produce an image from an object. Generally this also involves magnifying the image, but the main limitation on the maximum useful magnification is the resolution, which is defined as 'the smallest distance between two adjacent points which can be seen as separate'. The human eye can distinguish features separated by distances of ~0.2 mm. The resolving power of a light microscope is diffraction limited and is given by Equation (15.1):

$$r = \frac{0.61\lambda}{\mu \sin \alpha} \tag{15.1}$$

where λ is the wavelength of radiation, μ is the refractive index and α is the semi-angle subtended at the specimen.

Generally this corresponds to a resolution of about 200 nm, i.e. magnifications of around 1000×. Electron microscopy offers potentially improved resolution primarily due to the shorter wavelength of electrons compared to light. When electrons interact with a specimen a number of different signals can be produced, the most common of which are shown in Figure 15.1.

Historically, the first electron microscopes to be developed were transmission electron microscopes (TEM) in 1931; these electron optical systems being conceptually very similar to light microscopes (see Table 15.1). As well as imaging, TEM is used to obtain

Colloid Science: Principles, methods and applications, Second Edition Edited by Terence Cosgrove

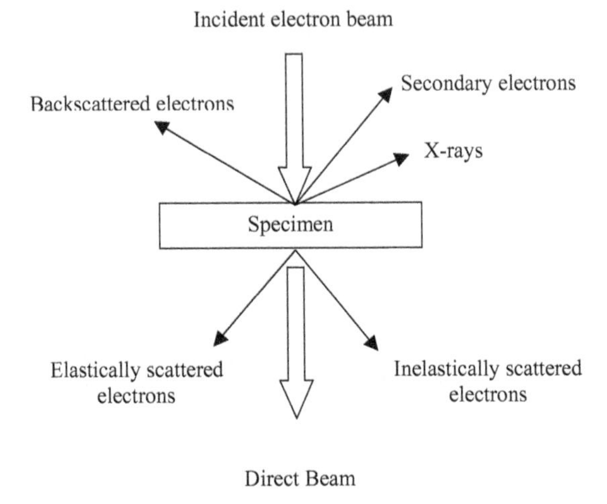

Figure 15.1 *Specimen electron interactions*

crystallographic information about samples using electron diffraction. The first commercial scanning electron microscope (SEM) was not produced until 1965. The two main types of signal used for imaging in SEM are the low energy secondary electrons (SE) and elastically scattered back-scattered electrons (BSE) (see Figure 15.2). Although the mechanism of image formation is very different to TEM, the images are easy to interpret, and the machines share a number of common components (electron gun, vacuum system, electromagnetic lenses).

The design of conventional electron microscopes imposes a number of restrictions on the nature of the specimens that can be imaged. The most obvious restriction is that in conventional electron microscopy (SEM or TEM) the specimens have to be vacuum stable (at room temperature and pressure the path length of electrons would only be a few mm). In addition the samples have to be stable to the electron beam (i.e. not thermally sensitive or photosensitive). These restrictions can be particularly important when imaging colloidal systems, and will be discussed further in the following sections.

Table 15.1 *A comparison of optical and electron microscopes*

Illumination	Electron beam	Light
Wavelength	0.0086 nm (20 keV) 0.0025 nm (200 keV)	750 nm visible
Construction	Vacuum	Atmosphere
Lens type	Electromagnetic	Glass
Aperture angle	35 mins	70°
Resolving power	0.2 nm	~200 nm
Magnification	$10-1\times10^6$; variable	10–200; fixed lens
Focussing	Electronic	Mechanical
Contrast	Scattering, absorption, diffraction, phase	Absorption, reflection, phase, polarisation

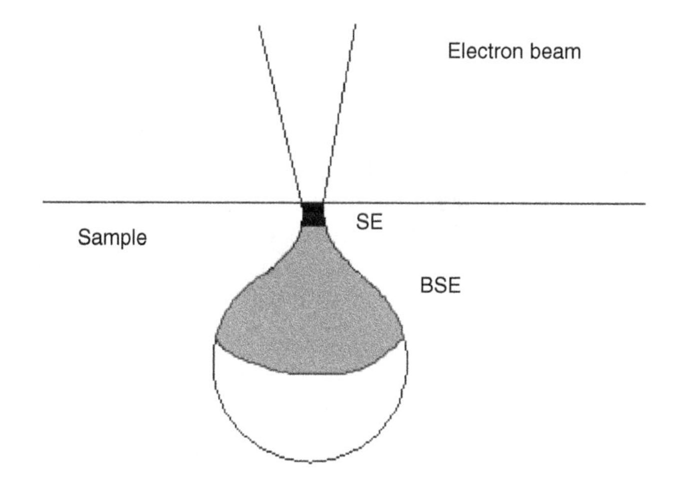

Figure 15.2 *Volume of interaction of electron beam with sample*

15.2 Conventional TEM

15.2.1 Background

In TEM the electron beam is passed through a thin specimen, and the transmitted image is magnified and focussed. Because electromagnetic lenses behave as thin lenses, electron optics can be treated similarly to light optics. However, for EM lenses, $\mu = 1$, and the electrons are deflected through very small angles so sin $\alpha = \alpha$. Therefore, $r = 0.61\lambda/\alpha$.

Resolution increases as wavelength decreases (increased accelerating voltages), and also as the objective lens aperture is made larger. For an instrument operated at an accelerating voltage of 50 keV ($\lambda = 0.0055$ nm) the theoretical resolution is $r = 0.003$ nm, which is subatomic.

In fact the actual resolution obtainable is typically of the order of 0.2–0.3 nm. The theoretical resolution is never obtained due to lens aberrations which are difficult to fully compensate for in electron optical systems. The two main sources which limit the resolution are chromatic aberrations and achromatic (spherical aberrations). Both of these types of aberration result in the electron beam being focussed in a range of positions along the 'optical' axis due to differences in the energy of the electrons (chromatic) or path length travelled (achromatic). Although chromatic aberration can be reduced using monochromatic sources, inelastic scattering means the electrons emerging from the specimen have a spread of energy. Thinner specimens and higher accelerating voltages reduce the number of inelastic scattering events, and enable higher resolution imaging. Achromatic aberrations can be reduced by using small objective apertures, i.e. selecting electrons close to the 'optical' axis. This also acts to increase contrast as electrons scattered through large angles do not contribute to the final image. However, the theoretical resolution is decreased when smaller objective apertures are used (α smaller).

Image contrast in the TEM results from three mechanisms, which can all contribute to the image: mass/thickness, diffraction and phase. In brightfield imaging the objective aperture is centred around the optical axis, and the size of the aperture determines the number of scattered electrons that contribute to the image. Thicker or higher density regions will scatter electrons more and therefore appear darker in the image. If the specimen is crystalline then the electrons may be diffracted, and the contrast will depend on crystal orientation and strongly diffracting regions will appear dark. Phase contrast arises when electrons of different phase are allowed through the objective aperture and contribute to image formation. As most scattering events result in a change in phase, most images contain some phase contrast as it is impossible to select a small enough objective aperture to exclude all scattered electrons. If diffracted electron beams are allowed through this aperture and interfere, a lattice image is produced which allows interplanar spacings of crystalline materials to be directly measured from the image.

15.2.2 Practical Aspects

The main instrumental variables in terms of the operating conditions of the TEM are the choice of accelerating voltage and objective aperture size.

Increasing accelerating voltage results in:

(i) increased specimen penetration
(ii) shorter electron wavelengths (better resolution)
(iii) reduced image contrast
(iv) reduced specimen damage.

Increasing objective aperture size results in:

(i) better theoretical resolution
(ii) reduced contrast
(iii) increased spherical aberration.

The actual conditions chosen are generally a compromise and depend on the nature of the specimen and the information required (contrast vs. resolution).

TEM is a very powerful technique for obtaining information on colloidal systems. It is routinely used for obtaining information on particle size, shape, dispersity and aggregation. It can also provide analysis of internal structure, chemical composition and crystallographic information. The most critical step in the analysis of colloidal systems is often the sample preparation. As mentioned previously, a number of general restrictions (vacuum, thermal and photostability) are placed on the specimen due to instrument design. For TEM analysis further limitations are imposed on specimen dimensions. The specimens are usually supported on 3 mm diameter copper mesh grids, covered with a thin film of carbon or a carbon-coated polymer film. Such support films are chosen because they are low atomic weight, and amorphous, so minimise information loss in the image. In addition the sample should generally be as thin as possible ($<1\ \mu m$) to allow transmission of electrons and to minimise beam damage.

A wide variety of techniques are available to prepare samples for analysis in the TEM (Table 15.2). However, all sample processing steps have the potential to introduce imaging artefacts. Below are some selected examples of TEM studies on colloidal particles which are used to highlight the range of information that can be obtained from simple dispersions of colloidal particles and the additional information that can be acquired by some of the indirect sample preparation methods.

Table 15.2 *Common methods of specimen preparation for TEM*

Small particles
- *Evaporation (Dispersions)*
- "Dusting" (Dry powders)
- Electrostatic deposition
- Freeze-Etching

Bulk materials
- *Ultramicrotomy*
- Ion-Beam, chemical, electrochemical thinning

General
- *Staining techniques*
- 'Decoration Techniques'
- Surface replication

15.2.3 Polymer Latex Particles

Suspensions of small particles can be dried directly onto grids for examination in the TEM. For particle sizing TEM is often used as a complementary technique, e.g. to light scattering. Image analysis software allows a range of measurements to be performed on the particles (Figure 15.3).

Assuming the aggregation of the particles is limited (e.g. by dilution of suspension), so that individual particles can be readily distinguished, statistically significant measurements on a number of particles can be performed relatively quickly. As well as aggregation other possible artefacts are particle shrinkage, on drying and examination under vacuum, and beam damage. The particles shown in Figure 15.3 are relatively large (~1100 nm) so the contrast in the projected image arises from absorption/inelastic scattering. This is fine for analysing external size and shape of these relatively stable particles. For more thermally sensitive particles this may result in 'melting' under exposure to the electron beam.

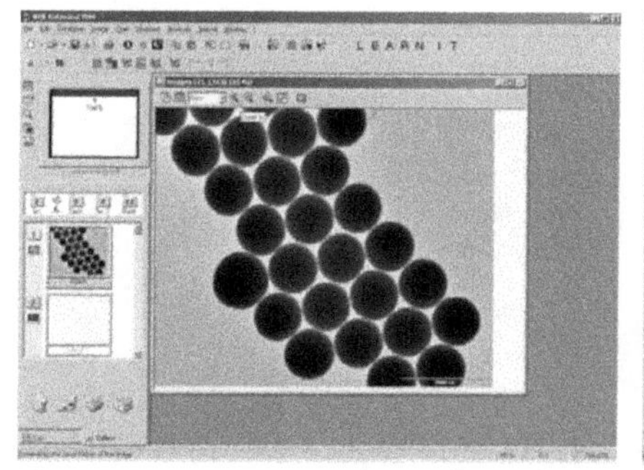

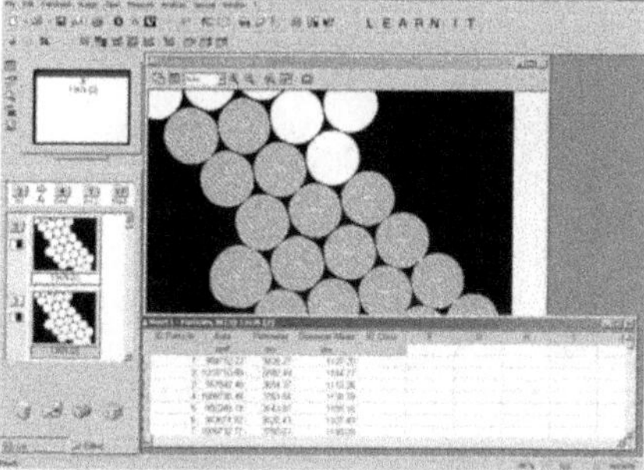

Figure 15.3 *Auto particle sizing of polymer latex particles*

15.2.4 Core/Shell Particles

TEM is particularly useful for directly imaging heterogeneous particles and allows the determination of information in addition to the external diameter. For example, Figure 15.4 shows a nano-structured coating assembled onto a polymer latex particle (1).

The shell is less dense than the core so is readily discernible in the projected image. Measurements of core diameter and shell thickness can be made, and in this particular example used to determine the increase in shell thickness as each new layer of shell material is sequentially added. However, after removal of the core, although the shells can still be imaged, it is difficult to measure the shell thickness directly. Cross-sections of the shells were prepared for TEM analysis by ultramicrotomy – a common technique for thinning bulk materials for TEM studies. The dry shell material was embedded in a polymer resin in a mould. Very thin slices (50–100 nm) of this sample were then cut using a diamond knife in an ultramicrotome. The slices were collected on a grid and images of the shells in cross-section were recorded on the TEM.

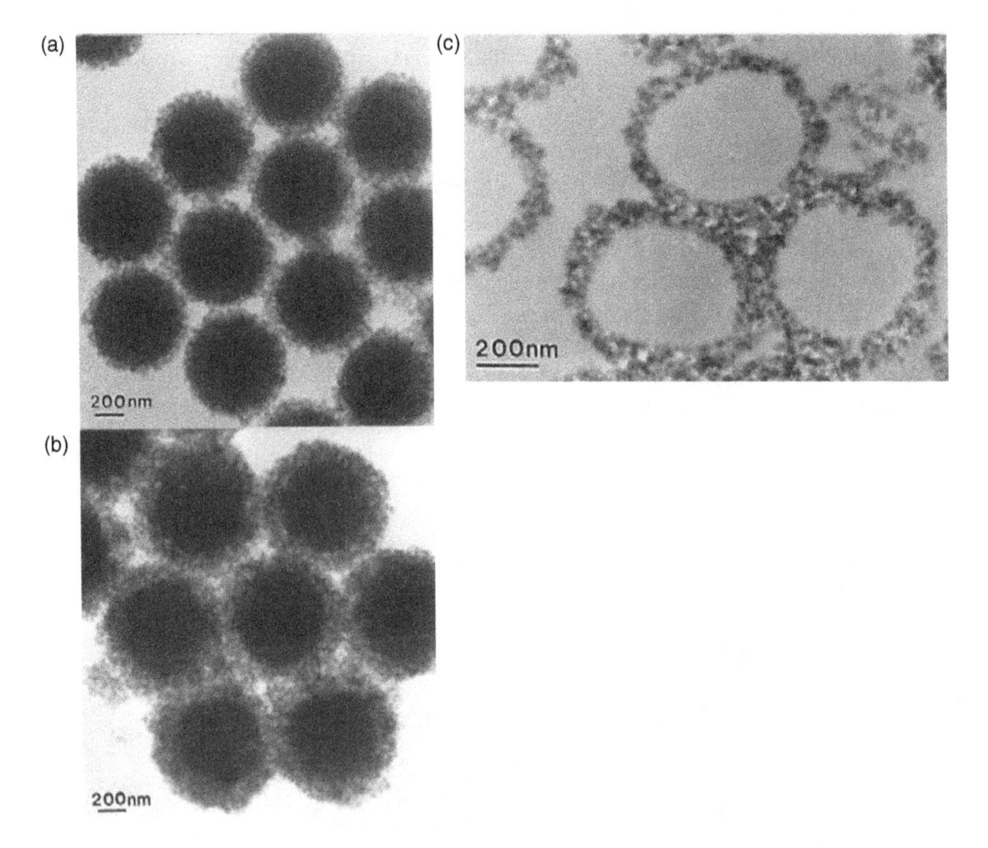

Figure 15.4 Layer by layer assembly of zeolite particles onto latex particles: (a) 3 layers, (b) 5 layers, (c) ultramicrotomed thin section of shell after core removal. (Reprinted with permission from Ref. (1). Copyright (2001) American Chemical Society)

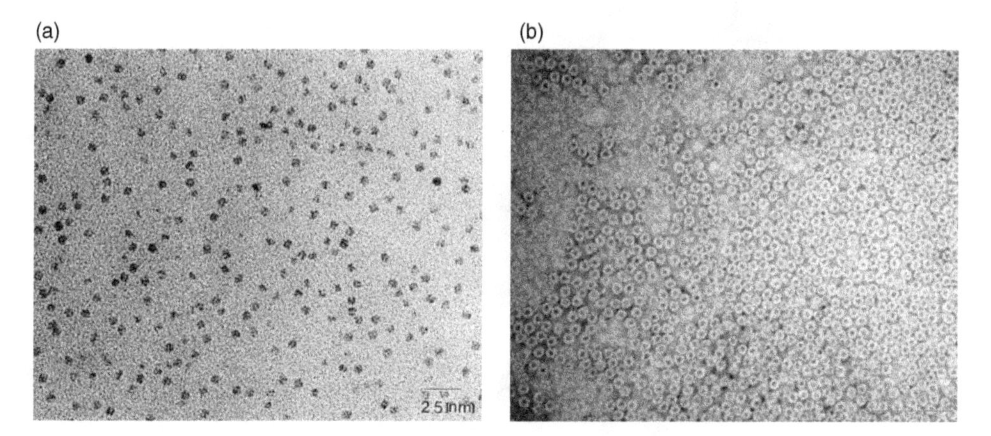

Figure 15.5 *TEM image of the iron storage protein ferritin: (a) image of iron oxide cores, (b) negatively stained sample to allow the protein shell to be imaged indirectly. Images courtesy of Mei Li, School of Chemistry, University of Bristol. Reprinted with permission. Copyright (2005) Mei Li*

As well as methods to make thin samples for TEM analysis it is often necessary to increase the contrast of material which cannot be imaged directly. The most common method for doing this is using stains, which were developed to improve the contrast of biological material in the TEM. The basic principle is that solutions of salts of heavy metals (inherently high contrast) can be applied to TEM samples to increase contrast. Positive stains act by interacting with specific functional groups (e.g. osmium tetroxide with -C=C-). Negative staining is an indirect method in which the stain solution is allowed to dry on the grid. The heavy metal salts concentrate around the low contrast material on the grid, allowing a 'negative' image to be obtained. For example, Figure 15.5 shows images of the iron storage protein ferritin. Normally only the iron oxide (ferrihydrite) core is visible when examined by TEM. If the sample is negatively stained with uranyl acetate, a halo around the dense cores can be discerned, which corresponds to the protein shell. This indirect imaging allows information on the core and shell diameter to be obtained.

However, again it must be remembered that any change in the physicochemical conditions (pH, concentration, temperature, ionic strength) during sample preparation can potentially change the nature of the colloidal system. For example, supramolecular aggregate structures of surfactants and lipids can be particularly susceptible to alteration during sample processing. For such labile systems the optimum technique is cryo TEM, where the thin film specimen is first rapidly frozen by plunge freezing in liquid ethane or propane and then the thin vitreous ice film is imaged at low temperature (liquid nitrogen) using a cryo TEM holder.

15.2.5 Internal Structure

As discussed above, one of the key advantages of TEM is the ability to image the internal structure of materials. TEM can be used to provide complementary information to other techniques such as surface area measurements for porous materials, or XRD for crystalline materials. However, the small sample volumes required for TEM means that when bulk

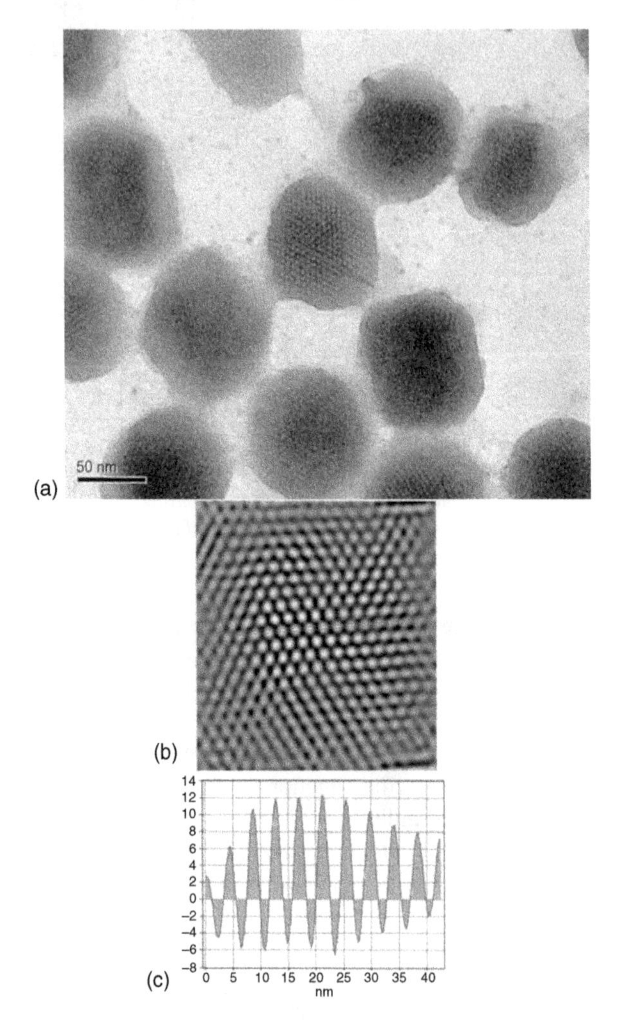

Figure 15.6 *(a) TEM image of colloidal particles of MCM-41, (b) filtered image and (c) corresponding linescan analysis of contrast variation*

analyses are not possible for comparison, characterisation can be provided by TEM analysis alone.

Over the last 10 years there has been considerable interest in meso-structured inorganic materials prepared using surfactant aggregate structures as templates. TEM analysis of such materials allows factors such as pore size, wall thickness, pore order, etc., to be determined directly. Figure 15.6 shows a TEM image of a colloidal dispersion of silica MCM-41 particles.

Image processing and analysis can be used to get accurate measurements of the pore–pore distance. For ceramic colloidal particles one common artefact is the sintering of aggregated particles, e.g. condensation reactions of surface silanol groups on silica particles. The related crystalline microporous zeolite materials are often extremely beam sensitive and

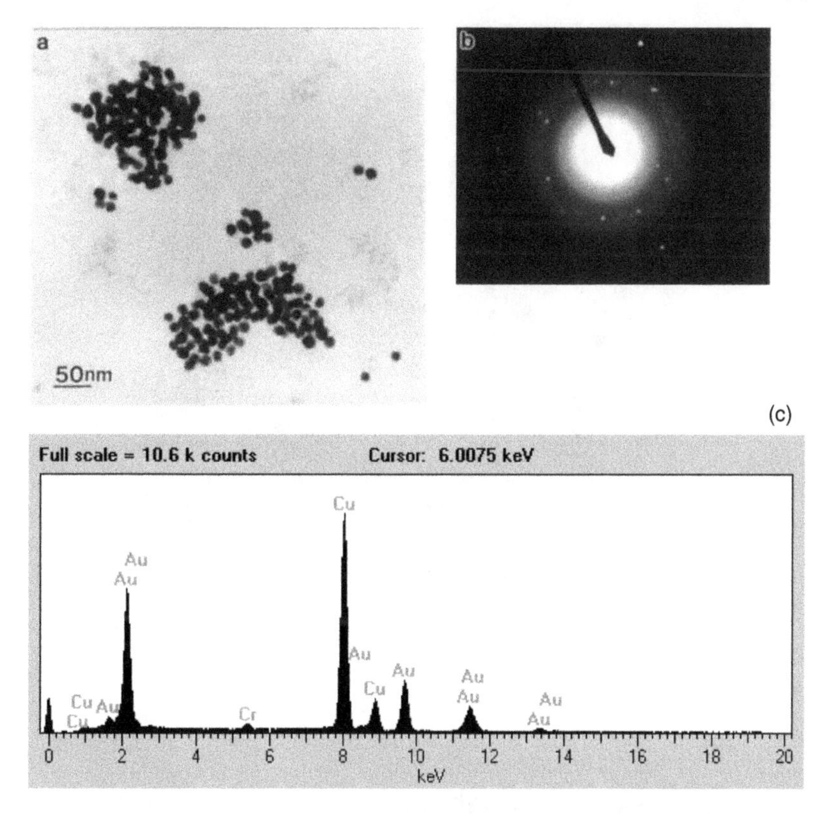

Figure 15.7 *(a) TEM characterisation of colloidal gold suspension, (b) corresponding electron diffraction pattern and (c) energy dispersive X-ray analysis spectrum (Cu peaks from support grid). (Reprinted with permission from Ref. (4). Copyright (2003) Macmillan Publishing Ltd)*

tend to lose crystallinity on prolonged beam exposure. With these and other thermally sensitive materials beam stability is improved by just imaging at low temperature, using a cryo TEM holder.

Crystalline colloidal materials are usually characterised by TEM analysis of a number of particles for size, shape, dispersity, composition and crystallinity (Figure 15.7).

Ultimately at high magnification the resolution of TEM allows lattice imaging of individual crystalline particles. The demand for high resolution TEM instruments has been driven in part by the explosion in interest in the synthesis and characterisation of nanostructured materials. Improvements in instrument specifications such as field emission guns, higher accelerating voltages, aberration corrected lenses and higher resolution digital cameras have led to HRTEM becoming a more routine characterisation technique.

Figure 15.8 shows a high-resolution image of gold nanoparticles ($\sim$3 nm). The fringes in the image correspond to the $\{200\}$ lattice planes of gold (spacing = 0.2039 nm). As well as the high-resolution imaging of individual particles it is also routinely possible to obtain diffraction and X-ray analysis from isolated nanoparticles using electron beams of nominal diameter < 10 nm. This is particularly useful when characterising particles of varying size, shape or composition. For example, Figure 15.9 shows a HRTEM image of a gold rod

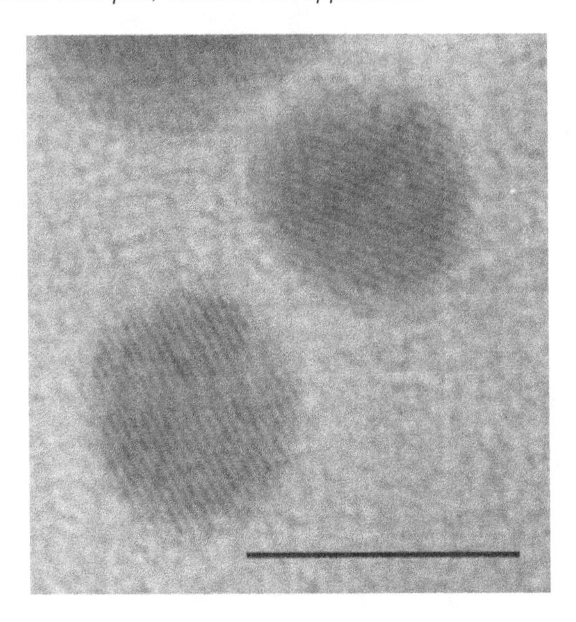

Figure 15.8 *HRTEM image of gold particles (scale bar 5 nm)*

Figure 15.9 *HRTEM image of gold nanorod viewed down the ⟨112⟩/⟨100⟩ zone showing continuous {111} fringes (d = 0.236 nm) parallel to the direction of elongation. The fringes are modulated in the central region of the twinned crystal into wider stripes due to double diffraction arising from the superposition of twin domains aligned along different zones (scale bar, 5 nm). (Reprinted with permission from Ref. (2). Copyright (2002) Royal Society of Chemistry)*

prepared by a seeded growth technique, from the particles shown in Figure 15.8 (2). From such images it was possible to determine that the rods were not single crystalline particles, but were in fact multiply twinned. Ultimately such insights allow for the improvement and optimisation of the synthesis to improve the yield of the desired anisotropic rod-like particles.

15.3 Conventional SEM

15.3.1 Background

In SEM the electron beam is scanned across the specimen surface point by point. The signal collected from each point is used to construct an image on the display, with the cathode ray tube beam and the column beam following a synchronised scanning pattern. Thus the displayed image is the variation in detected signal intensity as the column beam is scanned across the sample. The ultimate performance of the SEM is limited by the beam diameter. The lenses in the SEM do not magnify the image, they demagnify the beam. The condenser lens reduces the beam diameter from 50 μm to ~5 nm. The image is focused by adjusting the final lens such that the beam has the minimum diameter at the specimen surface. The magnification is given by the simple relationship between the width of specimen scanned relative to the width of displayed image.

15.3.2 Types of Signal

Secondary electrons are produced when incident beam electrons knock out loosely bound conduction electrons. Due to the low energy of secondary electrons (<50 eV) they can only escape if they are within ~10 nm of surface (Figure 15.2). The detected signal intensity depends on the angle between the beam and the specimen. These two factors mean that the secondary electron signal provides the highest resolution topographic information.

In contrast, the back-scattered signal is produced by elastically scattered electrons, deflected through angles between 0° and 180° by atoms within the specimen. Those scattered through greater than 90° can re-emerge from the specimen surface still with a high energy. Under similar operating conditions the signal will be produced from a larger volume than the secondary electron signal so will give lower resolution topographical information. However, scattering events are more likely with atoms of higher atomic weight (or if the incident electron has low energy) so the signal can be used to give qualitative compositional information in heterogeneous samples.

15.3.3 Practical Aspects

SEM is used to look at the surface structure of materials with a resolution <2 nm. Thus it can be used for determining particle size, shape and dispersity, (Figure 15.10) as well as chemical composition and distribution from the characteristic X-ray signal.

In the SEM, unlike the TEM, the sample dimensions are restricted purely by the physical size of the column rather than the size of lenses and apertures. Typically sample holders are in the region of 10–40 mm in diameter, and obviously the specimens can be a lot thicker. It is generally as easy to look at bulk specimens as it is thin films or dispersions. As well as the general stability criteria of the specimens that can be routinely imaged, non-conducting

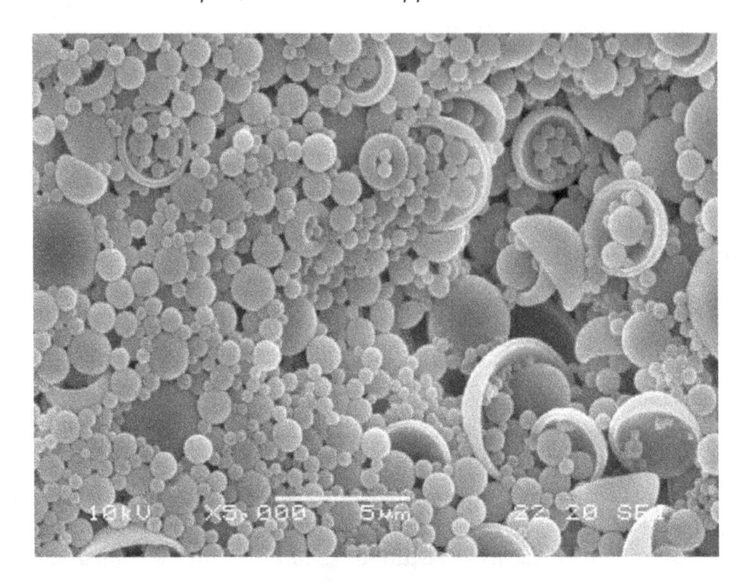

Figure 15.10 *SEM image of polydisperse latex sample*

specimens have to be coated in a thin conductive film (C, Au, Pt/Pd). This prevents charge build up on the specimen, and the associated image distortion (Figure 15.11).

The main instrumental operating conditions that can be varied are accelerating voltage, scan speed, spot size/beam current, aperture size, working distance and tilt.

The accelerating voltages used in the SEM are lower than for TEM, typically between 1 and 30 kV. The choice of voltage depends on the nature of the specimen material and the

Figure 15.11 *Powdered sample showing effects of specimen charging on image quality*

magnification range and image resolution required. Generally for high-resolution work small beam diameters are needed, thus high accelerating voltages are used to produce a good emitted signal for image formation. However, high accelerating voltages in SEM result in increased specimen penetration (e.g. Al 5 keV → 1 nm, 30 keV → 10 nm). Image information is produced from deeper within the specimen, so surface detail is lost and the chance of specimen damage is increased.

One of the main advantages of SEM over light microscopy is the increased depth of field (range of positions of object for which eye can detect no change in sharpness of image). For example, at an image magnification of 100×, the depth of field in the SEM is ~1 mm, compared with 1 μm in the optical microscope. A small aperture gives larger depth of field, as well as increased resolution. Short working distances are also used for high-resolution work, but this reduces the depth of field. For back-scattered electron imaging and X-ray analysis larger apertures and beam diameters are used to increase the beam current and hence yield of signal. Again, the increased beam current can lead to specimen damage (Figure 15.12) (3).

The scan speed is essentially increased to improve signal to noise ratio when recording images and focussing. For finding areas of interest a fast refresh rate is used (25 frames s^{-1}). The yield of secondary electrons detected can be improved by tilting the specimen (Figure 15.13), but this causes image distortion (however, particle sizes can still be determined from the images).

Another advantage of the SEM is that the characteristic X-rays detected can be plotted as an elemental composition image (Figure 15.14). Due to the larger volume from which the X-rays emerge from within the sample the resolution of these elemental maps is less than the corresponding SEM image. However, they can prove useful for determining the homogeneity of samples.

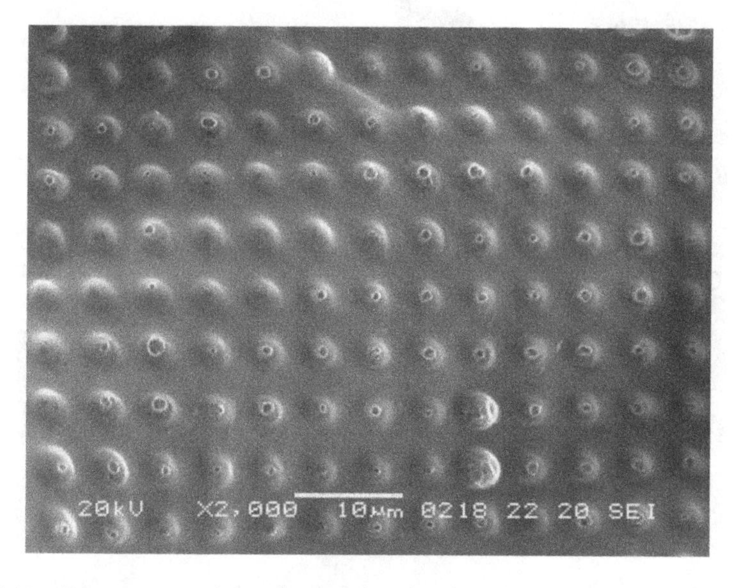

Figure 15.12 Damage to surface of a starch gel induced by high beam currents. (Image courtesy of Sean Davis. Reprinted with permission from Ref. (3). Copyright (2005) Sean Davis)

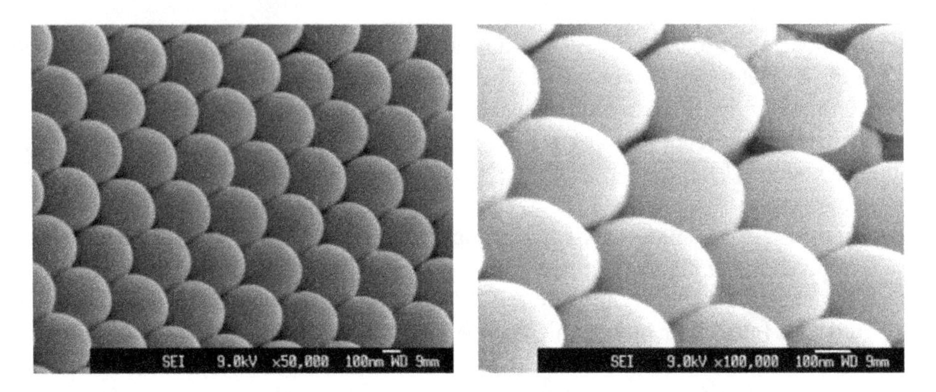

Figure 15.13 *Tilting spherical polymer latex specimen to improve signal results in image distortion*

In SEM the use of field emission guns (FEG) has resulted in improved resolution (Figure 15.15). With a resolving power of the order of a few nm, it is possible to routinely image individual particles within aggregate structures.

For example, the homogeneity of coverage of nanostructured thin films on solid substrates can be ascertained without all the associated sample preparation required for TEM analysis. Over the last decade a number of improvements to instrument design have been associated with developing non-conventional instruments which can image samples in their native state. Traditional methods of preparing hydrated samples for EM analysis

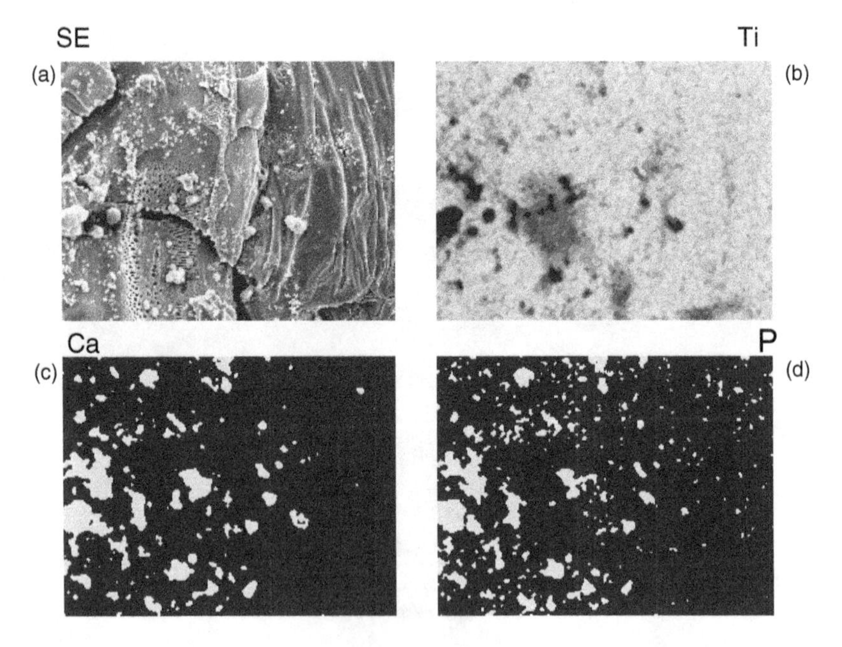

Figure 15.14 *Calcium phosphate precipitates grown on a titania substrate: (a) original image, (b) titanium, (c) calcium and (d) phosphorus images*

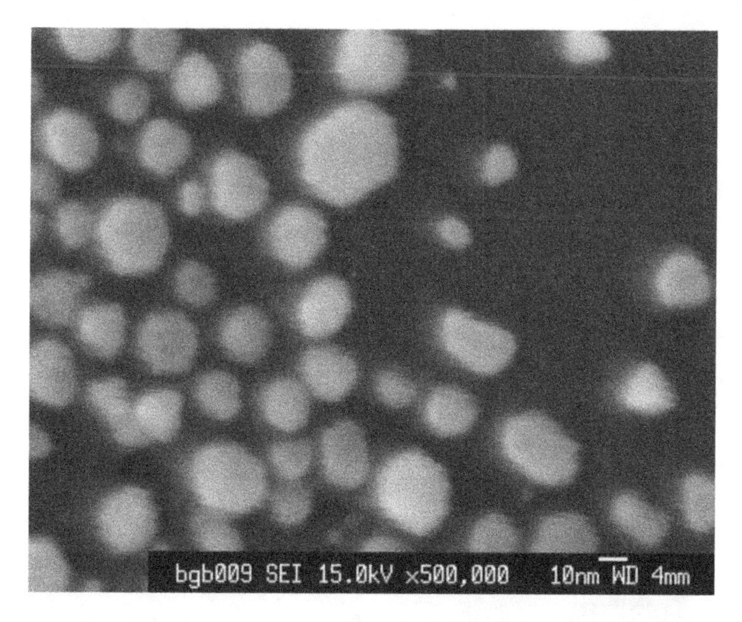

Figure 15.15 *FEG-SEM image of a resolution test specimen (gold on carbon)*

include freeze-drying and critical point drying. Although these techniques reduce structural damage caused by surface tension effects associated with air-drying wet samples, both have limitations. For example, the solvent exchange steps required to dehydrate samples prior to critical point drying (and also embedding material for sectioning) can result in some loss of structure or solubilisation of certain components. The technique of cryo SEM can minimise further the production of artefacts or loss of structure during specimen processing. Like cryo TEM, the first step is rapid freezing of the sample. The specimen is then transferred into a cooled specimen chamber, sputter coated and imaged.

The development of environmental SEMs has expanded further the range of information that can be obtained from samples (4). The gun is still maintained at high vacuum but a differential pumping system allows a low pressure of gas around the sample. Secondary electrons are used for image formation, and the resolution of these instruments is ~5 nm. The advantage of environmental chambers is that for insulators no conductive coating is required. In addition, if the gas is water hydrated specimens can be imaged. Controlling the temperature allows the state of hydration to be varied, thus dynamic processes such as aggregation and film formation can be studied (Figure 15.16).

Finally, a novel specimen holder has been developed to allow the imaging of samples in their hydrated state (5). The specimen capsules have a vacuum-resistant, electron-transparent membrane which also permits compositional analysis by EDX. Small volumes of suspension are applied to the membrane, and then the holder is sealed. Images are produced using high-energy back-scattered electrons (secondary electrons produced from the specimen are absorbed by the membrane). The detected signal is produced from particles close to or adsorbed on the membrane. Potential applications include imaging cells, emulsions, suspensions, creams, etc., with the obvious limitations of solvent compatibility and adherence to the membrane. An example is shown in Figure 15.17.

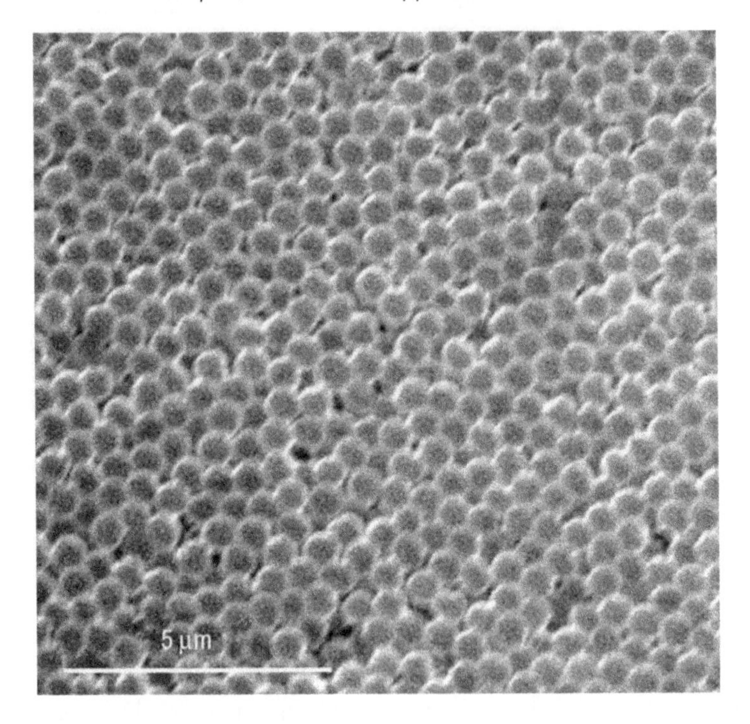

Figure 15.16 *ESEM image of partially dehydrated film forming latex particles prior to coalescence. (Reprinted with permission from Ref. (4). Copyright (2003) Macmillan Publishing Ltd)*

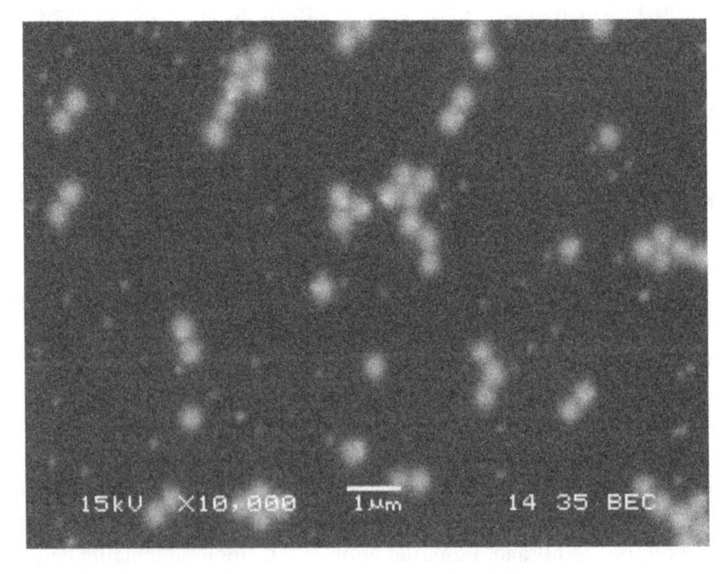

Figure 15.17 *BSE images of an aqueous suspension of 200 nm silica and 30 nm gold particles (Quantomix capsule)*

15.4 Summary

Although conventional SEM and TEM are well-established techniques, improvements in microscope design and new techniques continue to be developed. Currently a lot of these developments are being driven by the differing EM requirements of bio- and nanotechnologists. However, colloid scientists stand to benefit from these improvements as the range of techniques available for imaging soft matter and small particles increases.

The following texts are useful general references:

- *Electron Microscopy and Analysis*, P.J. Goodhew
- *The Operation of Transmission and Scanning Electron Microscopes*, D. Chescoe and P.J. Goodhew
- *Environmental Scanning Electron Microscopy*, Philips
- *The Principles and Practice of X-ray Microanalysis*, Oxford Instruments
- *A Guide to Scanning Microscope Observation*, Jeol
- www.matter.org.uk

References

(1) Davis, S. A., Breulmann, M., Rhodes, K. H., Zhang, B., Mann, S. (2001) Chem. Mater., 13: 3218–3226.
(2) Johnson, C. J., Dujardin, E., Davis, S. A., Murphy, C. J., Mann, S. (2002) J. Mater. Chem., 12: 1765–1770.
(3) Zhang, B., Davis, S. A., Mann, S. Unpublished results.
(4) Donald, A. M. (2003) Nature Materials, 2: 511.
(5) http://www.quantomix.com

16

Surface Forces

Wuge Briscoe

School of Chemistry, University of Bristol, UK

16.1 Introduction

Colloids are ubiquitous in industrial processes and products and in natural biological systems (c.f. Table 1.2 in Chapter 1), whereby colloidal particles are in close proximity to each other. The properties of these products, the efficacy of these processes and our understanding of many biological phenomena are largely determined by the interactions, called *surface forces*, between the colloidal particles in close range.

It has transpired in earlier chapters of this book that there are a number of different types of surface forces acting between colloidal particles. These include van der Waals forces and electric double layer forces (Chapter 3), and polymer-mediated surface forces (Chapters 8 and 9). In addition, we also often encounter in the literature hydrophobic interactions, structural forces, hydration forces, adhesion, capillary forces and so on. Such classifications are somewhat arbitrary and really only for our convenience of distinguishing between them. These surface forces between macroscopic bodies all originate from the forces on an atomic or molecular level.

16.1.1 Intermolecular Forces

There are four fundamental forces in nature: the *strong* and *weak* interactions between elementary particles, and the universally present *gravitational* and *electromagnetic forces*.

Colloid Science: Principles, methods and applications, Second Edition Edited by Terence Cosgrove

Intermolecular forces are electromagnetic forces, and can be further loosely divided into three categories according to their range.

1. *Coulomb forces* are the electrostatic interactions between permanent charges and dipoles, and have a long range.
2. *Polarisation forces*, giving rise to the van der Waals forces, arise from interactions between dipoles in atoms and molecules induced by nearby charges and permanent dipoles. van der Waals forces are considered long ranged on a molecular scale but short ranged on a colloidal scale.
3. Very short-ranged forces of *quantum mechanical* nature include *chemical bonds* and *steric* or *Born repulsions* due to Pauli's exclusion principle.

The exact expressions for these intermolecular forces are complex (1, 2), and the Lennard-Jones (L-J) potential is often used to describe the interaction energy or 'pair potential' $W(r)$ between two molecules a distance r apart,

$$W(r) = -\frac{C}{r^6} + \frac{B}{r^{12}} \tag{16.1}$$

where C and B are the constants respectively for the (negative) van der Waals attraction and the (positive) Born repulsion. Although semi-empirical, the L-J potential could well account for the interactions between two neutral molecules (not engaging in the chemical bonding process). For instance, we could split the constant C into three terms,

$$C = C_{\text{Keesom}} + C_{\text{Debye}} + C_{\text{London}} \tag{16.2}$$

to account for the three van der Waals components: the *Keesom* energy (interactions between two permanent dipoles), the *Debye* energy (interactions between a rotating dipole and an instantaneous dipole it induces) and the *London* dispersion energy (interactions between two instantaneous dipoles).

Many familiar macroscopic physical properties of matter can be appreciated by considering the intermolecular forces above. For instance, the *boiling point* of a (non-hydrogen bonding) liquid indicates the thermal energy required for the molecules to escape the van der Waals attraction between them (i.e. the first term in the L-J potential in Equation 16.1). Meanwhile, its melting point reflects the ability of its molecules to pack into a lattice, which depends largely on the size and shape of the molecules – these in turn are determined by the short-range intermolecular repulsion (the second term in the L-J potential in Equation 16.1). The same repulsion also gives rise to the excluded volume effect of the polymer monomers discussed in Section 7.6.4.

16.1.2 From Intermolecular Forces to Surface Forces

In theory, we could obtain the surface forces between colloidal particles by summing the intermolecular forces between all the constituent molecules in the system, including the intervening medium between the colloidal particles. If this is done, and as we know from our established experimental understanding, the intermolecular and surface forces differ in a number of aspects, here we highlight three most distinct features of surface forces: their range, the surface dominance and the confinement effect on the intervening medium.

16.1.2.1 Surface Forces are much Longer Ranged than Intermolecular Forces

Firstly, the range of surface forces is much longer than that of their originating intermolecular forces. To illustrate this, we consider only the van der Waals term in the L-J potential (Equation 16.1),

$$W(r) = -\frac{C}{r^6} \tag{16.3}$$

If the attraction between two atomic or molecular species at a distance σ (of some molecular dimension) is $W(\sigma) = -1/\sigma^6$, then at 2σ, its magnitude $W(2\sigma) = -1/(2\sigma)^6$ falls by a factor of $2^6 = 64$ already. In general, the intermolecular van der Waals forces are 'felt' in the range $r \sim 0.2$ nm to several nm.

We will compare this with that for two spherical colloidal particles of radii R_1 and R_2 respectively at a surface separation D apart (thus the centre-to-centre distance $c = R_1 + R_2 + D$), as illustrated in Figure 16.1(b). We are most interested in the case when the particles are in close proximity, i.e. $R_1, R_2 \gg D$. The total van der Waals interaction energy is obtained via the integration of the interactions between all the molecules over the volumes of the two colloidal particles,

$$W(D) = \int_{v_1} dv_1 \int_{v_2} dv_2 \rho_1 \rho_2 \left(\frac{-C}{r^6}\right) \tag{16.4}$$

where ρ_i and v_i are the number density of molecules and the volume of sphere i. To carry out this summation, we follow the approach of Hamaker (3) by first calculating the interaction between a point molecule P and the sphere R_1 at a surface separation D (Figure 16.1a).

Consider a thin shell (hatched in Figure 16.1) of thickness dr cut out by a sphere centred at P of radius r. The cut-out angle θ_0 is defined by

$$R_1^2 = b^2 + r^2 - 2R_1 r\cos\theta_0 \tag{16.5}$$

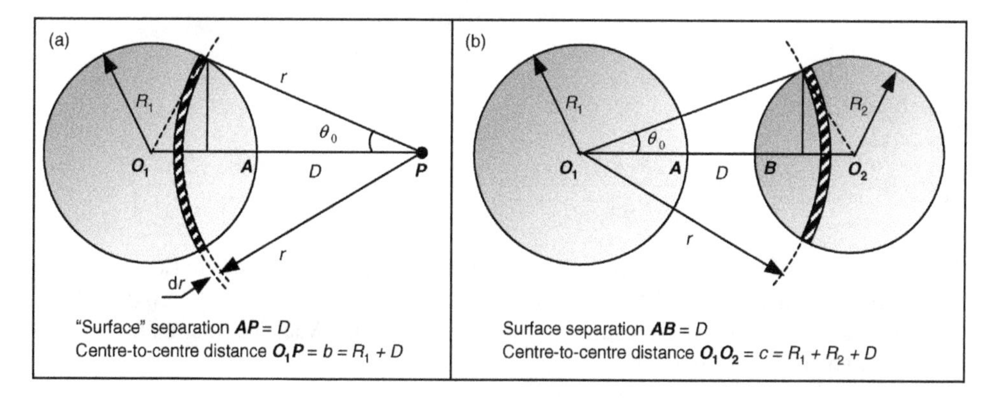

Figure 16.1 *(a) A molecule **P** interacting with a sphere of radius R_1 centred at **O_1**; (b) two spheres of radii R_1 and R_2 interacting with each other*

and the summation of the interaction between P and this thin shell volume is (c.f. Equation 16.3)

$$dW(r) = -\rho_1 dv_1 \frac{C}{r^6} \tag{16.6}$$

The volume of the thin shell dv_1 is, using Equation (16.5),

$$\begin{aligned} dv_1 &= \text{shell surface area} \times \text{shell thickness} \\ &= \left\{2\int_0^{\theta_0} \pi r^2 \sin\theta d\theta\right\} dr \\ &= \left\{\frac{\pi r}{b}\left[R_1^2 - b(b-r)^2\right]\right\} dr \end{aligned} \tag{16.7}$$

and the interaction energy between P and sphere R_1 is

$$W_1(D) = \int dW(r) = \int_D^{D+2R_1} \left(-\rho_1 dv_1 \frac{C}{r^6}\right) = -\int_D^{D+2R_1} \frac{\rho_1 C}{r^6}\frac{\pi r}{b}\left[R_1^2 - (b-r)^2\right] dr \tag{16.8}$$

Once this is done, the interactions between two spherical particles in Figure 16.1(b) can be obtained in a similar fashion by considering the interaction between sphere R_1 and a cut-out thin shell in sphere R_2 at r away from the centre O_1. From Equation (16.4), the sphere–sphere interaction energy is

$$\begin{aligned} W(D) &= \int_{v_1} dv_1 \int_{v_2} dv_2 \rho_1 \rho_2 \left(\frac{-C}{r^6}\right) \\ &= \int_{v_2} W_1(D)\rho_2 dv_2 \\ &= \int_{c-R_2}^{c+R_2} W_1(D)\frac{\pi r}{c}\left[R_2^2 - (c-r)^2\right] dr \end{aligned} \tag{16.9}$$

This finally gives

$$W(D) = -\frac{\pi^2 \rho_1 \rho_2 C}{6}\left[\frac{2R_1R_2}{c^2 - (R_1+R_2)^2} + \frac{2R_1R_2}{c^2 - (R_1-R_2)^2} + \ln\frac{c^2 - (R_1+R_2)^2}{c^2 - (R_1-R_2)^2}\right] \tag{16.10}$$

For two spheres of equal radius, $R_1 = R_2 = R$, the above equation reduces to Equation (3.1) in Chapter 3 (where $x = D/2R$); for two such colloidal particles in close proximity, i.e. $D \ll R$, it further simplifies to the familiar form for the van der Waals interaction free energy between two spheres of equal radius R,

$$W(D) = -\frac{AR}{12}\frac{1}{D} \tag{16.11}$$

where A is the now well-known *Hamaker constant*,

$$A = \pi^2 \rho_1 \rho_2 C \tag{16.12}$$

Comparing Equations (16.3) and (16.11), we recognise that, as a result of summation over the volumes of the two interacting colloidal particles, the inter-surface van der Waals interaction energy varies as $1/D$, much less rapidly than the $1/r^6$ decay between two molecules. This result holds generally for the argument of surface interactions, and as a rule of thumb, surface forces operate in the range from intimate contact to some 100 nm. In addition, we also note that the interaction in Equation (16.11) depends on the size (and indeed the geometry) of the particles, a point that is relevant when the measurement of surface forces is carried out and which we shall address shortly.

16.1.2.2 Colloidal Forces are Dominated by the 'Surface'

We see in Equation (16.11) that, in the limit of close proximity between two colloidal particles, the van der Waals interaction is a function of the *surface* separation, not of the centre-to-centre distance. In the case of the electric double layer force (c.f. Chapter 3), it arises from the surface charge either due to adsorption of charged species or due to dissociation of some ionisable surface groups (c.f. Chapter 2). Indeed, surface forces are dominated by the surface molecules. Hence, it is no coincidence that surfactants and polymers are commonly added to the system to mediate desired surface forces, as they readily anchor on the colloidal particle surface to form a surface layer, thereby modifying the surface properties of colloidal particles and hence their interactions.

This surface effect is also felt by the molecules of the intervening medium immediately adjacent to the surface. The van der Waals force field tends to densify them at the surface; surface charges universally present on colloidal particles in aqueous media tend to orient water molecules to form a tenacious yet fluid hydration layer; a hydrophobic colloid surface could induce nucleation of nanobubbles on the surface. All of these surface-induced effects could have direct and profound influence on the surface forces.

Considering above, it is thus pertinent we term the inter-colloidal particle forces *surface forces.*

16.1.2.3 Molecules of Intervening Media Experience Confinement by Surfaces

Because of their macroscopic size as compared with molecules, when colloidal particles come to a surface separation of a few molecular diameters, they create a nano-cavity. Under this condition, the confined molecules of the intervening medium can no longer be considered as a continuum – since the space between the confining walls is comparable with the molecular size – and the surface forces in this regime are very different to those at larger surface separations.

16.1.3 Why Measure Surface Forces?

We have carried out a summation of the van der Waals interaction between two spheres in Section 16.1.2.1. However, in practice, due to the lack of the prior knowledge of a system and the large number and many types of molecules involved, such a summation often cannot be easily performed. Measurement of surface forces is a direct way to find out about surface interactions experimentally, and it is clearly relevant to our understanding of colloidal stability, an issue that must be considered in every industrial process involving the colloidal state. By measuring the forces required to separate two surfaces from intimate contact, we could obtain the adhesion energy between the surfaces. In addition, given that surface forces are very sensitive to the surface condition, in particular to the structures of the surfactants and polymers on the surface, by studying the surface forces they mediate we can learn a great deal about these surface structures. For instance, a small volume fraction (a few percent) of polymer adsorbed on a surface is rather challenging for any scattering technique to 'see' (c.f. Chapter 8.4.1), but it would result in a detectable surface force when two such surface polymer layers begin to interact. Furthermore, the effect of confinement on the intervening medium can also be studied, in fact quite uniquely so, through surface force measurement. With the advancement of nanotechnology, there is a continuous drive for smaller and smaller components. The surface to volume ratio associated with this process ever increases, and so does the importance of the considerations of the surface interactions involved.

16.2 Forces and Energy; Size and Shape

Hitherto, the terms surface forces F, interactions and interaction free energy W have been used in an interchangeable manner, although they are indeed different but related to each other. Experimentally, we often measure the forces $F(D)$ between two bodies as a function of their surface separation D, which should be size and geometry dependent. Theoretically, it is most convenient to calculate the interaction free energy per unit area $W_a(D)$ between two parallel *planar surfaces*. In this section we will clarify the links between these terms and between interactions associated with different geometries.

16.2.1 Pressure, Force and Energy

Firstly, the surface force $F(D)$ between two colloidal particles at a separation D has its corresponding 'interaction free energy' $W(D)$. Here 'interaction' means it is the separation dependent components of the energy, and 'free' means that the system is at thermodynamic equilibrium. Thus, $W(D)$ is the work required to bring two colloidal particles from infinity (where there is no force) to the separation D, integrating the surface forces this work is done against from infinity to D, i.e.

$$W(D) = \int_{\infty}^{D} F(D)\mathrm{d}D \tag{16.13}$$

and conversely, the surface force is the separation variation (i.e. the negative gradient) of $W(D)$,

$$F(D) = -\frac{\mathrm{d}W(D)}{\mathrm{d}D} \tag{16.14}$$

The above relations hold similarly for the corresponding pair of quantities: pressure $P(D)$ (force per unit area) and interaction free energy per unit area $W_a(D)$,

$$W_a(D) = \int_{\infty}^{D} P(D)\mathrm{d}D \tag{16.15}$$

and

$$P(D) = -\frac{\mathrm{d}W_a(D)}{\mathrm{d}D} \tag{16.16}$$

16.2.2 The Derjaguin Approximation

The above $F(D)$–$W(D)$ and $P(D)$–$W_a(D)$ relations are formal, regardless of the size and geometry of the interacting colloidal particles. However, in practice we would like to relate $F(D)$ measured between *two spheres* (or *a sphere against a flat* in the case of AFM; or between *two crossed cylinders* in SFA) to $W_a(D)$ between two planar surfaces which is the quantity often calculated in theory. This would also enable us to compare our results from different experiments and from different experimental techniques employing different geometries. This important $F(D)$–$W_a(D)$ cross geometry comparison is facilitated by the well-known Derjaguin approximation, which will be presented below in an approach taken by Horn (4) that gives a clear dissemination of the approximation.

16.2.2.1 Four Approximations in the Derjaguin Approximation

Let us assume two spherical colloidal particles of radii R_1 and R_2 at a surface separation D apart, as illustrated in Figure 16.2. To obtain the total surface forces $F(D)$ between these two particles, we integrate the forces between a circular element strip of area $2\pi y dy$ on the upper surface and the lower surface,

$$F(D) = \int_{Z=D}^{Z=D+R_1} (2\pi y\mathrm{d}y) f(Z) \tag{16.17}$$

where $f(Z)$ is the force per unit area. There are indeed four approximations in deriving the $F(D)$–$W_a(D)$ relation in the Derjaguin approximiation.

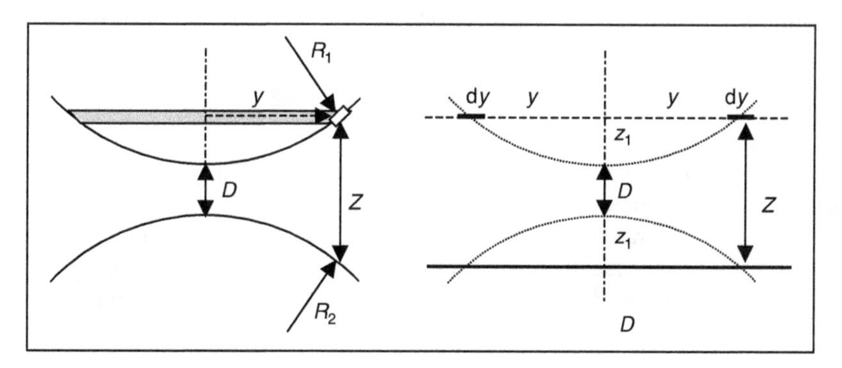

Figure 16.2 *The Derjaguin approximation*

The first one is the '*flat-for-curved*' approximation, for which we take three steps.

1. We divide the curved circular region into incremental area elements.
2. Replace each of these curved elements with a flat element parallel to the plane that is tangential to the point of the closest approach, so that the curved circular strip becomes a flat ring of width dy.
3. Replace the second surface (R_2) with a semi-infinite flat surface at the separation Z directly opposite to the flat area element – a reasonable approximation for R_1, $R_2 \gg D$.

The second approximation is the '*parabolic surface*' or *Chord Theorem* approximation, in which the surface curvature is assumed to be parabolic. This is reasoned from the relation,

$$R_1^2 = (R_1 - z_1)^2 + y^2 \tag{16.18}$$

and for $R_1 \gg D$ (and thus $R_1 \gg z_1$), this gives,

$$z_1 \approx \frac{y^2}{2R_1} \tag{16.19}$$

and similarly,

$$z_2 \approx \frac{y^2}{2R_2} \tag{16.20}$$

Thus,

$$Z = D + z_1 + z_2 \approx D + \frac{y^2}{2R_1} + \frac{y^2}{2R_2} \tag{16.21}$$

This gives,

$$\mathrm{d}z = \frac{y}{R}\mathrm{d}y \tag{16.22}$$

where

$$R = \frac{R_1 R_2}{R_1 + R_2} \tag{16.23}$$

Substitution of Equations (16.22) and (16.23) into Equation (16.17) leads to

$$F(D) = 2\pi R \int_{Z=D}^{Z=D+R_1} f(Z)\mathrm{d}Z \tag{16.24}$$

The third approximation is the '*range*' approximation, assuming that the range of the surface force is small compared with the size of the two spheres, R_1 and R_2. Thus, we can replace the upper integration limit in Equation (16.24) with infinity, and bearing in mind Equation (16.15) and the fact that $f(Z)$ is the force between a unit area of the upper surface and the entire lower surface (now approximated as a semi-infinite planar surface), we obtain

$$F(D) = 2\pi R \int_D^{\infty} f(Z)\mathrm{d}Z = 2\pi R W_{\mathrm{A}}(D) \tag{16.25}$$

where $W_A(D)$ is the interaction free energy between a unit area and a semi-infinite surface D apart.

The fourth approximation is the '*unit area*' approximation, where we assume that the interaction free energy between two unit areas $W_a(D)$ is very similar to $W_A(D)$, i.e.

$$W_a(D) \approx W_A(D) \tag{16.26}$$

a very good approximation for surfaces in close proximity. Hence, we finally arrive at the Derjaguin approximation,

$$F(D) = 2\pi R W_a(D) \tag{16.27}$$

16.2.2.2 Derjaguin's Approximation for Different Geometries

Equation (16.27) has been derived for two spheres, and can be easily adapted to other geometries. For two spheres of equal radius, $R_1 = R_2 = R$, it becomes

$$F(D) = \pi R W_a(D) \qquad \text{(for two sphere of equal radius } R\text{)} \tag{16.28}$$

For a sphere of radius $R_1 = R$ against a planar surface (as employed in colloidal atomic force microscopy (AFM)), we set $R_2 \gg R_1$ in Equation (16.23), and it gives

$$F(D) = 2\pi R W_a(D) \qquad \text{(for a sphere of radius } R \text{ against a planar surface)} \tag{16.29}$$

For two cylinders of radii R_1 and R_2 at D (closest approach) from each other with their axes crossed at an angle θ, it can be readily shown that

$$F(D) = 2\pi \frac{\sqrt{R_1 R_2}}{\sin\theta} W_a(D) = 2\pi R_c W_a(D) \qquad \text{(for two cross cylinders at an angle } \theta\text{)} \tag{16.30}$$

where R_c is the effective radius,

$$R_c = \frac{\sqrt{R_1 R_2}}{\sin\theta} \tag{16.31}$$

We would like to consider in on its own a special case where $R_1 = R_2 = R$ and $\theta = 90°$, that is, the two cylinders are of equal radius and crossed orthogonally as shown in Figure 16.3(a), since this geometry is employed in the surface force apparatus (SFA). Setting these parameters accordingly in Equation (16.30), we see that its Derjaguin approximation expression is identical to that for a sphere against a flat (Equation 16.29).

To prove the above geometrical equivalence, we need to establish that, for all pairs of points (e.g. A and B in Figure 16.3) directly opposite each other on two cylinder surfaces whose projections in the top view reside on a circle of radius r with $r^2 = x^2 + y^2$ (c.f. Figure 16.3c), the separation Z between them should be constant, just as in the case of a

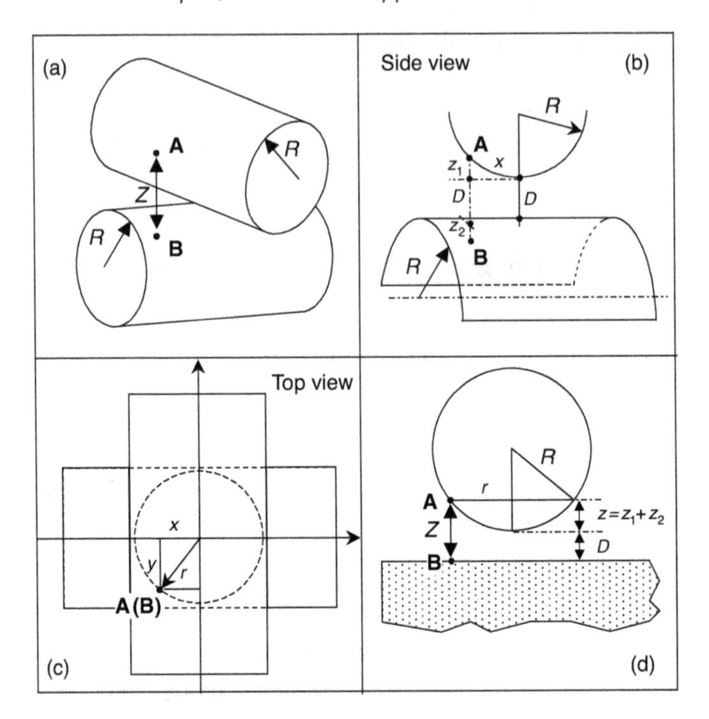

Figure 16.3 *Orthogonally crossed cylindrical geometry in an SFA (a)–(c) and its geometrical equivalence of a sphere vs. a planar surface (d)*

sphere against a flat plate (c.f. Figure 16.3d). Similarly to Equation (16.21), Z is defined as (c.f. the side view in Figure 16.3b)

$$Z = D + z_1 + z_2 \approx D + \frac{x^2}{2R} + \frac{y^2}{2R} = D + \frac{1}{2R}(x^2 + y^2) = D + \frac{1}{2R}r^2 \qquad (16.32)$$

which indeed is constant for a particular r value at a separation D, and this geometry is equivalent to a sphere of radius R at D away from a planar surface, as depicted in Figure 16.3(d). Thus,

$$F(D) = 2\pi R W_a(D) \qquad \text{(for two orthogonally crossed cyliners of equal radius } R\text{)} \qquad (16.33)$$

This is why we would plot the force–distance curve as F/R vs. D when reporting results from the colloidal AFM (a sphere against a flat) and SFA (two crossed cylinders), so that the data from these two techniques are directly comparable.

16.2.2.3 Limitations of Derjaguin's Approximation

The Derjaguin approximation holds remarkably well for colloidal particles in close range and it does so even when one or two of its four approximations are challenged. Precautions should, however, be exercised under some experimental conditions, and these include electrical double layer interactions between small colloids in non-polar media (see below) where the range of the surface forces is very long – comparable to the size of colloidal particles. Another situation involves the interpretation of the surface forces measured with a sharp tip which would invalidate the assumption $R \gg D$, which will in turn compromise the 'flat-for-curved' and 'parabolic surface' approximations above.

16.3 Surface Force Measurement Techniques

When carefully conducted and interpreted, direct force measurements can yield fruitful information on the solid–liquid interface (1, 5). An absolute force as small as some 10^{-12} N can be detected with great precision routinely using modern microbalance technology. However, in colloid science, the measured force between two surfaces can tell a 'meaningful story' only when the surface separation at which it is detected, and subsequently, its variation with the separation, can be established as well. Considerable effort has been dedicated to achieving this, and various techniques have been developed. Comprehensive reviews on the subject exist in the literature (e.g. the review by Claesson *et al.*, (6)), and only a brief account is given here, emphasising the underlying principles.

16.3.1 Optical Tweezers

In principle, the force is measured by gauging it with a known force, and different choices and implementations of the gauge in different techniques determine their sensitivity and suitability, and in turn, their applicability. When the technique of *optical tweezers* (Chapter 14) is adapted to force measurements (7, 8), this gauge is the thermal fluctuation energy kT, and it is particularly suitable for detecting the interaction between *particles of colloidal dimension* in pure water. In this case, a pair of dielectric colloidal particles are trapped and brought close to each other by two separate, tightly focused laser beams impinging upon them. The trapping occurs due to the fact that the induced dielectric dipole in the particle by the laser beam senses the gradient of the electromagnetic field intensity of the beam, and is drawn to the brightest region, i.e. the central axis of the beam (9–11). Once the laser beams are switched off, the particles are released from the optical traps and fluctuate about their equilibrium positions. The probability $p_n(D)$ of finding the particles at a centre-to-centre separation D is related to the total interaction free energy $W_{total}(D)$ between them as gauged by kT,

$$p_n(D) = \Omega \exp\left[-\frac{W_{total}(D)}{kT}\right] \tag{16.34}$$

where Ω is a constant. The probability function $p_n(D)$ can be established by switching on and off (or *blinking*) the laser beams numerous times, and taking the snapshots of the trajectories of the particles with different initial separations using the method of *digital video microscopy* facilitated by a rigorous image processing algorithm (12). The limited resolution of the particle separation determination, currently ± 50 nm, is due to its reliance on digital video microscopy, but the blinking optical tweezers remain as one of the few techniques available for direct force measurements between colloidal particles (a few hundred nm to a few μm in diameter).

16.3.2 Total Internal Reflection Microscopy (TIRM)

This limitation in the separation determination is overcome in the *total internal reflection microscopy* (TIRM) as developed by Prieve *et al.* (13), which also employs kT as the gauge for the interaction free energy between, in this case, *a colloidal particle and a plate* in an aqueous medium. A colloidal particle ($3 \sim 30\,\mu m$ in diameter) is allowed to settle above a transparent plate due to its gravity, and to fluctuate about its equilibrium position. The transient separation at which the particle appears can be sampled with 1 nm resolution from the scattered light intensity by the particle when it is illuminated with an evanescent wave, which is produced when a laser beam is incident upon the plate–liquid interface with an angle of incidence greater than the critical angle. The total interaction free energy between the particle and the plate still obeys Equation (16.34), but now with an additional gravity component, which can be determined with ease from the linear part of the interaction free energy profile. $p_n(D)$ in this case is established from the probability of the observation for different light intensities, if a large enough number of observations is made. TIRM enjoys both sensitivity in the magnitude of the force detectable and relatively high resolution in the separation determination, and it is particularly suitable for probing the long-range tail of weak interactions. However, it suffers a compromise in that the separation range of the interaction accessible to the technique is limited by how close the particle can settle to the plate, that is, by the gravitational force. This may be circumvented to a certain extent by exerting an additional variable optical force on the particle, thus expanding the range of the interaction that can be probed.

16.3.3 Atomic Force Microscope (AFM)

In the territory of the detection of colloidal forces, TIRM encounters competition from the *atomic force microscope* (AFM) equipped with a colloidal probe (14). In this case, AFM gauges the interaction between a colloidal particle attached to the tip of its cantilever and an approaching planar substrate, with the deflection of the cantilever spring. The deflection can be determined by a number of means, most commonly the laser optical technique which claims a sub-Ångstrom resolution. At any particular separation D, the total interaction free energy and the surface force can be obtained through

$$\begin{aligned} 2\pi R W_{\text{total}}(D) &= F(D) \\ &= 2\pi R[W(D) - W(\infty)] \\ &= K[\Delta x(D) - \Delta x(\infty)] \\ &= K\Delta d(D) \end{aligned} \tag{16.35}$$

where K is the spring constant, R the radius of the colloidal particle, Δx the deflection of the AFM cantilever spring, and $\Delta d(D)$ the deviation of the spring deflection from that if the surface force is absent. In practice, the infinity is chosen to be a large enough separation such that the surface interaction can be regarded as zero. The spring constant K for an AFM can be as weak as $0.5\,N\,m^{-1}$, thus forces as small as $10^{-12}\,N$ can be detected. However, the radius of the colloidal probe R is normally very small, i.e. of some μm, which then brings the detection limit in $W_{\text{total}}(D)$ to a modest $10^{-6}\,J\,m^{-2}$. The advantage of AFM over TIRM is that it can have access to the full separation range of the interaction, and is suited to detect the strong force when the surfaces are close to contact. However, it shares the drawback with all the above-mentioned techniques that the true intimate contact between the surfaces cannot be established.

16.3.4 Surface Force Apparatus (SFA)

Also using the deflection of cantilever springs as the gauge, the interferometric *surface force apparatus* (SFA), as originally configured (15, 16), measures *surface forces between macroscopic bodies*, one of which is suspended on the force-measuring cantilever springs. The employment of an interferometry technique called fringes of equal chromatic order (FECO) (17, 18) determines the surface separation with 0.2 nm resolution but requires the use of the thin, smooth, transparent material as the model substrate. Most often this is mica, although alternative materials have been explored (19–22). This restriction has led to the development of alternative versions of SFA to enable a broader range of materials to be studied (6, 23–27), in which the surface separation is monitored by other means. However, the versatility and effectiveness of the interferometry technique in the SFA should be appreciated. Among all the techniques available, it is the only one that affords the unequivocal determination of true intimate contact between surfaces as well as allowing the direct observation of the surface condition during the measurement, and enables the investigation of the optical properties of the medium confined between the substrates (28, 29). In addition, it can yield information that is valuable in contact mechanics (30) and surface topography (31, 32). The spring constant K used in SFA can be varied over many orders of magnitude, i.e. $10^2 \sim 10^5$ $\mathrm{N\,m^{-1}}$, with a common value of $\sim 150\ \mathrm{N\,m^{-1}}$ which is towards the low end of the range. This translates to 10^{-7} N and $10^{-5}\ \mathrm{J\,m^{-2}}$ in the force and interaction free energy detection limits respectively, given the typical radius of curvature R is 1 cm.

First developed by Tabor and Winterton (33) and Israelachvili and Tabor (34), a number of versions of the SFA now exist in the world. Figure 16.4 shows the essential components of a version with sensitive friction measurement capability developed by Klein (35) (this is sometimes called the surface force balance, SFB, by Klein). As schematically shown in Figure 16.4, in this version of SFA, measurement is often made between μm thick mica surfaces (a) mounted in a crossed cylindrical geometry immersed in a liquid (b) contained in a boat (d). The normal force F is obtained from the deflection of a pair of cantilever springs (c) carrying the bottom surface. The top surface is mounted on a sectored piezoelectric ceramic tube (e), which is suspended via a rigid cradle (f) on a pair of vertical springs (g). The deflection of the vertical springs, as attained from the displacement of a polished stainless steel flag (h) gauged by a capacitance probe (i), gives the lateral or friction force F_S. Separation D between, and the geometry of, the surfaces can be monitored interferometrically by observing the FECO fringes focused into a scanning spectrometer (not shown) with an objective (j) by shining a beam of collimated white light through the surfaces.

16.3.5 Other Techniques

There exist a variety of other techniques, for instance, the *evanescent wave light scattering microscopy* (EVLSM) (26), the *light lever instrument for force evaluation* (LLIFE) (41), and the *measurement and analysis of surface interaction forces* (MASIF) (27). Their operating principles are largely in line with what has been outlined in the foregoing

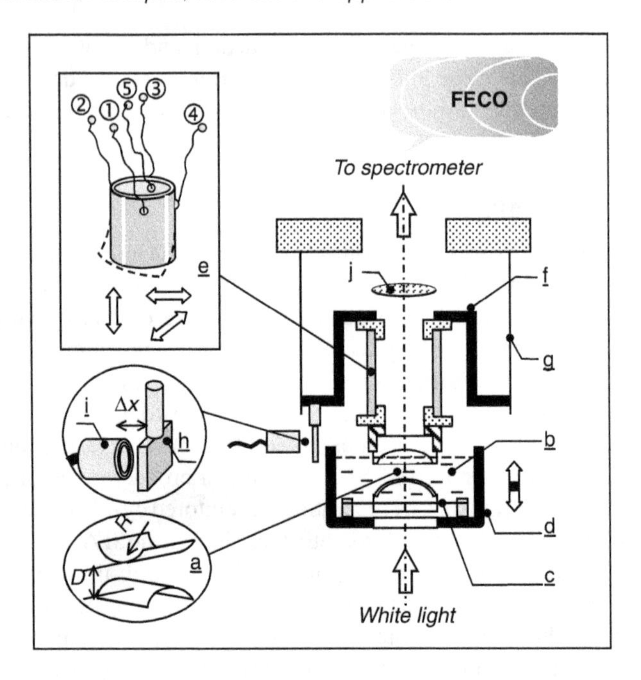

Figure 16.4 Key components of a version of the surface force apparatus with sensitive friction measurement capability, sometimes called a surface force balance

discussions, although the implementations may bear some distinct features. The capabilities of these techniques have sometimes been compared with one another, in terms of their detection limits in force or interaction free energy (13, 42). This comparison, however, is simplistic; instead, it should be appreciated that various techniques are suitable for different systems and are capable of yielding information that is complementary.

Derjaguin *et al.* (36–40) have constructed a beam balance and were the first to measure the van der Waals force between macroscopic bodies in air and vacuum. However, the application of the balance in other media or to detect other interactions has never been reported in the literature. It is thus difficult to bring its advantages into comparison with other techniques.

16.4 Different Types of Surface Forces

All originating from intermolecular forces as discussed in Section 16.1 above, surface forces have been traditionally divided into different types. Our knowledge of some of these surface forces is quite well established, such that we have quite a few tricks up our sleeves to control them in order to achieve desired inter-colloidal interactions. Others remain to be fully understood. Comprehensive reviews on surface forces exist (1, 4, 5) to which interested readers are referred. Certain aspects of van der Waals forces and electric double layer forces have been discussed in Chapter 3, and polymer-mediated surface forces discussed in Chapter 9. Here we only present a brief review of different types of surface forces commonly encountered in dealing with colloidal suspensions.

16.4.1 van der Waals Forces

The van der Waals force between two colloidal particles can be obtained by summing all the interactions between atomic and molecular dipoles in the particles. There are two approaches to doing this. The first one due to Hamaker (43), as conducted in Section 16.1, is called pair-wise addition, simply adding together all the contributions from constituent atoms and molecules. Once this is done, we find the van der Waals interaction free energy per unit area $W_a(D)$ between two flat surfaces D apart is

$$W_a(D) = -\frac{A}{12\pi}\frac{1}{D^2} \tag{16.36}$$

and the inter-particle force between two spherical colloids of radius R is obtained through the Derjaguin approximation (Equation 16.17)

$$F(D) = -\frac{AR}{12}\frac{1}{D^2} \tag{16.37}$$

Note that Equation (16.37) could also be obtained from the $F(D)$–$W(D)$ relation (Equation 16.14) by differentiating the van der Waals interaction free energy between two spheres obtained in Equation (16.11) with respect to D. The Hamaker constant A in the above equations, given by Equation (16.12), is related to the intermolecular pair potential coefficient C, which in turn depends on the polarisabilities, permanent dipole moment and ionisation energy of the interacting molecules.

The Hamaker pair-wise addition approach, however, does not take into account the many-body effect, that is, the dipole field of one molecule is influenced by its neighbouring molecules. The alternative Lifshitz approach addresses this by considering each interacting body (colloid) as a dielectric continuum, characterised by a frequency/wavelength dependent dielectric constant ε (or refractive index n). The summation is made of the fluctuation modes of the electromagnetic field as two surfaces approach each other. The energy associated with this summation is the van der Waals interaction free energy. Its detail is sophisticated and beyond the scope and purpose of this chapter. It suffices to say that as a result, Equation (16.37) is still valid, but instead of Equation (16.12), the Hamaker constant A is computed from the wavelength-dependent refractive indices of the interacting surfaces and the intervening medium.

In practice, the values of the Hamaker constant for different materials across different media are given in text books and they are of the order $(0.4–40) \times 10^{-20}$ J, (e.g. 1, 2) which we can use to compute the van der Waals force between interacting colloidal particles. It is useful to know several features of the van der Waals force. It is ubiquitous, and is always attractive – that is the Hamaker constant is positive – between colloidal particles of similar material, regardless of the intervening medium. However, the van der Waals force between dissimilar materials (1 and 2) can be repulsive, if the value of refractive index of the medium (3) is intermediate between those of material 1 and 2, i.e. if $n_1 < n_3 < n_2$. An example for this positive van der Waals force is realised in the interaction between gold and polytetrafluoroethylene (PTFE) in cyclohexane.

16.4.2 Electric Double Layer Forces in a Polar Liquid

Most colloidal particles acquire a surface charge in a polar liquid, often negative, via a number of mechanisms, as discussed in Chapter 2. A key prerequisite for the presence of this surface charge is the high dielectric constant ε_r of a polar medium which reduces the Coulomb energy between oppositely charged ions. The Coulomb energy favours combining the surface charge and the opposite charges (called *counter-ions*) to maintain charge neutrality, whereas the entropic thermal agitation of the counter-ions favours smearing them out throughout the polar medium. The balance between these two effects results in the formation of the electric double layer: the surface charge layer and the diffuse layer of opposite charges. A detailed description of the sophisticated Helmholtz–Gouy–Chapman model of this double layer is given in Chapter 2, whereas here for surface forces we are mainly concerned with the diffuse layer (i.e. the outer Gouy–Chapman layer).

The thickness of this double layer is measured by the Debye length κ^{-1},

$$\kappa^{-1} = \left(\frac{e^2 \sum \rho_i z_i^2}{\varepsilon_0 \varepsilon_r kT} \right)^{-1/2} \tag{16.38}$$

where $e = -1.609 \times 10^{-19}$ C is the electronic charge, ρ_i is the number density of ion species i and z_i its valence, $\varepsilon_0 = 8.854 \times 10^{-12}\ \mathrm{C^2\,J^{-1}\,m^{-1}}$ is the dielectric permittivity of a vacuum, ε_r is the dielectric constant of the medium, $k = 1.381 \times 10^{-23}\ \mathrm{J\,K^{-1}}$ is the Boltzmann constant and T is the absolute temperature in K. For example, $\kappa^{-1} = 9.6$ nm in 1 mM NaCl solution (c.f. Table 2.1), and for pure water at pH = 7, it approaches 1 μm, although water is never that pure, with its pH ~ 5.5 due to CO_2 solubility.

Quantitatively, the interaction between the surface charge and the counter-ions is described by the Poisson equation, which relates the electrical potential $\psi(z)$ due to the surface charge at any position z away from the surface and the charge density $\rho(z)$ at this position. Concurrently, the distribution of the ions away from the surface is described by the Boltzmann distribution which relates $\rho(z)$ in the vicinity of the surface to the charge density in the bulk liquid. Equating the electrical potentials in the Poisson and Boltzmann equations leads to the Poisson–Boltzmann equation, whose solution reveals that the electrical potential $\psi(z)$ decays exponentially away from the surface, with its decay length κ^{-1}.

When two similarly charged colloidal particles are brought into close proximity, so that the diffuse layers of their electrical double layers overlap, it results in the ionic species between the surfaces getting crowded, leading to a repulsion which is entropic in its origin. The complete solution to this expression for the repulsion needs to be obtained numerically, and it can be simplified by a linearisation in the *weak overlap approximation* for large surface separations ($D > \kappa^{-1}$). The linearised interaction free energy per unit area for two planar surfaces, for symmetric $z:z$ electrolytes, is

$$W_a(D) = \left(64\, kT\kappa^{-1} \sum \rho_i\right) \left[\tanh\left(\frac{ze\psi_0}{4kT}\right)\right]^2 \exp(-\kappa D) \tag{16.39}$$

where ψ_0 is the surface potential and is related to the surface charge density σ_s through

$$\sigma_s = \left(4ze\kappa^{-1} \sum \rho_i\right) \sinh\left(\frac{ze\psi_0}{2kT}\right) \tag{16.40}$$

For low ψ_0 values, Equation (16.39) can be further simplified to

$$W_a(D) \approx 2\varepsilon_0\varepsilon_r\kappa^2\psi_0^2\exp(-\kappa D) = \frac{2\sigma_s^2}{\kappa\varepsilon_0\varepsilon_r}\exp(-\kappa D), \quad \text{(per unit area for planar surfaces)} \tag{16.41}$$

which is applicable for all electrolytes regardless of their valence. Subsequently, the electric double layer force *F*(*D*) between two spheres can be obtained through the Derjaguin approximation.

A number of assumptions have been made in arriving at the above expressions in the original Gouy–Chapman theory. These include:

- the ions are point charges, i.e. without any physical size
- the intervening medium is a structureless continuum, characterised only by its dielectric constant
- the distribution of the surface charge is uniform.

These assumptions, clearly not all fulfilled in practical colloidal interactions, work quite well for large surface separations $D > \kappa^{-1}$, but start to break down at smaller *D*, where an accurate description of the interaction should resort to numerical solution of the Poisson–Boltzmann equation.

A particular relevant issue is related to the adsorption of counter-ions on the surface at small *D*. If there is no such adsorption, it is termed the boundary condition of constant charge density. Alternatively, counter-ions could adsorb on the surface to maintain a constant surface potential boundary condition, which always results in a lower repulsion than the constant charge density boundary condition. In practice, it is likely for the electrical double layer interaction to lie between these two limits.

16.4.3 The DLVO Theory

The DLVO theory of colloidal stability, named after the two Russian scientists (Derjaguin and Landau) and the two Dutch scientists (Verwey and Overbeek) who developed it around the 1940s, is a cornerstone of colloid science. It is based on the assumption that the total force between colloidal particles is obtained by adding together the van der Waals and electrical double layer forces between them. Chapter 3 has described this theory and how it is employed in considering colloidal stability. We will not reproduce it here, except noting that it is very useful in predicting the correct trends, despite a number of non-DLVO surface forces that could also be operating between colloidal particles. An example of the DLVO forces measured between mica surfaces in pure water is shown in the inset of Figure 16.5. Some of these non-DLVO forces will be described below.

16.4.4 Non-DLVO Forces

It has now been realised that it is commonplace rather than a rarity that surface forces of a non-DLVO nature act between colloidal particles. As we have discussed in Sections 16.1.2.2 and 16.1.2.3, a surface could interact with the molecules of the intervening medium and modify their distribution adjacent to the surface in a number of ways. In general, their range

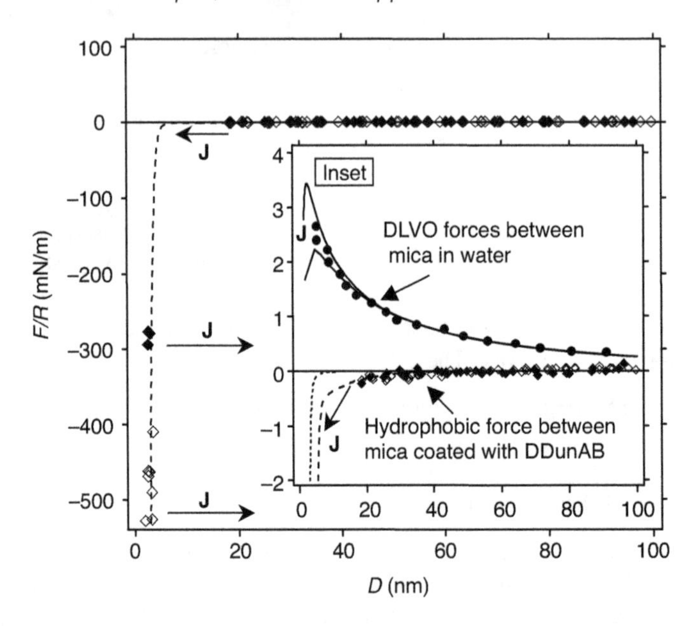

Figure 16.5 *Normalized surface forces F/R as a function of surface separation D between mica surfaces with adsorbed DDunAB surfactant in pure water (◆ and ◇), measured when they are brought into contact. On a magnified scale in the inset, the DLVO interaction between bare mica in pure water (●) is shown for comparison, with upper and lower solid curves assuming constant surface charge density and surface potential respectively. The dotted curve is the van der Waals interaction, using a Hamaker constant of 4×10^{-20} J for hydrocarbon surfaces across water. The dashed curve is an empirical double exponential fit to the data as a guide for eyes. Letter **J** indicates 'jumps' when the gradient of the attractive force exceeds that of the spring constant in the SFA*

is confined within the surface separation of several molecular dimensions, except for the hydrophobic force; they can be repulsive, attractive or oscillatory and their magnitude could far exceed the DLVO forces, thus playing an important role in considering overall surface forces. A consensus for a 'name' for these non-DLVO forces is yet to be reached. Israelachvili (1) has termed them collectively solvation forces with three different sub-categories (oscillatory, hydration and hydrophobic forces); Derjaguin (44) has referred to hydration forces as structural forces; whereas, Horn (5) has elected to distinguish between them more clearly, a convention we follow here.

16.4.4.1 Oscillatory Structural Forces

These forces have a geometric origin. As the molecules of the medium are confined to a surface separation D of a few molecular dimensions σ, they are induced to form ordered quasi-discrete layers. When the surface separation is equal to an integral number of molecular dimensions, i.e. $D = m\sigma$ (where m is an integer), the molecules pack comfortably and efficiently in the gap, with each m corresponding to an energy minimum and thus an attractive force of increasing magnitude as m decreases. As the molecular layers are being squeezed out, i.e. D does not allow them to pack into integral number of layers, the corresponding energy maxima result in repulsive forces between the surfaces, increasing in magnitude as m decreases. Adding together, this leads to an oscillatory structural force, alternating between attractive minima and repulsive maxima, with the spacing of the oscillation $\sim\sigma$ and its amplitude diminishing as D is greater than several σ.

An approximate theoretical expression for this oscillatory structural force (45) is given in terms of pressure $P(D)$ experienced by the surfaces D apart,

$$P(D) \approx -kT\rho_s(\infty)\cos\left(2\pi\frac{D}{\sigma}\right)\exp\left(-\frac{D}{\sigma}\right) \tag{16.42}$$

where $\rho_s(\infty)$ is the density of the molecules of the intervening medium, and can be approximated as (for close packing)

$$\rho_s(\infty) \approx \frac{\sqrt{2}}{\sigma^3} \tag{16.43}$$

The above equation basically states that the period of the oscillations is $\sim\sigma$ and the peak-to-peak amplitude of the oscillations decays with D exponentially with the decay length also $\sim\sigma$. Using the P(D)–W_a(D) relation (Equation 16.15), we integrate the above equation to get the interaction free energy per unit area between two planar surfaces:

$$W_a(D) \approx \frac{\sigma kT\rho_s(\infty)}{1+4\pi^2}\left[\cos\left(2\pi\frac{D}{\sigma}\right) - 2\pi\sin\left(2\pi\frac{D}{\sigma}\right)\right]\exp(-D/\sigma) \tag{16.44}$$

As first reported by Horn and Israelachvili (46) (see Figure 16.6), such structural forces have been observed in experiments on polar liquids (such as water) and a range of simple

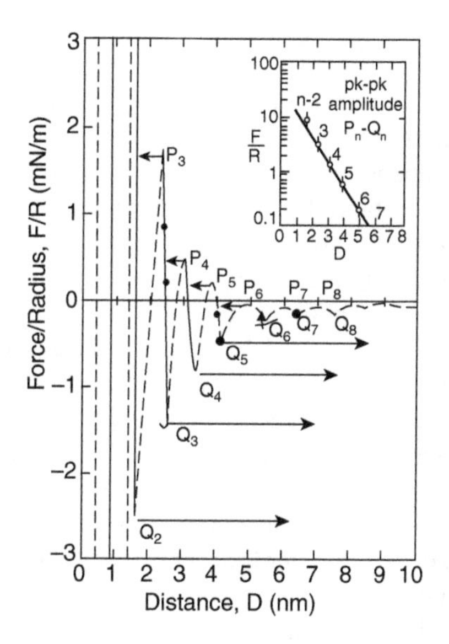

Figure 16.6 Oscillatory structural forces due to layering of OMCTS between mica surfaces measured with an SFA. The oscillation period is ~1.0 nm, comparable to the diameter of the OMCTS molecule. The inset shows the peak-to-peak amplitude with a decay length of 1.0 nm. (Figure provided by J. N. Israelachvili, UCSB)

liquids under confinement, and these include spherical molecules (octomethylcyclotetrasiloxane, OMCTS) and linear chain molecules (alkanes) using the SFA. The range within which it is observable is of up to 5–10 σ, and its magnitude could well exceed the van der Waals force in this range. Its manifestation depends highly on the shape of the molecules being confined and the smoothness of the confining surfaces. For instance, the lack of symmetry and regularity in branched alkanes means that they cannot pack easily into ordered layering structures, and thus they do not mediate the oscillatory forces upon confinement. Furthermore, a few Å of surface roughness would also disrupt the packing order and smear out the oscillations.

One might argue that the interacting surfaces in real life are rarely molecularly smooth, and thus the oscillatory forces are irrelevant to practical applications. However, surface asperities invariably deform under applied pressure, particularly in the case of soft surfaces, to form locally flattened, intimate contacts where such oscillatory structural forces would play an important role. Moreover, it has been suggested that the out-of-plane layering of the confined molecules, as revealed by the structural forces, could lead to a quasi-phase transition of the liquid to a solid-like state in the case of simple liquids, corresponding to the attractive minima in the oscillations, where the effective viscosity of the liquid could dramatically increase by many orders of magnitude. This is very relevant to oily lubricants found in many engineering applications. Interestingly and intriguingly, when similarly confined, water molecules do not seem to undergo this 'solidification' process due to its expansive freezing property (i.e. ice floats on water), that is, densification does not lead to solidification for water. This unique property of water, and the issue of the fluidity of water molecules, remains the subject of intensive research and enthusiastic debate. We will return to this in the following section.

16.4.4.2 Hydration Forces

In the case of the structural force above, the surfaces have largely served as inert confining boundaries for the intervening medium. However, the surfaces themselves can also become solvated. In the case of aqueous media, repulsive hydration forces arise between two surfaces when water molecules bind strongly to surface groups. These surface groups could be ionic or hydrogen bonding groups, which would orient water molecules around them to form hydration layers. The strength of the hydration force depends largely on the energy required to dehydrate these hydrophilic groups.

In the case of many hydrophilic colloidal particles and clays, monovalent or multivalent cations could bind to the usually negatively charged surface, carrying with them a full hydration shell of water molecules and leading to hydration repulsion when the surfaces approach the separation where the hydration sheaths become restricted. For monovalent cations, as their radius decreases from Cs^+ to Li^+, the hydration number n_H (number of water molecules that tenaciously bind to the ions to form the primary hydration layer) increases from 1–2 to 5–6 due to stronger Coulomb interactions with water molecules, and their resultant hydration radii are all comparable at ~0.33–0.38 nm. (However, different methods of measurement give different values for the hydration number. For example, n_H for Li^+ can vary from 2 to 6.) For divalent cations (such as Be^{2+}, Mg^{2+} and Ca^{2+}), their bare ionic radii are generally smaller due to the double cationic charge losing an extra electron, and this electron-nakedness again leads to stronger hydrogen bonding interactions with

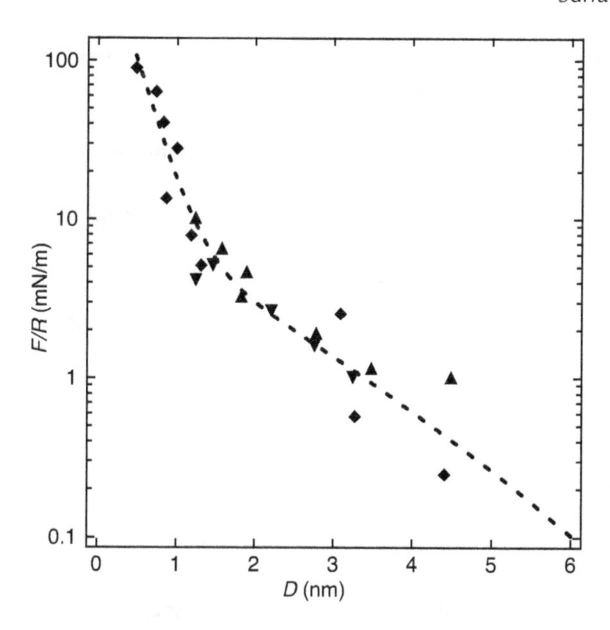

Figure 16.7 *Hydration force between mica surfaces across 60 mM NaCl solution measured with an SFA. The dashed curve shows the hydration force, fitted using Equation (16.45), with* $W_0 = 100\,\mathrm{mJ\,m^{-2}}$ *and* $\lambda_0 = 0.25\,\mathrm{nm}$*, added to the longer ranged double layer force (data from S. Perkin, UCL)*

surrounding water molecules, and in turn larger hydration radii compared with monovalent cations (by ~0.1 nm) and a more hydrated state with $n_H = 4$–6. Empirically, the repulsive hydration energy (per unit area) follows an exponential decay:

$$W_a^H(D) = W_0 \exp(-D/\lambda_0) \tag{16.45}$$

where typical values for 1 : 1 electrolytes are $W_0 = 3$–$30\,\mathrm{mJ\,m^{-2}}$ and the decay length $\lambda_0 \sim 0.6$–1.1 nm. The effective range of hydration forces is ~3 nm as measured experimentally, which is some twice that of the oscillatory structural forces above. Figure 16.7 shows an example of the hydration force between mica surfaces, in this case in 60 mM NaCl solution measured with an SFA, and the hydration component of the repulsion (added to the double layer force) can be fitted using Equation (16.45), with $W_0 \sim 100\,\mathrm{mJ\,m^{-2}}$ and $\lambda_0 = 0.25\,\mathrm{nm}$ in this case.

In addition to adsorbed cations, other surface hydrophilic groups such as hydroxyl groups, quaternary ammonia groups, and sugar and zwitterionic groups (e.g. present on biological cell membranes) can also facilitate hydration repulsion. The presence of hydration forces could offer explanations for a range of phenomena, such as swelling of certain clays and surfactant soap films, repulsion between biological membranes and colloidal stability of silica dispersions in high salt. However, the nature and origin of hydration forces remain to be fully understood, particularly on a theoretical level.

Another topical issue related to hydration forces is the fluidity of the water molecules in the primary hydration layers of the surface hydrophilic groups. Though it is difficult to strip these water molecules, there is evidence that they retain high fluidity even under high compression, due to high exchange rate with bulk water molecules. This exchange rate or lifetime of primary hydration molecules depends very much on the valency of the cations, and ranges from 10^{-9}–10^{-8} s for monovalent cations to 10^{-1}–1 s for Al^{3+} and even many orders of magnitude slower for other ions, such as Cr^{3+}. Both the repulsive hydration forces and the fluidity of bound primary water molecules associated with multivalent cations remain to be fully explored both experimentally and theoretically.

16.4.4.3 Hydrophobic Forces

First reported between two mica surfaces coated with cationic surfactant monolayers in water using an SFA, hydrophobic forces now have been observed between a wide range of hydrophobic surfaces prepared in different ways. These include surfaces bearing hydrogenated or fluorinated surfactant monolayers by adsorption or Langmuir–Blodgett deposition, chemically modified with silane layers, plasma polymerisation from vapour phase on surfaces, or spin coating of hydrophobic polymer films on a solid. The experimental observations of hydrophobic forces remain vastly varied in range and magnitude, but it is widely perceived that hydrophobic forces are very long ranged – measurable at D up to ~10–250 nm and much stronger than the van der Waals attraction. The hydrophobic interaction free energy per unit area between two planar surfaces sometimes can be described empirically by a double exponential expression,

$$W_{a}^{Hph}(D) = -2C_1 \exp(-D/\lambda_1) - 2C_2 \exp(-D/\lambda_2) \tag{16.46}$$

where the typical values for first pre-exponential factor $C_1 = 10$–$50\,\mathrm{mJ\,m^{-2}}$ and the first decay length $\lambda_1 \sim 1$–3 nm can describe shorter range components of the hydrophobic forces from a number of experiments; whereas, the values for the second pre-exponential factor C_2 and decay length λ_2 vary vastly, indicating the particular variability of the longer range components of hydrophobic forces. An example of the hydrophobic force measured between mica surfaces in water with adsorbed monolayers of di-chained cationic surfactant $[(CH_3(CH_2)_{10})2N^+(CH_3)_2Br^-]$ (DDunDAB) is shown in Figure 16.5, and is given as a contrast to the DLVO force between bare mica in water. In this case, the hydrophobic force is not as long ranged as some reported values in literature, but it is much longer ranged than the van der Waals force (dotted curve in the inset), detectable up to $D \sim 30$ nm.

The origin of hydrophobic forces remains to be established. Earlier suggestions have included possible charge correlation effects due to patchy and mobile charges on surfaces, and extended water network propagating into the bulk. Other experimental observations have pointed to the presence of nanobubbles (47), and there are indications that the charge correlation effects could be gaining some currency again (48). It is likely that a number of mechanisms corroborate or operate in different experimental conditions.

16.4.4.4 Surface Forces in Contact: Adhesion and Capillary Force

When two surfaces come into intimate contact ($D = D_0$) in a vapour, for example in the case of two spherical colloidal particles of equal radius R shown in Figure 16.8, the interaction energy per unit area in Equation (16.28) becomes

$$F(D_0) = \pi R W_a(D_0) = 2\pi R \gamma_s \tag{16.47}$$

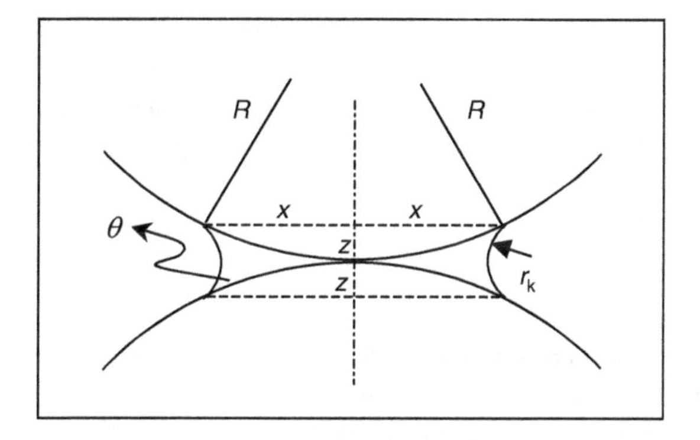

Figure 16.8 *Capillary of Kelvin radius r_k forms between two colloidal particles of equal radius R in contact, when the liquid condensed from vapour wets the surface, i.e. $\theta < 90°$*

where $\gamma_S = \frac{1}{2}W_a(D_0)$ is the surface energy of the solid in vacuum and $D_0 \approx \sigma/2.5$ is taken as the contact separation of the order of a fraction of the molecular dimension. In general, under such adhesion, the solid surfaces would become elastically deformed, a subject dealt with in the theory of contact mechanics. We will not discuss it in detail here.

If a vapour is present, and if its condensed liquid of surface energy γ_L wets the surface (i.e. the contact angle $\theta < 90°$), a meniscus of liquid would form around the annulus of the contact area, characterised by the *Kelvin radius* r_k, as illustrated in Figure 16.8,

$$r_k = \frac{\gamma_L V}{N_A kT \log(p/p_s)} \tag{16.48}$$

where V is the molar volume for the condensing liquid, $N_A = 6.022 \times 10^{23}\,\text{mol}^{-1}$ is the Avogadro constant and p/p_s (i.e. the partial pressure over the saturation pressure at T) is the relative vapour pressure of the liquid. For water at 20 C, $\gamma V/N_A kT = 0.54$ nm. Since $p/p_s < 1$, $r_k < 0$, that is, the meniscus is concave. For instance, for $p/p_s = 0.5$, $r_k \approx -1.6$ nm.

A negative r_k also means that the Laplace pressure ΔP in the liquid due to the curvature of the meniscus is negative too, i.e.

$$\Delta P = \frac{\gamma_L}{r_k} < 0 \tag{16.49}$$

which means the surfaces experience an attraction due to the condensed liquid meniscus. We wish to estimate the contribution of this attractive force F_k and compare it with the adhesion force from Equation (16.47). The Laplace pressure is acting on an area $\pi x^2 \approx 2\pi R z$, and we also have from the meniscus geometry $2z \approx 2r_k \cos\theta$. Thus the attractive force due to the Laplace pressure is

$$F_k = \pi x^2 \Delta P = 2\pi R \gamma_L \cos\theta \tag{16.50}$$

We also note that γ_S in Equation (16.47) should be replaced with γ_{SL} (the solid–liquid interfacial energy) due to the capillary condensation. Thus the total adhesive force in the presence of the meniscus is

$$F_{ad} = F_k + F(D_0, \gamma_{SL}) = 2\pi R(\gamma_L \cos\theta + \gamma_{SL}) = 2\pi R \gamma_{SV} \tag{16.51}$$

Often $\gamma_L \cos\theta > \gamma_{SL}$, which means the adhesion force is largely determined by the capillary force and in turn by the surface energy of the condensing liquid.

Capillary condensation could also occur when the two spheres are immersed in a liquid (1) with trace amounts of another immiscible liquid (2), e.g. an oil with a small amount of water. Then Equation (16.48) becomes

$$r_k = \frac{\gamma_{12} V}{N_A kT \log(c/c_s)} \tag{16.52}$$

where c/c_s is the ratio between the concentration of the condensing liquid and its saturation concentration (or solubility), and γ_{12} is the interfacial tension between the two liquids. Now the adhesion force is dominated by the capillary force $F_k = 2\pi R \gamma_{12} \cos\theta$.

Such capillary forces are relevant to the colloidal stability in non-polar media, powder processing and, of course, sand castle building.

16.4.5 Neutral Polymer-mediated Surface Forces

As we have seen, in general, the surface separation between colloidal particles where the colloidal interactions become important lies in the range ~0–100 nm. This is comparable to the length scale of the radius of gyration R_g of typical polymers in a good solvent. Once added to a colloidal system, the polymer-mediated interaction will normally dominate other types of surface forces. Chapter 9 has explored how surface forces and colloidal stability could be mediated and affected by adding polymers to the colloidal system. It is found that the key parameter to be considered there is whether or not the polymer adsorbs favourably to colloids. If the polymer does so at a high surface density or coverage, it leads to repulsion between polymer-coated colloids. On the hand, if the surface coverage is low, bridging attraction could occur as a polymer chain finds itself adsorbed on more than one colloidal particle. In the case that the polymer does not adsorb onto the colloid surface, depletion attraction may manifest when the surface separation becomes comparable to R_g due to the osmotic pressure imbalance inside and outside the gap between the colloidal particles.

An effective strategy to anchor polymers on a colloidal particle surface at a high density is to end anchor the polymer chains on the surface to form a brush. Considerable experimental and theoretical efforts have been made to measure such polymer-brush-mediated surface force, so we will examine it here briefly. For a brush to form, the solvent must be a good solvent so that the chain is stretched out from the surface rather than adsorbing or collapsing on it. In addition, the polymer chain density must be high so that the spacing s between the polymer chains is smaller than R_g.

Figure 16.9 shows schematically such a polymer brush with a uniform brush equilibrium height L_0, each chain with N number of (neutral) monomers of size a and an end anchoring energy of αkT per chain. (Note that the brush thickness is denoted as δ_H in Chapter 8.6.) The volume per brush chain is thus $V_{chain} = s^2 L$ for any brush height L, and volume fraction of monomers in the brush is $\phi = Na^3/Ls^2$. The monomer volume fraction is such that it falls into the semi-dilute regime, i.e. $\phi^* \ll \phi \ll 1$, with $\phi^* \sim N^{-4/5}$ the threshold volume fraction where polymer chains start to overlap. We will set out to find firstly the equilibrium brush height L_0 on a single surface and the equilibrium free energy associated with the brush by considering the energetic balance between the osmotic repulsion and elastic stretching energies. Secondly, we will bring two such brush-bearing planar surfaces to a separation

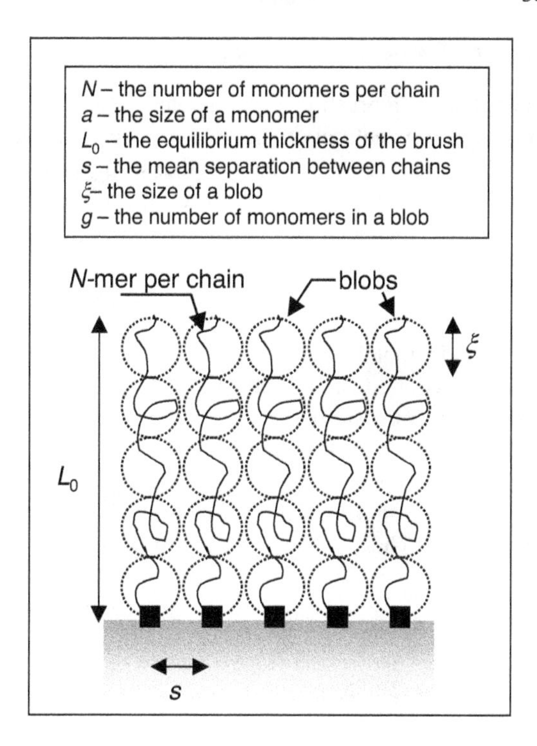

Figure 16.9 *A polymer brush*

D so that the brushes are confined to $D < 2L_0$. The D-dependent free energy change will give us the interaction free energy per unit area $W_a(D)$. This will be carried out following the approach by Alexander (49) and de Gennes (50). In doing so, we typically ignore some prefactors of order unity to the expressions we use.

We will present our derivation in terms of energy per chain, W^{chain}; and given that the area per chain is s^2, it relates to energy per unit area as $W_a(D) = W^{\text{chain}}/s^2$. There are two components in W^{chain} due respectively to osmotic repulsion and chain elastic stretching, i.e.

$$W^{\text{chain}} = W^{\text{chain}}_{\text{osm}} + W^{\text{chain}}_{\text{stretch}} \tag{16.53}$$

16.4.5.1 Osmotic Pressure Π_{osm} in Polymer Brush

Firstly, the osmotic repulsion between monomers in a polymer brush arises because the monomers do not like to be crowded together in a chain, and the osmotic term favours stretching the chain away from the surface. In a semi-dilute regime, the monomers osmotic pressure in the brush scales as $\Pi_{\text{osm}} \cong kT\phi^{9/4}/a^3$, and thus the associated energy per chain is

$$W^{\text{chain}}_{\text{osm}} = kTV_{\text{chain}}\Pi_{\text{osm}} \cong kT(s^2L)\left(\frac{kT}{a^3}\phi^{9/4}\right) \cong kTN\left(\frac{Na^3}{s^2L}\right)^{5/4} \tag{16.54}$$

16.4.5.2 Elastic Stretching Energy of a Brush Chain

Secondly, the polymer chain in a brush is highly stretched, and associated elastic energy favours restoring the chain to its natural configuration, which, in a semi-dilute regime, is characterised by an end-to-end size $R(\phi)$. The effective spring constant of such a chain is $\sim kT/R^2(\phi)$, so that the elastic restoring energy for a brush thickness L is

$$W_{\text{stretch}}^{\text{chain}} \cong \frac{kT}{R^2(\phi)} L^2 \tag{16.55}$$

To obtain $R(\phi)$, we consider that the polymer chain is made up of independent spheres of size ξ, called blobs, shown as dotted circles in the figure (see also Chapter 8, Section 8.3.4), each with g number of monomers. Inside one blob, g monomers behave like a Flory chain, so that $\xi = ag^{3/5} = a\phi^{-3/4}$ according to the standard scaling relation and there are N/g number of the blobs in each brush chain. However, these blobs do not interact with each other in the semi-dilute regime, such that it is as if we have an ideal chain of size $R(\phi)$ consisting of N/g blobs as its 'blob monomers'. Thus, we have, as for an ideal chain,

$$R^2(\phi) = \frac{N}{g}\xi^2 \cong Na^2\phi^{-1/4}, \qquad (\phi^* \ll \phi \ll 1) \tag{16.56}$$

and substitution into Equation (16.55) gives

$$W_{\text{stretch}}^{\text{chain}} \cong kT\frac{L^2}{Na^2}\phi^{1/4} = kTL^{7/4}N^{-3/4}a^{-5/4}s^{-1/2} \tag{16.57}$$

16.4.5.3 Equilibrium Brush Thickness L_0 and Free Energy per Chain W_0^{chain}

Adding together the osmotic and elastic stretching terms in Equations (16.55) and (16.57), we obtain the total energy per chain,

$$W^{\text{chain}} = W_{\text{osm}}^{\text{chain}} + W_{\text{stretch}}^{\text{chain}} \cong kT\left(N\left(\frac{Na^3}{s^2L}\right)^{5/4} + \left[L^{7/4}N^{-3/4}a^{-5/4}s^{-1/2}\right]\right) \tag{16.58}$$

and minimising it with respect to L we obtain the equilibrium brush thickness,

$$L_0 = Na\left(\frac{a}{s}\right)^{2/3} \tag{16.59}$$

(Note that s is related to the grafting density σ in Chapter 8, Section 8.6, as $\sigma = 1/s^2$, and thus from Equation (16.59) we have $L_0 \sim N\sigma^{1/3}$, in agreement with the result shown in Figure 8.36.)

We see that the brush thickness scales with N, as compared to $N^{0.6}$ for a free chain in a good solvent. It means the polymer chain takes up a very extended conformation in a brush. The free energy per chain is obtained by inserting L_0 above into Equation (16.59),

$$W_0^{\text{chain}} = kTN\left(\frac{a}{s}\right)^{5/3} \tag{16.60}$$

16.4.5.4 *Interaction Free Energy per Unit Area $W_a(D)$ Between Brushes: the Alexander–de Gennes Theory*

When two brushes (on two planar surfaces of separation D) are compressed against each other so that $D < 2L_0$, they tend to interdigitate only weakly, with the intedigitation depth d_p estimated as

$$d_p \cong \left(\frac{2L_0}{D}\right)^{1/3} s \tag{16.61}$$

We see that the denser the brush (smaller s), the less the interdigitation. Hence, the interaction free energy due to compressing two chains is approximately twice that of a single chain compressed to $D/2$. The monomer fraction of the compressed brushes is

$$\phi(D) = \frac{2Na^3}{s^2 D} \tag{16.62}$$

and substitution of $\phi(D)$ and L_0 into Equations (16.55) and (16.57) gives the interaction free energy per chain, and we divide it by the area per chain s^2 to finally obtain the interaction free energy per unit area $W_a(D)$ between two polymer brushes compressed to D whose equilibrium thickness is L_0 and chain spacing is s,

$$W_a(D) = \frac{2kTL_0}{s^3}\left[\frac{4c_1}{5}\left(\frac{2L_0}{D}\right)^{5/4} + \frac{4c_2}{7}\left(\frac{D}{2L_0}\right)^{7/4} - \left(\frac{4c_1}{5} + \frac{4c_2}{7}\right)\right] \tag{16.63}$$

where c_1 and c_2 are prefactors of order unity. Equation (16.63) and the preceding derivations leading to it are often referred to as the Alexander–de Gennes theory for polymer brush interactions. The first term in the square brackets in Equation (16.63) is the osmotic repulsion which dominates under high compression, the second term is the entropy gain as the chain is pushed back from its stretched conformation, and the last term is added to ensure $W_a(D) = 0$ for $D \geq 2L_0$. An expression for the pressure between two such brushes is given in Equation (9.3) in Chapter 9 (omitting the prefactors), and integration of Equation (9.3) using the $P(D)$–$W_a(D)$ relation (c.f. Equation 16.15) will also arrive at Equation (16.63).

Experimentally, the interactions between polymer brushes as described above have been verified using the SFA (51), and an example is given in Figure 16.10 (52), in which the experimentally obtained force between two 50 nm thick polystyrene brushes end adsorbed on mica in toluene agrees closely with the fit (solid curve in Figure 16.10) using Equation (16.63).

16.4.6 Surface Forces in Surfactant Solutions

Surfactant molecules readily adsorb on solid surfaces, and the resulting surface structures – their morphology, thickness and density, depend on the headgroup, molecular architecture and concentration of the surfactant as well as a number of solution parameters such as pH and electrolyte concentrations. For example, at very dilute surfactant concentrations, monolayers or partial monolayers would form on a hydrophilic or charged surface, rendering it hydrophobic. As the surfactant concentration increases, partial bilayers or

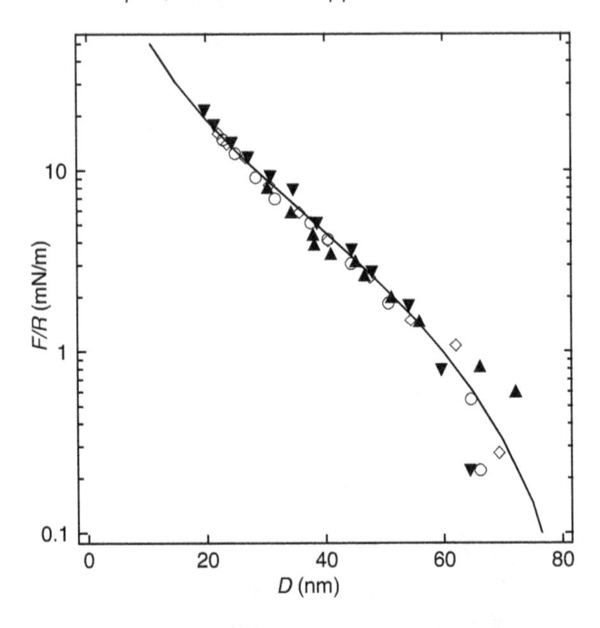

Figure 16.10 *Surface forces (symbols) between two polystyrene brushes in toluene end-anchored on mica via a zwitterionic group. The MW of the brush chain is 65 kDa. The solid curve is a fit to the Alexander–de Gennes theory Equation (16.63), with* $L_0 = 43$ *nm and* $s = 7.5$ *nm*

bilayer-like surface aggregates such as cylindrical micelles or hemi-micelles would form. The surface coverage approaches saturation as the surfactant concentration reaches the critical surface micellisation concentration, which is typically much lower than the bulk cmc.

The presence of surfactants dramatically alters the surface force between solid surfaces. As the above process takes place, the surface forces exhibit a plethora of behaviours. If charged, the headgroup density, and thus the surface charge density, varies as the surfactant concentration varies, in turn affecting the long range double layer force. The presence of surfactant molecules in the bulk also alters the Debye decay length of the double layer force. Surfactant headgroups tend to be highly hydrated, and this gives rise to an additional hydration repulsion at short range. A large number of related surface force experiments have been carried out in surfactant solutions. For instance, the first observation of the hydrophobic interaction has been made between surfactant monolayers in the SFA, as is the hydrophobic force shown in Figure 16.5. However, it is fair to comment that a complete picture of surface forces in surfactant solutions is yet to emerge.

16.5 Recent Examples of Surface Force Measurement

Direct measurement of surface forces, particularly using the SFA, has greatly enhanced our knowledge of intermolecular and inter-surface interactions, and it remains at the frontier of current research. Some of the future challenges will be outlined in the next section. Here we describe some of recent results obtained using the SFA.

16.5.1 Counter-Ion Only (CIO) Electric Double Layer Interactions in a Non-Polar Liquid

Due to its very low dielectric constant, e.g. around 2 to 4, the Coulomb attraction between the cation and anion in an electrolyte in a non-polar liquid is strengthened by a factor of 20 to 40 in comparison to that in water. Consequently, the dissociation of electrolyte, and in turn the ionic concentration, is minimal, and the validity of the charging mechanisms, which are operative in the aqueous medium, becomes questionable. It is for this reason that it has been somewhat controversial to ask whether the solid can acquire a surface charge and if the electrical double layer interaction plays a role in colloidal stability in the non-polar liquid (53, 54). Controversy had persisted as to whether it even exists in non-polar media. Previously believed to be undetectable, such a force in a non-polar liquid is very difficult to measure directly due to its gradual decay over a large range and its weak magnitude.

In a recent study (55, 56), modifications of an SFA have been carried out to enhance its capacity, with which an attempt has been successfully made to measure an electrical double layer interaction between two mica surfaces immersed in decane with an added surfactant, sodium di-2-ethylhexyl-sulfosuccinate (AOT) at mM concentrations (see Figure 16.11). The interaction is long ranged and weaker than that in water by one order of magnitude. This, together with observations from light scattering and FTIR experiments, has allowed the proposition of a charging mechanism at the solid–non-polar liquid interface. The charging occurs when the ions on the surface are transferred into the water cores of

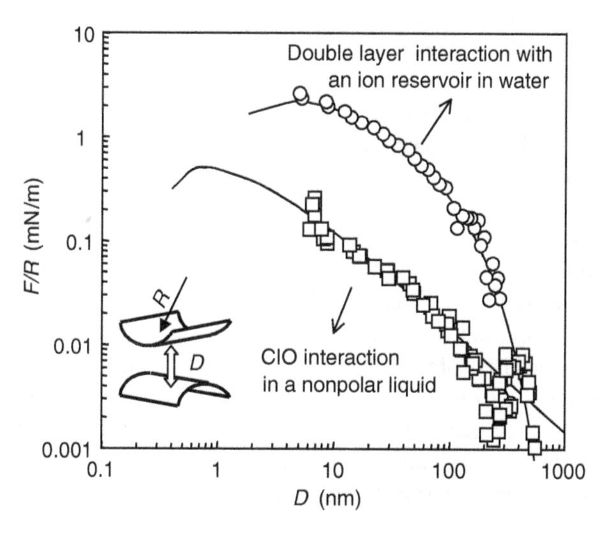

Figure 16.11 Comparison of the measured double layer interactions in a non-polar liquid (empty squares) and in pure water (empty circles) respectively with the counter-ion only (CIO) theory (power-law decay; lower curve) and the DLVO theory (exponential decay; upper curve). Fitting the measured interaction in the non-polar liquid with our theory gives a surface charge density of 10^{-3} C m^{-2} at the mica–decane interface. The measured interaction is the strongest possible CIO interaction in a non-polar liquid, but it is still much weaker than that in water. It is the first time such a CIO interaction has been detected in a non-polar liquid

the inverse micelles formed by the surfactant molecules and subsequently carried away from the surface.

Can we describe such a force in non-polar media using the DLVO theory? In a related study, it was discovered that a different theoretical analysis is required to describe the double layer force in non-polar media. It is called the *counter-ion only* (CIO) double layer, as in this case almost all the ionic species come from solids. Accordingly, a counter-ion only theory (CIO) has been developed, taking a *constrained total entropy approach* (57), and a pleasing agreement was found with the experimental data (see Figure 16.10) (55). Asymptotically, the CIO interaction decays as a power law, in contrast to the exponential decay of double layer forces in water. This serves to question the appropriateness of borrowing the aqueous DLVO theory to treat the double layer interaction in non-polar liquids, a practice widely adopted.

16.5.2 Interactions between Surface-grown Biomimetic Polymer Brushes in Aqueous Media

Conventionally, polymer brushes are formed with a grafting-to approach, in which non-adsorbing polymer chains are end-anchored on a surface via a functional group. That anchoring energy is typically a few kT per chain in the case of physical adsorption, and such a relatively low sticking energy limits the density and thickness of the brush we could access. The brush obtained is also not robust enough to sustain high compression, as they get uprooted from the surface. The polymer chains in the grafting-to approach are also chemically anchored (see Chapter 8, Section 8.6), although such brush formation can be kinetically slow due to the chains arriving later being sterically hindered by the chains that are already there. An alternative approach is the grafting-from brush, in which polymer brushes are grown directly from activated surface sites, providing strong anchorage and chemically tuneable brush density. In particular, surface-initiated atom transfer radical polymerisation (SI-ATRP) has been used to obtain surface-grown polymer brushes of tailored architecture and molecular weight. However, direct measurement of such SI-ATRP brushes had not been previously performed mainly due to the experimental difficulties to construct such brushes suitable for SFA's stringent measurement conditions.

Very recently, an attempt has been made to successfully grow a biomimetic polyzwitter-ionic polymer brush, poly[2-(methacryloyloxy)ethyl phosphorylcholine] (pMPC), from mica, and the surface force between such brushes in pure water is shown in Figure 16.12 (58, 59). Thanks to the robustness of these brushes, surface force measurement under very high compression could be performed, where a noticeable discrepancy with the Alexander–de Gennes theory (dashed segment of the fitted curve) is observed, possibly due to higher order terms in the expression for the osmotic pressure becoming dominant under high compression, and thus high volume fraction of monomers.

16.5.3 Boundary Lubrication Under Water

We have focused on the normal surface force in this chapter. Friction between surfaces in contact is also tremendously important to many technological and engineering applications. Early scientific investigations on friction could be traced back to da Vinci (AD 1452–1519) and it remains the subject of intensive research today. Here an example is given of a recent experiment on aqueous boundary lubrication using the version of SFA shown in Figure 16.4.

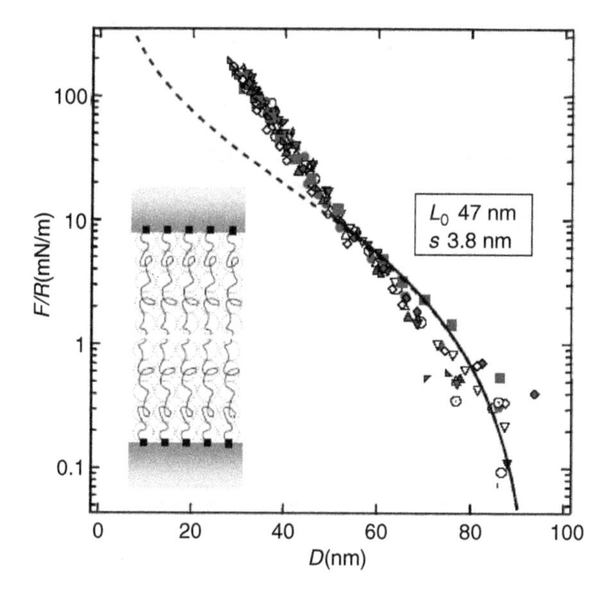

Figure 16.12 *Surface forces between two surface grown polymer brushes, pMPC, in water. The dashed and solid curve is the fit using the Alexander–de Gennes theory (Equation 16.63), and a significant deviation of the experimental result (symbols) from the theory is observed for* $D < \sim 50\,\text{nm}$ *(dashed curve) when the polymer brushes are under high compression*

In classic boundary lubrication in air or oil media, monolayers of surfactants are anchored on the rubbing solid surfaces to reduce friction and wear, and its molecular mechanism is well understood: rubbing between the underlying substrates is largely replaced by sliding between the hydrocarbon tails of the anchored surfactant boundary layers, as the van der Waals bonds between the tails are the weakest link with respect to shear. Surfactant molecules are also ubiquitous in aqueous media and may adsorb readily onto solid surfaces to form various surface aggregates. How would they participate in the tribological process in aqueous media?

In a recent study (60, 61), friction has been measured between two mica surfaces bearing surfactant monolayers of a di-chained cationic surfactant, $[(CH_3(CH_2)_{10})2N^+(CH_3)2Br^-]$. Figure 16.13 shows kinetic friction F_s against shear velocity V_s, as the surfaces are made to slide past each other when they are in a strongly adhesive contact under water, with an adhesion energy of $\sim$40 mJ m^{-2}. The inset shows the corresponding shear stress $\sigma_0 = F_s/A_0$ where $A_0 \sim 1000\,\mu\text{m}^2$ is the contact area between the surfaces. The measured friction and shear stress under water (empty circles; ○) are much lower than those in dry air from previous measurements (indicated by the hatched regions), by up to some two orders of magnitude. To explain the observation of this reduction in friction, it is proposed that the aqueous boundary lubrication mechanism has a different molecular origin.

Water can penetrate into the monolayers and hydrate the surfactant head groups, forming 'molecular water puddles' at the surface. This hydration significantly enhances the lateral mobility of surfactant molecules on the surface, and also promotes possible structural changes in the surfactant layers. This mechanism suggests that, unlike in air or

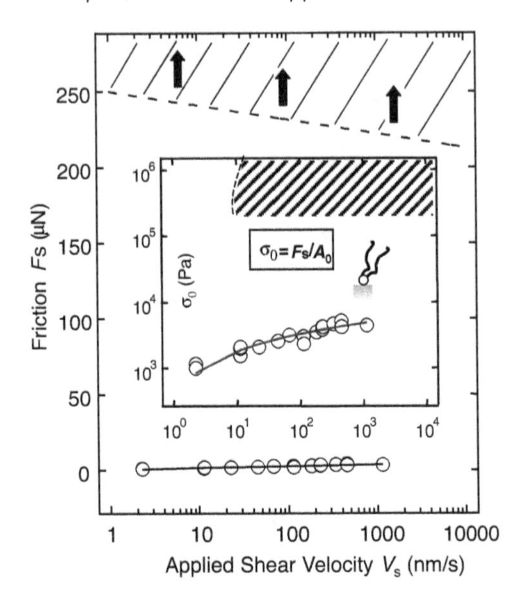

Figure 16.13 *Friction F_s and shear stress σ_0 (inset) between two mica surfaces bearing surfactant monolayers under water as a function of applied shear velocity V_s*

oil media, lubrication under water is facilitated by the water molecules tenaciously held around the charged surfactant head groups. Thus, sliding mostly takes place at the mica surface where the hydration layer is located, instead of the interface between the hydrocarbon tails.

16.6 Future Challenges

While the DLVO forces are well understood, we have also accumulated significant experimental data to help us in tailoring surface forces by adding polymers and surfactants to the system. From a surface force measurement viewpoint, most of the equilibrium forces have been measured and appropriate theories have been put forward to analyse experimental results. However, several types of forces need further investigations. These include first of all hydrophobic forces, particularly regarding the postulated mechanisms of nanobubbles and patchy charges. In addition, the experiment on the electrical double layer force in a non-polar liquid highlighted in this chapter remains one of very few such measurements, and further experiments are required to fully explore parameters relevant to solid charging in non-polar media. Another area that is lacking experimental data is the surface force measurement between polyelectrolyte brushes (62, 63) for which model surfaces required for the SFA measurement are difficult to obtain. These charged brushes are expected to behave rather differently from the neutral brushes hitherto studied, particularly their postulated response to added multivalent ions. Similarly, further systematic studies on hydration forces and surface forces in different surfactant solutions will also add to the completeness of our current understanding.

A new area that deserves focused effort is the surface force mediated in ionic liquids, whose 'green' credentials have brought them into the spotlight recently. Despite an early comprehensive surface force study in an ionic liquid (64) (called 'molten salts' at the time, before they gained their current fame), surface measurements have been few and far between since. Given the unique nature and the ready availability of a vast number of possible molecular architectures, we expect fruitful results from future studies in this area.

As compared with the equilibrium forces, surface forces involving non-equilibrium effects are even less well studied and understood. These include friction forces mediated by polymer and surfactant surface structures, particularly in aqueous media and hydrodynamic forces involving deformable surfaces (65).

References

(1) Israelachvili, J. N. (1991) Intermolecular and Surface Forces. Academic Press, London.
(2) Mahanty, J., Ninham, B. W. (1976) Dispersion Forces. Academic Press, New York.
(3) Hamaker, H. C. (1937) Physica 4: 1058–1072.
(4) Horn, R. G. (1995) in Ceramic Processing, Terpstra, R. A., Pex, P. P. A. C., de Vries, A. H. (eds.). Chapman & Hall, London, vol. 3, pp. 58–101.
(5) Horn, R. G. (1990) J. Am. Ceramic Soc., 73: 1117–1135.
(6) Claesson, P. M., Ederth, T., Bergeron, V., Rutland, M. W. (1996) Adv. Colloid Interface Sci., 67: 119–183.
(7) Crocker, J. C., Grier, D. G. (1994) Phys. Rev. Lett., 73: 352–355.
(8) Grier, D. G. (1998) Nature, 393: 621–623.
(9) Ashkin, A. (1970) Phys. Rev. Lett., 24: 156–159.
(10) Ashkin, A. (1980) Science, 210: 1081–1088.
(11) Grier, D. G. (1997) Curr. Opin. Colloid Interface Sci., 2: 264–270.
(12) Crocker, J. C., Grier, D. G. (1994) Phys. Rev. Lett., 73: 352–355.
(13) Prieve, D. C. (1999) Adv. Colloid Interface Sci., 82: 93–125.
(14) Ducker, W. A., Senden, T. J., Pashley, R. M. (1991) Nature, 353: 239–241.
(15) Tabor, D., Winterton, R. H. S. (1969) Proc. R. Soc. Lond. A, A312: 435–450.
(16) Israelachvili, J. N., Adams, G. E. (1976) Nature, 262: 774–776.
(17) Tolansky, S. (1966) in An Introduction to Interferometry. Longmans, London, vol. 14, pp. 173–196.
(18) Tolansky, S. (1970) Multiple-beam Interference Microscopy of Metals. Academic Press, London and New York.
(19) Horn, R. G., Clarke, D. R., Clarkson, M. T. (1988) J. Mater. Res., 3: 413–416.
(20) Horn, R. G., Smith, D. T., Haller, W. (1989) Chem. Phys. Lett., 162: 404–408.
(21) Parker, J. L., Cho, D. L., Claesson, P. M. (1989) J. Phys. Chem., 93: 6121–6125.
(22) Horn, R. G., Smith, D. T. (1992) Science, 256: 362–364.
(23) Tonck, A., Georges, J. M., Loubet, J. L. (1988) J. Colloid Interface Sci., 126: 150–163.
(24) Crassous, J., Charlaix, E., Gayvallet, H., Loube, J.-L. (1993) Langmuir, 9: 1995–1998.
(25) Belouschek, P., Maier, S. (1986) Prog. Colloid Polym. Sci., 72: 43–50.
(26) Tanimoto, S., Matsuoka, H., Yamauchi, H., Yamaoka, H. (1999) Colloid Polym. Sci., 277: 130–135.
(27) Parker, J. L. (1992) Langmuir, 8: 551–556.
(28) Mächtle, P., Muller, C., Helm, C. A. (1994) J. Phys. II Fr., 4: 481–500.
(29) Kékicheff, P., Spalla, O. (1994) Langmuir, 10: 1584–1591.
(30) Horn, R. G., Israelachvili, J. N., Pribac, F. (1987) J. Colloid Interface Sci., 115: 480–492.
(31) Levins, J. M., Vanderlick, T. K. (1993) J. Colloid Interface Sci., 158: 223–227.
(32) Heuberger, M., Luengo, G., Israelachvili, J. N. (1997) Langmuir, 13: 3839–3848.
(33) Tabor, D., Winterton, R. H. (1968) Nature, 219: 1120–1121.

(34) Israelachvili, J. N., Tabor, D. (1972) Proc. R. Soc. Lond. A., 331: 19–38.
(35) Klein, J., Kumacheva, E. (1998) J. Chem. Phys., 108: 6996–7009.
(36) Abrikossova, I. I., Derjaguin, B. V. (1957) in Electrical Phenomena and Solid/Liquid Interface, Schulman, J. H. (ed.). Butterworths Scientific Publications, London, pp. 398–405.
(37) Derjaguin, B. V., Titijevskaia, A. S., Abrikossova, I. I., Malkina, A. D. (1954) Disc. Faraday Soc., 18: 24–41.
(38) Derjaguin, B. V., Abrikossova, I. I., Lifshitz, E. M. (1956) Quart. Rev., Chem. Soc., 10: 295–329.
(39) Gauthier-Manuel, B., Gallinet, J.-P. (1995) J. Colloid Interface Sci., 175: 476–483.
(40) Klein, J. (1983) J. Chem. Soc., Faraday Trans. I, 19: 99–118.
(41) Pashley, R. M., Karaman, M. E., Craig, V. S. J., Kohonen, M. M. (1998) Colloids Surf. A, 144: 1–8.
(42) Cappella, B., Dietler, G. (1999) Surf. Sci. Rep., 34: 1–104.
(43) Hamaker, H. C. (1937) Physica, 4: 1058–1072.
(44) Derjagui, B. V., Churaev, N. V. (1974) J. Colloid Interface Sci., 49: 249–255.
(45) Tarazona, P., Vicente, L. (1985) Mol. Phys., 56: 557–572.
(46) Horn, R. G., Israelachvili, J. N. (1981) J. Chem. Phys., 75: 1400–1411.
(47) Tyrrell, J. W. G., Attard, P. (2002) Langmuir, 18: 160–167.
(48) Perkin, S., Kampf, N., Klein, J. (2006) Phys. Rev. Lett., 96.
(49) Alexander, S. (1977) J. Phys., 38: 983–987.
(50) de Gennes, P.-G. (1987) Adv. Colloid Interface Sci., 27: 189.
(51) Taunton, H. J., Toprakcioglu, C., Fetters, L., Klein, J. (1990) Macromolecules, 23: 571–580.
(52) Dunlop, I. E., Briscoe, W. H., Titmuss, S., Sakellariou, G., Hadjichristidis, N., Klein, J. (2004) Macromol. Chem. Phys., 205: 2443–2450.
(53) Osmond, D. W. J. (1966) Disc. Faraday Soc., 42: 247.
(54) Albers, W., Overbeek, J. Th. G. (1959) J. Colloid Sci., 14: 510–518.
(55) Briscoe, W. H., Horn, R. G. (2002) Langmuir, 18: 3945–3956.
(56) Briscoe, W. H., Horn, R. G. (2004) Prog. Colloid Polym. Sci., 123: 147–151.
(57) Briscoe, W. H., Attard, P. (2002) J. Chem. Phys., 117: 5452–5464.
(58) Chen, M., Briscoe, W. H., Armes, S. P., Cohen, H., Klein, J. (2007) Chem. Phys. Chem., 8: 1303–1306 (cover picture).
(59) Chen, M., Briscoe, W. H., Armes, S. P., Klein, J. (2009) Science, 323: 1698–1701.
(60) Briscoe, W. H., Titmuss, S., Tiberg, F., Thomas, R. K., McGillivray, D. J., Klein, J. (2006) Nature, 444: 191–194.
(61) Briscoe, W. H., Klein, J. (2007) J. Adhesion, 83: 705–722.
(62) Dunlop, I. E., Briscoe, W. H., Titmuss, S., Jacobs, R. M. J., Osborne, V. L., Edmondson, S., Huck, W. T. S., Klein, J. (2009) J. Phys. Chem. B, 113: 3947–3956.
(63) Liberelle, B., Giasson, S. (2008) Langmuir, 24: 1550–1559.
(64) Horn, R. G., Evans, D. F., Ninham, B. W. (1988) J. Phys. Chem., 92: 3531–3537.
(65) Connor, J. N., Horn, R. G. (2001) Langmuir, 17: 7194–7197.

Index

Colloid Science: Principles methods and applications, Second Edition Edited by Terence Cosgrove

www.ingramcontent.com/pod-product-compliance
Lightning Source LLC
Chambersburg PA
CBHW080940260125
20788CB00015BA/161

* 9 7 8 1 4 4 4 3 2 0 2 0 6 *